Lecture Notes in Mathematics 1475

S. Busenberg M. Martelli (Eds.)

Delay Differential Equations and Dynamical Systems

Proceedings of a Conference in honor of Kenneth Cooke held in Claremont, California, Jan. 13–16, 1990

Springer-Verlag
Berlin Heidelberg New York
London Paris Tokyo
Hong Kong Barcelona
Budapest

Editors

Stavros Busenberg
Department of Mathematics
Harvey Mudd College
Claremont, CA 91711, USA

Mario Martelli
Department of Mathematics
California State University
Fullerton, CA 92634, USA

Mathematics Subject Classification (1985): 34C, D, E, G, K; 35B, K; 39A11

ISBN 3-540-54120-9 Springer-Verlag Berlin Heidelberg New York
ISBN 0-387-54120-9 Springer-Verlag New York Berlin Heidelberg

Printed in Germany

Printing and binding: Druckhaus Beltz, Hemsbach/Bergstr.
2146/3140-543210 - Printed on acid-free paper

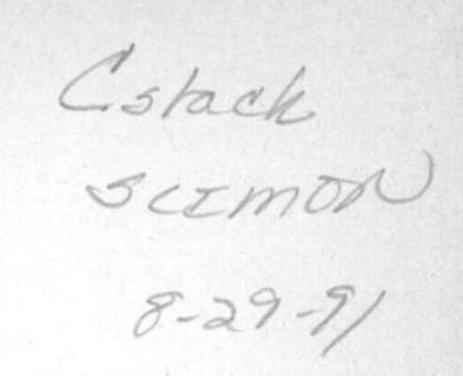

Preface

The area of differential equations has played a central role in the development of mathematics and its applications for nearly four centuries, yet it still displays an unabated vitality. The field involves vigorous cross-disciplinary interactions, and research papers appear in a variety of mathematics and science journals. Such breadth and fertility create difficulties for those who wish to keep abreast of recent developments. Consequently, it is beneficial to have occasions which bring together experts in this field in a forum that encourages the exchange of ideas and leads to timely publication of new results. Such an occasion occurred during January 13 to 16, 1990 when some two hundred research workers participated in the International Conference on Differential Equations and Applications to Biology and Population Dynamics which was held in Claremont. The occasion was particularly noteworthy because it provided a venue for the celebration of the 65th birthday of Kenneth Cooke, a seminal worker who has made numerous pioneering contributions in delay differential equations and population dynamics.

This volume contains a selection of papers on delay differential equations and dynamical systems which were presented at the conference and accepted after peer review. A companion volume in the Biomathematics Lecture Notes series of Springer contains papers devoted to applications in biology and population dynamics. The areas of these volumes have close and fruitful interactions, and Kenneth Cooke has been one of the most artful and original practitioners in this interdisciplinary research work.

The contributions in this volume are collected in two groups, the first consisting of survey articles and the second of research papers. The three survey articles are by Kenneth Cooke and Joseph Wiener who review the recently opened area of differential equations with piecewise continuous arguments; by Jack Hale who discusses a fascinating array of results in the stability of delay differential equations viewed as dynamical systems; and by Paul Waltman who presents an overview of useful new results on persistence in dynamical systems. The research contributions part of the volume consists of nineteen papers which present new results in delay differential equations and dynamical systems. The papers are united by the common thread of the underlying topic but, as is characteristic of this field, employ a wide array of deep mathematical theories and techniques. These include methods from linear and nonlinear functional analysis, a number of topological and topological degree techniques, as well as asymptotic and other classical analysis methods. Many of these mathematical techniques were originally created in order to address problems arising in the field of differential equations and are still being stimulated by challenges from this field.

The research conference which lead to this volume was supported by the National Science Foundation through grant number MCS-8912391 and by Harvey Mudd College which hosted the meeting. A large number of individuals gave invaluable help during all phases of the conference. Our special gratitude goes to those who helped prepare the manuscripts for publication: Sue Cook, Jeffrey

McClelland, Beth Nyerges, Barbara Schade and David Williamson. The editorial staff of Springer gave us constant support and was always ready to respond to our requests for help and information. Finally, the many researchers who contributed time and expertise in refereeing the papers provided an invaluable service to the members of the mathematical community who will use this volume. Our sincere gratitude goes to all, both named and unnamed, who have helped in this mathematical endeavor.

Stavros Busenberg and Mario Martelli
Claremont, California

Contents

A Survey of Differential Equations with Piecewise Continuous Arguments

Kenneth L. Cooke[1] *and Joseph Wiener*[2]

[1] Mathematics Department Pomona College, Claremont, CA 91711
[2] Department of Mathematics The University of Texas-Pan American Edinburg, Texas, 78539

1 Introduction

The general theory and basic results for functional differential equations have by now been thoroughly explored and are available in the famous book of Hale [19] and subsequent articles by many authors. Nevertheless, there is still need for investigation of special equations, which may have interesting properties and provide insight into the more abstract theory. In this review article, we shall describe some of the work that has been done by us and others over the last few years on the differential equations that we call equations with piecewise continuous arguments, or EPCA. Our attention was directed to these equations by an article of A.D. Myshkis [23], who observed that a substantial theory did not exist for differential equations with lagging arguments that are piecewise constant or piecewise continuous. Since that time several authors have investigated equations of this type. The purpose of this article is to give a brief survey of the present status of this research and to point out some directions for further study.

A typical EPCA is of the form

$$x'(t) = f(t, x(t), x(h(t))) \ , \tag{1}$$

where the argument $h(t)$ has intervals of constancy. For example, in [9] equations with $h(t) = [t]$, $[t-n]$, $t - n[t]$ were investigated, where n is a positive integer and $[\ ]$ denotes the greatest integer function. Note that $h(t)$ is discontinuous in these cases, and although the equation fits within the general paradigm of delay differential or functional differential equation, the delays are discontinuous functions. Also note that the equation is non-autonomous, since the delays vary with t. Moreover, as we show below, the solutions are determined by a finite set of initial data, rather than by an initial function as in the case of general functional differential equations.

Before delving into a general theoretical discussion, we mention an example, the equation

$$x'(t) = ax(t)(1 - x([t])) \ . \tag{2}$$

This equation is analogous to the famous logistic differential equation, but t in one argument has been replaced by $[t]$. As we shall explain in further context in Sect. 6, the equation has solutions that display complicated dynamics. It is a first order equation, and is one of the simplest examples of a differential equation with chaotic dynamics. It seems likely that other simple nonlinear EPCA's may display other interesting behavior.

This paper is organized in the following way. In Sect. 2, we attempt to introduce most of the kinds of EPCA that have been considered, including linear and nonlinear, retarded, advanced, neutral, and mixed. Basic definitions of solution will be given, connections with difference equations will be shown, and theorems on existence, uniqueness, representation, and stability will in some typical cases be given. In Sect. 3, we derive specific conditions for stability as it depends on the "delay" for a class of scalar equations. In Sect. 4, we survey some of the work that has been done on oscillation and existence of periodic solutions.

The numerical approximation of differential equations can give rise to EPCA in a natural way, although it is unusual to take this point of view. For example, the simple Euler scheme for an ordinary differential system $x'(t) = f(x(t))$ has the form $x_{n+1} - x_n = hf(x_n)$, where $x_n = x(nh)$ and h is the step size. This is equivalent to the EPCA

$$x'(t) = f(x([t/h]h)) \ . \tag{3}$$

In Sect. 5, we take this point of view and present some theorems on the approximation of differential delay equations by EPCA.

Another potential application of EPCAs is in the stabilization of hybrid control systems with feedback delay. By a hybrid system we mean one with a continuous plant and with a discrete (sampled) controller. Some of these systems may be described by EPCA. This problem is presently being investigated by Mr. Gregg Turner at the Claremont Graduate School.

Section 6 contains additional comments, including a description of equations of alternating type, and more information concerning the above chaotic equation.

2 Existence, uniqueness, representation, stability

An equation in which $x'(t)$ is given by a function of x evaluated at t and at arguments $[t], \ldots, [t-k]$ where k is a non-negative integer may be called of retarded or delay type. If the arguments are t and $[t+1], \ldots, [t+k]$, the equation is of advanced type. If both these types of argument appear in the equation, it may be called of mixed type. If the derivative of highest order appears at t and at another point, the equation is generally said to be of neutral type. The equations may, of course, be linear or nonlinear. Equations of retarded type have been studied by Cooke and Wiener [9], [12], Aftabizedeh, et. al. [6] and Gyori and Ladas [16]. Equations of advanced, neutral, and mixed types were investigated by Shah and Wiener [25], Cooke and Wiener [10], and others. Equations of alternating type have also been investigated and these will be described later in Sect. 6.

We now describe how an initial value problem may be posed and solved for the following linear equation of mixed type, since it provides a simple framework for understanding more complicated problems.

$$x'(t) = Ax(t) + \sum_{j=-N}^{N} A_j x\,[[t+j]] \ , \tag{4}$$

$$x(j) = c_j, \quad -N \le j \le N-1 \ , \tag{5}$$

where $[\cdot]$ designates the greatest-integer function, A and A_j are constant $r \times r$ - matrices, and x and c_j are r-vectors. Following [10], we say that a solution of (4)-(5) on $(-\infty, \infty)$ is a vector $x(t)$ that satisfies the conditions: (i) $x(t)$ is continuous on $(-\infty, \infty)$; (ii) the derivative $x'(t)$ exists everywhere, with the possible exception of the points $[t]$, where one-sided derivatives exist; (iii) Equations (4)-(5) are satisfied on each interval $[n, n+1)$, with integer n.

In contrast to the situation with general functional differential equations, the fact that (4)-(5) contains both retarded and advanced arguments does not pose particular difficulties. We shall now prove Theorem 1, which generalizes Theorem 2.4 in [10]. First, we define some notation. Let

$$M_0(t) = e^{At} + [e^{At} - I]A^{-1}A_0 \ , \tag{6}$$

$$M_j(t) = [e^{At} - I]A^{-1}A_j, \quad j = \pm 1, \pm 2, \ldots, \pm N \tag{7}$$

$$B_1 = M_1(1) - I, \quad B_j = M_j(1), \ j \neq 1 \tag{8}$$

Theorem 1. *If the matrices A and $B_{\pm N}$ defined in* (8) *are nonsingular, then problem* (4)-(5) *has a unique solution on* $(0, \infty)$. *This solution cannot grow to infinity faster than exponentially.*

Proof. Let $x_n(t)$ be a solution of (4)-(5) on the interval $[n, n+1)$. If we let $c_n = x(n)$, for integer n, then we have the equation

$$x_n'(t) = Ax_n(t) + \sum_{j=-N}^{N} A_j c_{n+j} \tag{9}$$

with the solution

$$x_n(t) = e^{A(t-n)}c_n + \left[e^{A(t-n)} - I\right] A^{-1} \sum_{j=-N}^{N} A_j c_{n+j}$$

which can be written, by virtue of (6)-(7), as

$$x_n(t) = \sum_{j=-N}^{N} M_j(t-n)c_{n+j} \ . \tag{10}$$

From (10) we see that it suffices to know the constants c_n in order to determine $x(t)$. Taking into account that $x_n(n+1) = x_{n+1}(n+1) = c_{n+1}$, we obtain

$$c_{n+1} = \sum_{j=-N}^{N} M_j(1) c_{n+j}, \quad n \geq 0 . \tag{11}$$

This equation takes the form

$$\sum_{j=-N}^{N} B_j c_{n+j} = 0 . \tag{12}$$

Its particular solution is sought as $x_n = \lambda^n k$, where k is a nonzero vector, then

$$\det \left[\sum_{j=0}^{2N} B_{-N+j} \lambda^j \right] = 0 . \tag{13}$$

Equation (13) has $2Nr$ nontrivial solutions if $\det B_{-N} \neq 0$ and $\det B_N \neq 0$. Assuming that these roots are simple we write the general solution of (12)

$$c_n = \sum_{j=1}^{2Nr} \lambda_j^n k_j , \tag{14}$$

with constant vectors k_j each of which depends on the corresponding value λ_j and contains one arbitrary scalar factor. These factors can be found from the initial conditions in (5). Letting $n = -N, \ldots, N-1$ and $c_n = x(n)$ in (4)-(5), we get a system of equations with Vandermonde's determinant $\det[\lambda_j^i]$ which is different from zero. Hence, the unknown vectors k_j are uniquely determined by initial conditions (5). If some roots of (13) are multiple, then

$$c_n = \sum_{j=1}^{n} \lambda_j^n p_j(n), \qquad m \leq 2Nr$$

where the components of the vectors $p_j(n)$ are polynomials of degree not exceeding $m-1$. Substituting the vectors c_n in (10) yields the solution of problem (4) on $n \leq t \leq n+1$. This concludes the proof.

Remark. Another way to interpret Theorem 2.1 is as follows. We see that the values $c_n = x(n)$ satisfy an autonomous difference equation, and so there is an underlying discrete dynamical system that determines the solution $x(t)$ for real t.

Remark. The requirement that B_{-N} be nonsingular is nonessential. If B_{-N} is singular, solution (10) on $(0, \infty)$ depends on, at most $2N-1$ initial conditions $x(j) = c_j$, $-(N-1) \leq j \leq N-1$.

If B_{-N} is nonsingular, the solution of problem (4)-(5) has a unique backward continuation on $(-\infty, 0)$. From (10) and (14) it follows that the solution $x = 0$ of (4) is globally asymptotically stable as $t \to +\infty$ if and only if the roots of (13) satisfy the inequalities

$$|\lambda_j| < 1 \quad j = 1, \ldots, 2Nr \ . \tag{15}$$

Theorem 2. *The problem*

$$x'(t) = Ax(t) + \sum_{j=-N}^{N} A_j x\,[[t+j]] + f(t), \quad N \geq 2 \tag{16}$$

$$x(j) = c_j, \quad j = 0, 1, \ldots, 2N-1 \tag{17}$$

has a unique solution on $[2N-1, \infty)$ *if the matrices* A *and* $B_{\pm N}$ *are nonsingular and the vector* $f(t)$ *is locally integrable on* $[0, \infty)$.

The proof has been given in [10] and employs the formula

$$x_n(t) = \sum_{j=-N}^{N} M_j(t-n) c_{n+j} + \int_n^t e^{A(t-s)} f(s) ds \tag{18}$$

for the general solution of (16) on the interval $n \leq t < n+1$, where the $M_j(t)$ are defined in (6)-(7).

The following results have also been established in [10].

Theorem 3. *All solutions of (16) are bounded on* $[2N-1, \infty)$ *if* $f(t)$ *is bounded on* $[0, \infty)$ *and inequalities (15) hold true.*

Theorem 4. *All solutions of (16) tend to zero as* $t \to +\infty$ *if the roots of (13) satisfy (15) and* $\lim f(t) = 0$ *as* $t \to +\infty$. *The solutions of (16) cannot grow faster than exponentially if* $f(t)$ *has the same property.*

If the coefficients of (4) are variable matrices of $t \in [0, \infty)$, its solution $x_n(t)$ on $n \leq t < n+1$ satisfying the condition $x_n(n) = c_n$ is given by the expression

$$x_n(t) = U(t) \left[U^{-1}(n) c_n + \int_n^t U^{-1}(s) \sum_{j=-N}^{N} A_j(s) c_{n+j} ds \right] , \tag{19}$$

where $U(t)$ is the solution of the matrix equation

$$U'(t) = A(t)U(t), \quad U(0) = I \ . \tag{20}$$

In [10], this representation was derived and used to obtain sufficient conditions in order that solutions tend to zero as t approaches ∞.

In concluding this section, we point out that existence-uniqueness theorems for nonlinear EPCA have been proved in [9,10,28]. Logistic equations with piecewise constants arguments have been explored in [7,13]. The study of differential inequalities with piecewise constant argument and their application to EPCA was initiated in [1,3]. Continued fractions appeared useful in the theoretical and numerical analysis of EPCA [29].

3 Stability as a function of delay

In this section, we consider stability of the null solution of equations of the form

$$x'(t) = ax(t) + \sum_{j=0}^{N} a_j x\left(\left[\frac{t}{\alpha r}\right]\alpha r - j\alpha r\right) \tag{21}$$

where a and a_j are real constants, $r > 0$, and α is a positive parameter. Special cases of (21) may arise in hybrid systems where there is a time delay r in applying feedback control. For the case $r = \alpha = 1$, this equation was studied in [9]. We now direct attention to the way in which stability depends on αr, as well as on the coefficients.

Let $x_n(t)$ denote a solution on the interval $[n\alpha r, (n+1)\alpha r)$. On this interval, we have

$$x_n'(t) = ax_n(t) + \sum_{j=0}^{N} a_j x_n((n-j)\alpha r) \tag{22}$$

which has the solution

$$x_n(t) = x_n(n\alpha r)e^{a(t-n\alpha r)} + (e^{a(t-n\alpha r)} - 1)a^{-1}\sum_{j=0}^{N} a_j x_n((n-j)\alpha r) \tag{23}$$

Let $x(n\alpha r) = c_n$. Then for continuity at $(n+1)\alpha r$ we obtain

$$c_{n+1} = c_n e^{a\alpha r} + (e^{a\alpha r} - 1)a^{-1}\sum_{j=0}^{N} a_j c_{n-j} \tag{24}$$

Stability for this difference equation is governed by the characteristic equation

$$f(\lambda, \alpha r, A) \equiv \lambda^{N+1} - B_0(\alpha)\lambda^N - \ldots - B_N(\alpha) = 0 \tag{25}$$

where

$$B_j(\alpha) = (e^{\alpha r a} - 1)a^{-1}a_j \quad (j = 1, \ldots, N).$$

$$B_0(\alpha) = e^{\alpha r a} + (e^{\alpha r a} - 1)a^{-1}a_0.$$

We now prove the following theorem

Theorem 5. *Let $r > 0$ and assume*

$$(H_1) \qquad a + \sum_{j=0}^{N} a_j < 0.$$

Then there exists a maximal interval $(0, \alpha_0)$, with $0 < \alpha_0 \leq \infty$, such that all roots of $f(\lambda, \alpha r, A)$ lie in $|\lambda| < 1$ for $\alpha \in (0, \alpha_0)$, and therefore the zero solution of (21) is asymptotically stable.

Proof. Since $B_j(0) = 0$, $B_0(0) = 1$, (25) has, at $\alpha = 0$, an N-fold root at $\lambda = 0$ and a simple root at $\lambda = 1$. Consider the root $\lambda(\alpha)$ such that $\lambda(0) = 1$. This root depends continuously on α and is a differentiable function of α for small α. We have

$$(N+1)\lambda^N \frac{d\lambda}{d\alpha} - N\lambda^{n-1}B_0(\alpha)\frac{d\lambda}{d\alpha} - B_0'(\alpha)\lambda^N - \sum_{j=1}^{N} B_j'(\alpha)\lambda^{N-j} - \sum_{j=1}^{N}(N-j)\lambda^{N-j-1}B_j(\alpha)\frac{d\lambda}{d\alpha} = 0.$$

Now

$$B_j'(\alpha) = ra_j e^{\alpha ra}, \qquad B_0'(\alpha) = r(a+a_0)e^{\alpha ra},$$
$$B_j'(0) = ra_j, \qquad B_0'(0) = r(a+a_0).$$

Setting $\alpha = 0, \lambda = 1$, we obtain

$$\frac{d\lambda}{d\alpha}|_{\alpha=0} = r\left[a + \sum_{j=0}^{N} a_j\right].$$

By hypothesis (H_1), this is negative. Consequently, $\lambda(\alpha) < 1$ and $\lambda(\alpha)$ is real for all sufficiently small positive α. Thus all roots of $f(\lambda, \alpha r, A) = 0$ lie in $|\lambda| < 1$ for sufficiently small positive α. Since B_0 and B_j depend continuously on α, so do the roots, and the existence of the maximal α_0 follows.

Corollary 6. *Assume that* (H_1) *holds and also*

$$(H_2) \qquad f(\lambda, \alpha r, A) \neq 0, \quad \text{for } |\lambda| = 1, \quad 0 < \alpha < \infty\ .$$

Then (21) is asymptotically stable for every positive α.

Indeed, since (H_1) holds, there exists α_0 such that all roots of $f(\lambda, \alpha r, A) = 0$ lie inside the unit circle for $0 < \alpha < \alpha_0$. By (H_2), no root reaches or crosses the unit circle as α increases, and hence there is asymptotic stability for $0 < \alpha < \infty$.

Corollary 7. *Assume that* (H_1) *holds and also*

$$(H_3) \qquad f(\lambda, \alpha r, A) \neq 0, \quad \text{for } |\lambda| = 1.\ \ 0 < \alpha \leq 1.$$

Then (21) is asymptotically stable for $0 < \alpha \leq 1$, *and, in particular, for (21) with* $\alpha = 1$.

In [12], the "first-order" equation with $N = 0$ was examined and the stability region in the (a, a_0) parameter space was precisely described and was compared with the stability region for the first order differential-difference equation with constant lag r. The "second-order" equation with $N = 1$ was also investigated and a set of (a, a_0, a_1) found for which there is asymptotic stability for every positive r.

4 Oscillatory and periodic solutions

The oscillatory and asymptotic behavior of solutions of differential equations with deviating argument has been the subject of many recent investigations. Of particular importance, however, has been the study of oscillations which are caused by the argument deviations and which do not appear in the corresponding ordinary differential equation. Although some oscillatory properties of EPCA were mentioned in [9,10,25,27], the first consistent attempt in this direction was made in [5]. For scalar equations with constant coefficients, examination of the related difference equation can yield results. The following theorem was proved in [25].

Theorem 8. *In each interval* $(n, n+1)$ *with integral endpoints the solution of the equation*

$$x'(t) = ax(t) + a_0 x([t]) + a_1 x([t+1])$$

with the condition $x(0) = c_0 \neq 0$ *has precisely one zero*

$$t_n = n + \frac{1}{a} \ln \frac{a_0 + a_1 e^a}{a + a_0 + a_1}$$

if

$$\left[a_0 + \frac{ae^a}{e^a - 1}\right]\left[a_1 - \frac{a}{e^a - 1}\right] > 0. \tag{26}$$

If (26) is not satisfied and $a_0 \neq -ae^a/(e^a - 1)$, *the solution has no zero in* $[0, \infty)$.

The paper [5] deals with the oscillatory properties of solutions of the scalar first order EPCA

$$x'(t) + a(t)x(t) + p(t)x([t]) = 0 \ , \tag{27}$$

and

$$x'(t) + a(t)x(t) + q(t)x([t+1]) = 0 \ , \tag{28}$$

where $a(t), p(t)$, and $q(t)$ are continuous on $[0, \infty)$. Sufficient conditions under which equations (27) and (28) have oscillatory solutions are given. These conditions are the "best possible" in the sense that when a, p, and q are constants, these conditions reduce to

$$p > \frac{a}{(e^a - 1)} \text{ and } q < \frac{-ae^a}{(e^a - 1)}$$

which are necessary and sufficient conditions. The proof of these conditions depends on establishing the following results on differential inequalities.

Theorem 9. *Consider the delay differential inequality*

$$x'(t) + a(t)x(t) + p(t)x([t]) \leq 0 \ , \tag{29}$$

where $a(t)$ *and* $p(t)$ *are continuous on* $[0, \infty)$. *Assume that*

$$\limsup_{n \to \infty} \int_n^{n+1} p(t) \exp\left[\int_n^t a(s)ds\right] dt > 1 \ . \tag{30}$$

Then (29) has no eventually positive solution.

Theorem 10. *If (30) is satisfied, the delay differential inequality*

$$x'(t) + a(t)x(t) + p(t)x([t]) \geq 0$$

has no eventually negative solution.

From these theorems it follows that subject to (30), (27) has oscillatory solutions only. Note that when $a(t)$ and $p(t)$ are constants (30) reduces to

$$p > \frac{a}{(e^a - 1)}$$

which is sharp. In the same fashion, the condition

$$\liminf_{n\to\infty} \int_n^{n+1} q(t) \exp\left[-\int_t^{n+1} a(s)ds\right] dt < -1 \ . \tag{31}$$

is sufficient to show that (28) has oscillatory solutions only. If $a(t)$ and $q(t)$ are constants, (31) becomes

$$q < \frac{-ae^a}{(e^a - 1)}$$

which is sharp, according to (26).

Oscillatory and periodic solutions of the equation

$$x'(t) + a(t)x(t) + b(t)x([t-1]) = 0 \ , \tag{32}$$

where $a(t)$ and $b(t)$ are continuous on $[0, \infty)$, have been studied in [6]. In particular, suppose that $b(t) > 0$, and for $t \geq 0$

$$\limsup_{n\to\infty} \int_n^{n+1} b(t) \exp\left[\int_{n-1}^{t} a(s)ds\right] dt > 1 \ ,$$

then (32) has oscillatory solutions only. Another condition of this kind is obtained in the following:

Theorem 11. *Assume that*

$$\liminf_{n\to\infty} \left[\exp\left[\int_n^{n+1} a(s)ds\right]\right] \cdot \liminf_{n\to\infty} \left[\int_n^{n+1} b(t) \exp\left[\int_n^{t} a(s)ds\right] ds\right] > \frac{1}{4} \ . \tag{33}$$

Then (32) has oscillatory solutions only.

Similar results for higher order equations have not yet been found.

If $a(t)$ and $b(t)$ are constants, i.e.

$$x'(t) + ax(t) + bx([t-1]) = 0 \ , \tag{34}$$

then the solution of (34) with the initial conditions $x(-1) = c_{-1}$, $x(0) = c_0$ can be written as

$$x(t) = c_n e^{-a(t-n)} + \frac{b}{a}(e^{-a(t-n)} - 1)c_{n-1} \ ,$$

for $t \in [n, n+1), n = 0, 1, 2, \ldots$, and

$$c_n = \frac{[\lambda_1^{n+1}(c_0 - \lambda_2 c_{-1}) - (c_0 - \lambda_1 c_{-1})\lambda_2^{n+1}]}{\lambda_1 - \lambda_2} ,$$

where λ_1, λ_2 are the roots of

$$\lambda^2 - e^{-a}\lambda + \frac{b}{a}(1 - e^{-a}) = 0 .$$

If these roots are real, assume $\lambda_1 > \lambda_2$. Equation (34) has no oscillatory solution if either of the following hypotheses holds true:

1. $b < 0$ and $c_0 - \lambda_2 c_{-1} \neq 0$, or
2. $0 < b \leq \frac{ae^{-a}}{4(e^a - 1)}$.

Observe that any one of these conditions implies λ_1 and λ_2 are real. Furthermore, if $b < 0$ and $c_0 = \lambda_2 c_{-1}$, then (34) has oscillatory solutions. Hence, we conclude that a necessary and sufficient condition for the solutions of (34) to be oscillatory is either

$$b > \frac{ae^{-a}}{4(e^a - 1)} \tag{35}$$

or

$$b < 0 \text{ and } c_0 = \lambda_2 c_{-1} . \tag{36}$$

This proves that (33) is the "best possible" in the sense that when a and b are constants, it reduces to (35)-(36) which is sharp. If either

$$\frac{ae^{-a}}{4(e^a - 1)} < b < \frac{ae^a}{e^a - 1}$$

or

$$-a < b < 0 \text{ and } c_0 = \lambda_2 c_{-1}$$

takes place, then every oscillatory solution of (34) tends to zero as $t \to +\infty$.

Equation (34) exhibits unusually interesting properties concerning the existence of periodic solutions. Some results in this direction may be found in [6].

Theorem 12. *Assume $b > 0$, and let k be a positive integer. Then every oscillatory solution of (34) is periodic of period k if and only if*

$$b = \frac{ae^a}{e^a - 1} \text{ and } a = -\ln\left[2\cos\frac{2\pi m}{k}\right], \tag{37}$$

where m and k are relatively prime and $m = 1, 2, \ldots, [\frac{k-1}{4}]$.

Theorem 13. *Let $b < 0$ and $c_0 = \lambda_2 c_{-1}$. Then every oscillatory solution is periodic of period 2 if and only if*

$$b = -\frac{a(e^a + 1)}{e^a - 1} .$$

Theorem 14. *For given c_0 and c_{-1}, the set of all equations of type (34) having periodic solutions is countable.*

The above results were obtained with the implicit assumption $a \neq 0$. If $a = 0$, then (37) becomes

$$b = \lim_{a \to 0} \frac{ae^a}{e^a - 1} = 1 \ .$$

In this case (34) with $a = 0$ and $b = 1$ has periodic solutions of period 6.

In conclusion, we list a few properties of (4)-(5) with constant scalar coefficients concerning the existence of oscillatory solutions.

Theorem 15. *Suppose that $x(t)$ is a solution of (4) such that $x(n) \neq 0$ and $x(n+1) \neq 0$. Then $x(t)$ has a zero in $(n, n+1)$ if and only if $x(n)$ and $x(n+1)$ have different signs, and this zero is unique.*

Proof. If $x(n+1)/x(n) < 0$, it is clear that $x(t)$ has a zero in $(n, n+1)$. Assuming the existence of several zeros implies that between any two consecutive zeros there is a point ξ such that $x'(\xi) = 0$. On the given interval (4) becomes

$$x'(t) = ax(t) + \sum_{j=-N}^{N} a_j c_{n+j} \ ,$$

and differentiating this relation successively yields

$$x^{(k)}(\xi) = 0, \quad k \geq 2 \ .$$

Since $x(t)$ is analytic in $(n, n+1)$, it is constant which is impossible. On the other hand, if $x(n+1)/x(n) > 0$ and we assume that $x(t_n) = 0$ for some $t_n \in (n, n+1)$, then we must conclude that either $x'(t_n) = 0$ or that $x(t)$ has another zero in $(n, n+1)$. In both cases, $x(t)$ is constant.

Theorem 16. *If all roots of (13) are positive, equation (4) has no nontrivial oscillatory solution.*

Proof. We have

$$\frac{x(n+1)}{x(n)} = \frac{c_{n+1}}{c_n} = \frac{\sum_{j=1}^{2N} k_j \lambda_j^{n+1}}{\sum_{j=1}^{2N} k_j \lambda_j^{n}} \ ,$$

and since not all of the coefficients k_j equal zero, then

$$\lim_{n \to \infty} (c_{n+1}/c_n) = \lambda_m > 0$$

where λ_m is a root of (13). Hence, $c_{n+1}/c_n > 0$ for large n, and the proof follows from Theorem 15.

Theorem 17. *If all roots of (13) are negative, every solution of (4) is oscillatory and has a unique zero in each interval $[n, n+1)$, for large n.*

Oscillatory properties of n-dimensional systems (4) have been studied in [18] and [21], where it is shown that every solution of (4) oscillates (componentwise) if and only if its characteristic equation has no positive roots. Oscillatory properties of some classes of nonlinear EPCA may be found in [16], where it is shown that under appropriate hypotheses a nonlinear EPCA oscillates if and only if an associated linear equation oscillates. Stability and oscillation of neutral EPCA with both constant and piecewise constant delays have been investigated in [24]. These properties for a second order EPCA alternately of retarded and advanced type were also explored in [22]. The characteristic equation for linear EPCA with both constant and piecewise constant delays was discussed in [14]. The forthcoming book [17] contains a chapter devoted to EPCA. The study of oscillations in systems of EPCA was originated for the first time in [30].

5 Approximation of equations with discrete delay

Equations with piecewise constant delay can be used to approximate delay differential equations that contain discrete delays. For example, consider the scalar equation, with one constant delay τ,

$$x'(t) = -p(t)x(t-\tau), \quad t \geq 0 \tag{38}$$

with the intial condition

$$x(t) = \phi(t), \quad -\tau \leq t \leq 0 \tag{39}$$

where ϕ is a given function in $C = C([-\tau, 0], \mathbb{R})$. Let k be any positive integer and let $h = \tau/k$. We may approximate this problem by the EPCA

$$y'(t) = -p(t)y\left(\left[\frac{t}{h} - [\frac{\tau}{h}]\right] h\right) \tag{40}$$

with initial condition

$$y(nh) = \phi(nh), \qquad n = -k \ldots, 0 \ . \tag{41}$$

Moreover, by using the methods already described we find that $y(nh) = c_n$ satisfies the difference equation

$$c_{n+1} - c_n = -\int_{nh}^{(n+1)h} p(u) du c_{n-k} \tag{42}$$

$$c_n = \phi(nh), \qquad n = -k, \ldots, 0 \ . \tag{43}$$

It was proved in [15] that the solution of (40), (41) provides a uniformly good approximation to the solution of problem (38), (39) on any compact interval $[0, T]$, $T > 0$. This result was extended in [8] to convergence on $t \in (0, \infty)$, in case the zero solution of (38) is exponentially stable. Specifically, the following theorem was proved. Let $x(t_0, \phi)$ denote the solution of (38) with initial condition set on $[t_0 - \tau, t_0]$. Assume that there exist constants L and α such that for every $t_0 \geq 0$ and continuous ϕ

$$|x(t_0, \phi)(t)| \leq L\|\phi\|e^{-\alpha(t-t_0)}, \quad t \geq t_0 \tag{44}$$

where $\|\phi\|$ denotes the supremum norm of ϕ.

Theorem 18. *(a) For every $h = \tau/k$ ($k \geq 1$) and every ϕ in C, there exist constants $k, M_1(\phi, h) \geq 0$, and $M_2(\phi) \geq 0$ such that $M_1(\phi, h)$ tends to zero as $h \to 0$, and such that the solutions $x(\phi)(t)$ of (38), (39) and $y(t)$ of (40), (41) satisfy*

$$|x(\phi)(t) - y(\phi)(t)| \leq \{M_1(\phi, h) + htM_2(\phi)\}e^{-(\alpha - Kh)t} \tag{45}$$

for $t \geq 4\tau$.

(b) Assume the additional hypothesis that ϕ is differentiable on $[-\tau, 0]$ and satisfies the consistency condition $\phi'(0-) = -p(0)\phi(-\tau)$. Then there exists a constant $M_3(\phi) > 0$ such that for $t \geq 0$

$$|x(\phi)(t) - y(\phi)(t)| \leq \{M_3(\phi) + tM_2(\phi)\}he^{-(\alpha - Kh)t} \ . \tag{46}$$

Thus, if h is small enough, $y(t)$ provides a uniform approximation on $[0, \infty)$ and has an exponential decay rate very close to that of $x(t)$.

This result generalizes easily to the case of any finite number of discrete delays $\tau_1, \ldots, \tau_m$, and to systems of equations.

6 Concluding remarks

Differential equations of the form

$$x'(t) = f\left(x(t), x([t + \frac{1}{2}])\right) \tag{47}$$

have stimulated considerable interest and have been studied in [2,11,20,28]. In these equations, the argument deviation $T(t) = t - \left[t + \frac{1}{2}\right]$ changes its sign in each interval $n - \frac{1}{2} < t < n + \frac{1}{2}$ (n integer). Indeed, $T(t) < 0$ for $n - \frac{1}{2} < t < n$ and $T(t) > 0$ for $n < t < n + \frac{1}{2}$, which means that the equation is alternately of advanced and retarded type. This complicates the asymptotic behavior of the solutions, generates two essentially different conditions for oscillations in each interval $\left[n - \frac{1}{2}, n + \frac{1}{2}\right]$, and leads to interesting properties of periodic solutions.

In Sect. 1, we mentioned the equation

$$x'(t) = ax(t)(1 - x([t])), \quad x(0) = c_0 \ , \tag{48}$$

where $a > 0$, $c_0 > 0$. This equation may be regarded as a semi-discretization of the ordinary logistic equation, but its solutions display a much greater variety of dynamics. In fact, if we let $c_n = x(n)$ for integer n, the continuity of $x(t)$ at each integer implies the discrete difference equation

$$c_n = c_{n-1}e^{a(1-c_{n-1})}, \quad n = 1, 2, \ldots \ . \tag{49}$$

Since the function $f(x) = xe^{a(1-x)}$ is a C^1 unimodal map on $[0, \infty)$, it is not difficult to establish that there are period-doubling bifurcations, and the whole Sarkovskii sequence of periodic solutions and chaotic behavior [7]. It is intriguing to note that, in contrast, the semi-discretization

$$x'(t) = ax(t)(1 - x([t + \frac{1}{2}])) \tag{50}$$

has the property that the equilibrium $x(t) \equiv 1$ is asymptotically stable for all $a > 0$.

EPCA have only been studied for a few years, and it is not yet clear how important they may become in theory or application. At the least, they provide us with a new class of hereditary and anticipatory differential equations and with the challenge of working out the properties of these equations. The problems studied so far are closely related to ordinary difference equations and indeed have stimulated new work on these. In each of the areas touched on above – existence, asymptotic behavior, periodic and oscillating solutions, and approximation – there appears to be ample opportunity for extending the known results.

References

1. Aftabizedeh, A.R., Wiener, J. (1988): Oscillatory and periodic solutions for systems of two first order linear differential equations with piecewise constant argment. Applicable Analysis. **26**, 327-333
2. Aftabizedeh, A.R., Wiener, J. (1987): Oscillatory and periodic solutions of advanced differential equations with piecewise constant argument. In Nonlinear Analysis and Applications. V. Lakshmikantham, ed. Marcel Dekker Inc., N.Y. and Basel, 31–38
3. Aftabizedeh, A.R., Wiener, J. (1987): Differential inequalities for delay differential equations with piecewise constant argument. Applied Mathematics and Computation **24**, 183–194
4. Aftabizedeh, A.R., Wiener, J. (1986): Oscillatory and periodic solutions of an equation alternately of retarded and advanced type. Applicable Analysis **23**, 219–231
5. Aftabizadeh, A.R., Wiener, J. (1985): Oscillatory properties of first-order linear functional differential equations. Applicable Analysis **20**, 165–187
6. Aftabizedeh, A.R., Wiener, J., Xu, J-M. (1987): Oscillatory and periodic solutions of delay differential equations with piecewise constant argument. Proc. Amer. Math. Soc. **99**, 673–679
7. Carvalho, L.A.V., Cooke, K.L. (1988): A nonlinear equation with piecewise continuous argument. Differential and Integral Equations **1**, 359–367
8. Cooke, K.L., Gyori, I. (1990): Numerical approximation of the solutions of delay differential equations on an infinite interval using piecewise constant arguments. Inst. Math. and Its Applications, IMA Preprint Series # 633
9. Cooke, K.L., Wiener, J. (1984): Retarded differential equations with piecewise constant delays. J. Math. Anal. Appl. **99**, 265–297
10. Cooke, K.L., Wiener, J. (1987): Neutral differential equations with piecewise constant argument. Bollettino Unione Matematica Italiana **7**, 321–346

11. Cooke, K.L., Wiener, J. (1987): An equation alternately of retarded and advanced type. Proc. Amer. Math. Soc. **99**, 726–732
12. Cooke, K.L., Wiener, J. (1986): Stability regions for linear equations with piecewise continuous delay. Comp. & Math. with Appls. **12A(6)**, 695-701
13. Gopalsamy, K., Kulenovic, M.R.S., Ladas, G.: On a logistic equation with piecewise constant arguments. Preprint
14. Grove, E.A., Gyori, I., Ladas, G.: On the characterisitc equation for equations with piecewise constant arguments. Preprint
15. Gyori, I.: On approximation of the solutions of delay differential equations by using piecewise constant arguments. Intern. J. Math. Math. Sci. (in press)
16. Gyori, I., Ladas, L. (1989): Linearized oscillations for equations with piecewise constant argument. Differential and Integral Equations **2**, 123–131
17. Gyori, I., Ladas, G.: Oscillation Theory of Delay Differential Equations with Applications. Oxford University Press (to appear)
18. Gyori, I., Ladas, G., Pakula, L.: Conditions for oscillation of difference equations with application to equations with piecewise constant arguments. Preprint
19. Hale, J.K. (1977): Theory of Functional Differential Equations. Springer-Verlag, New York-Heidelberg-Berlin
20. Huang, Y.K. (1989): On a system of differential equations alternately of advanced and delay type. In Differential Equations and Applications I A.R. Aftabizadeh, Ed. Ohio University Press, Athens, 455–465
21. Ladas, G. (1989): Oscillations of equations with piecewise constant mixed arguments. In Differential Equations nad Applications. II Aftabizadeh, Ed. Ohio University Press. Athens, 64–69
22. Ladas, G., Partheniadis, E.C., Schinas, J.: Oscillation and stability of second order differential equations with piecewise constant argument. Preprint
23. Myshkis, A.D. (1977): On certain problems in the theory of differential equations with deviating argument. Uspekhi Mat. Nauk. **32**, 173–202
24. Partheniadis, E.C. (1988): Stability and oscillation of neutral delay differential equations with piecewise constant argument. Differential and Integral Eqautions **1**, 459–472
25. Shah, S.M., Wiener, J. (1983): Advanced differential equations with piecewise constant argument deviations. Internat. J. Math. & Math.. Sci **6(4)**, 671-703
26. Wiener, J. (1983): Differential equations with piecewise constant delays. In Trends in the Theory and Practice of Nonlinear Differential Equations. V. Lakshmikantham, Ed. Marcel Dekker, New York, 547–552
27. Wiener, J. (1984): Pointwise initial-value problems for functional differential equations. In Differential Equations. I.W. Knowles and R.T. Lewis, Eds. North-Holland, New York, 571–580
28. Wiener, J., Aftabizadeh, A.R. (1988): Differential equations alternately of retarded and advanced type. J. Math. Anal. Appl. **129**, 243–255
29. Wiener, J., Shah, S.M. (1987): Continued fractions arising in a class of functional differential equations. Journ. Math. Phys. Sci. **21**, 527–543
30. Wiener, J., Cooke, K.L. (1989): Oscillations in systems of differential equations with piecewise constant argument. J. Math. Anal. Appl. **137**, 221-239

Dynamics and Delays

Jack K. Hale

Center for Dynamical Systems and Nonlinear Studies, Georgia Institute of Technology, Atlanta, Georgia 30332

To Kenneth Cooke on his 65th *birthday*

1 Introduction

Delay differential equations or more generally functional differential equations have been studied rather extensively in the past thirty years and are used as models to describe many physical and biological systems. In spite of this fact, there are few examples for which one can describe how the flow defined by the equation changes with the delays. One of these is the class of equations with one delay in which the vector field has negative feedback. In this case, Mallet-Paret (1988) has shown that there is a Morse decomposition of the attractor with the basic sets being given by the number of zeroes of solutions on a delay interval. Further development of these ideas has led to many interesting results on the existence of periodic solutions for large delays. See, for example, Chow and Mallet-Paret (1983), Mallet-Paret and Nussbaum (1986), Chow, Lin and Mallet-Paret (1989), Cao (1989) and the references therein.

When there is more than one delay in the equation, there is very little information about the behavior of solutions. Even the local theory is not complete. The purpose of this paper is to point out some of the recent results dealing with several delays in order to illustrate present and future directions of research.

2 Dynamics on center manifolds

For a given $r > 0$, we let $C = C([-r,0], \mathbb{R}^n)$. If $x : [-r, \alpha) \to \mathbb{R}^n$ is a given continuous function, we denote by $x_t \in C, t \in [0, \alpha)$ the function $x_t(\theta) = x(t+\theta)$ for $\theta \in [-r, 0]$. For a given neighborhood $U \subset C$ of the origin and a given function $f \in C^k(U, \mathbb{R}^n)$ and a given continous linear map $L : C \to \mathbb{R}^n$, we consider the functional differential equation

$$\dot{x} = Lx_t + f(x_t). \tag{1}$$

The eigenvalues of the linear equation

$$\dot{x} = Lx_t \tag{2}$$

are defined to be the solutions of the characteristic equation

$$\det(\lambda I - Le^{\lambda \cdot} I) = 0. \tag{3}$$

We suppose that there are exactly p eigenvalues $\lambda_1, \ldots, \lambda_p$ of (2) on the imaginary axis. If we suppose also that $\|f\|_{C^k(U,\mathbb{R}^n)} < \delta$ with δ sufficiently small, then there is a center manifold $M(L, f)$ of (1). The flow on the center manifold is given by an ordinary differential equation (ODE). In case $n = 1$, the flow on the center manifold is given by p^{th}-order scalar ODE.

Theorem 2.1 (scalar FDE). *The flow on the center manifold M(L,f) is given by a pth-order ODE*

$$y^{(p)} + a_1 y^{(p-1)} + \cdots + a_p y = g_{L,f}(y, y^{(1)}, \ldots, y^{(p-1)}), \tag{4}$$

where $g_{L,0} = 0$ and the eigenvalues of the left hand side of (4) are $\lambda_1, \ldots, \lambda_p$.

The next result states that all polynomial vector fields on center manifolds can be realized by delay differential equations with $p - 1$ delays.

Theorem 2.2 (Scalar delay equation). *For any polynomial $G(z_1, \ldots, z_p)$ of degree q, there are a neighborhood $U \subset C$ of zero, a positive constant δ, real numbers $a_1, \ldots, a_p$, postive constants $r_1, \ldots, r_{p-1}$, and a function $f \in C^{q+1}(U, \mathbb{R})$ with $\|f\|_{C^{q+1}(U,\mathbb{R}^n)} < \delta$ such that the flow on the center manifold of the delay differential equation*

$$\dot{x}(t) = a_1 x(t) + a_2 x(t - r_1) + \cdots + a_p x(t - r_{p-1}) + f(x(t), x(t - r_1), \ldots, x(t - r_{p-1})) \tag{5}$$

in U is given by the ODE

$$y^{(p)} + a_1 y^{(p-1)} + \cdots + a_p y = G(y, y^{(1)}, \ldots, y^{(p-1)}). \tag{6}$$

The proofs of these results as well as those in Remarks 2.3 and 2.4 below may be found in Hale (1985), (1986). Theorem 2.1 is a consequence of elementary properties of control systems treating the function f as the control parameter. The proof of Theorem 2.2 makes use of the Implicit Function Theorem and the linear independence of the eigenfunctions. In Hale (1985), (1986), it is asserted that the function G in Theorem 2.2 may be an arbitrary C^q-function. A careful examination of those papers (as pointed out to the author by P. Poláčik) shows that the proof there holds only for a G polynomial. The general case remains open.

From Theorem 2.2, we see that each flow in the ODE (6) with a polynomial G can be realized by delay differential equations with $p-1$ delays. Thus, complicated dynamics can be expected with several delays and complicated behavior of solutions can be observed in a local neighborhood of an equilibrium point.

Theorem 2.2 asserts that there exist $p-1$ delays. We remark that the delays can be chosen almost arbitrarily; that is, there are few restrictions on the delays and the restrictions that are imposed depend upon the nature of the eigenfunctions associated to the eigenvalues. In the examples, this will be made precise.

Remark 2.3. Is is not known if the number of delays in Theorem 2.2 is optimal. The optimal number must depend upon the eigenvalues on the imaginary axis. For example, if all of the eigenvalues are equal to zero, then we must have $p-1$ delays in order to have the linear part of the delay differential equation to have p zero eigenvalues. If the eigenvalues are distinct and equal to $0, i\omega_j, -i\omega_j, j = 1, \ldots, m$, then we need only m delays to have a linear part with these eigenvalues. On the other hand, it is not known if this number is sufficient to take care of the nonlinear terms in the flow on the center manifold.

Remark 2.4. For $n > 1$, that is, systems of equations, using ideas from control theory with the function f as the control parameter, we can show that Theorems 2.1 and 2.2 remain valid for some special L and f. For general L and f, it is not known how to relate in a systematic way the flow on center manifolds for FDE to that of delay differential equations.

Remark 2.5. Theorems 2.1 and 2.2 remain valid for the neutral FDE

$$\frac{d}{dt}Dx_t = Lx_t + f(x_t)$$

where D is a stable linear operator.

Let us illustrate these results with a few examples.

Example 2.6. Suppose that (2) has a double eigenvalue 0 and no other eigenvalues on the imaginary axis. We can realize this situation with a linear delay differential equation with one delay:

$$\dot{x}(t) = x(t) - x(t-1)$$

Since Theorem 2.2 states that we need only one delay to reproduce all polynomial flows

$$y'' = G(y, y'),$$

we may rescale time to permit the choice of the delay as 1. If we choose

$$G(z_1, z_2) = \lambda_1 z_1 + \lambda_2 z_2 + \alpha z_1^2 + \beta z_1 z_2,$$

then there is a function f such that the flow on the center manifold of the equation

$$\dot{x}(t) = x(t) - x(t-1) + f(x(t), x(t-1))$$

is given by the ODE

$$\dot{z}_1 = z_2, \qquad \dot{z}_2 = \lambda_1 z_1 + \lambda_2 z_2 + \alpha z_1^2 + \beta z_1 z_2.$$

This is the Bogdanov-Takens bifurcation for $\lambda_1 = \lambda_2 = 0$.

Example 2.7. We now consider an equation with two delays, one of which can be taken to be 1 without loss of generality:

$$\dot{x}(t) = -a_0 x(t) - b_0 x(t-1) - c_0 x(t-r) + f(x(t), x(t-1), x(t-r)), \tag{7}$$

where a_0, b_0, c_0, r are chosen so that the only eigenvalues on the imaginary axis are $0, i\omega, -i\omega$ with $\omega > 0$. For a given $\omega > 0$, it is easy to verify that the coefficients a_0, b_0, c_0 are uniquely determined provided that $\sin(r-1)\omega \neq 0$.

Let $\Phi = (\varphi_1, \varphi_2, \varphi_3)$ be a real basis for the eigenfunctions corresponding to the eigenvalues on the imaginary axis. As shown in Hale (1985), if the vectors $\Phi(0), \Phi(-1), \Phi(-r)$ are linearly independent, then it is possible to reproduce any polynomial flow

$$y''' + \omega^2 y' = G(y, y', y'') \tag{8}$$

on the center manifold. It is easily verified that the condition of linear independence is equivalent to

$$\sin(r-1)\omega + \sin\omega - \sin r\omega \neq 0$$

which is satisfied except for a discrete set of values r.

In appropriate coordinates, the normal form for (8) up through terms of order three is given by

$$\begin{aligned} \dot{\theta} &= 1 \\ \dot{\rho} &= \lambda\rho + a\rho y + d\rho^3 \\ \dot{y} &= \beta y + b y^2 + c\rho^2, \end{aligned}$$

which is a famous singularity for $\lambda = \beta = 0$ (see, for example, Chow and Hale 1982).

3 Hopf bifurcation with respect to delays

In this section, we consider the problem of the creation of periodic orbits from an equilibrium point by varying the delays following the approach in Hale (1979). We obtain a bifurcation function which is C^{k-1} in the delays if the vector field is C^k. The stability properties of the periodic orbits are determined from this bifucation function.

Suppose $\alpha \in \Omega \subset \mathbb{R}^k$ is a parameter and consider the equation

$$\dot{x} = L(\alpha)x_t + f(\alpha, x_t) \tag{9}$$

in C, where $L(\alpha)\varphi$ is continuous and linear in φ and the function $f(\alpha, \varphi)$ is C^2 in φ, $f(\alpha, 0), D_\varphi(\alpha, 0) = 0$. We need the following hypothesis:

(H_1) *The characteristic matrix of the linear part of (9)*

$$\Delta(\alpha,\lambda) = \lambda I - L(\alpha)e^{\lambda\cdot}I$$

is a C^1 function of α and there are simple eigenvalues $\lambda_0 = i\nu_0, \bar{\lambda}_0 = -i\nu_0, \nu_0 > 0$, of $\Delta(\alpha_0,\lambda)$ and all other eigenvalues λ_j satisfy $\lambda_j \neq m\lambda_0$ for any integer m.

The next result is almost an immediate consequence of the Implicit Function Theorem.

Lemma 3.1 *Under hypothesis* (H_1), *there is a $\delta > 0$ such that, for $|\alpha - \alpha_0| < \delta$, there is a simple eigenvalue $\lambda(\alpha)$ of $\Delta(\alpha,\lambda)$ which is a C^1-function,* $\mathrm{Im}\lambda(\alpha) > 0$, *and $\lambda(\alpha_0) = i\nu_0$.*

From this lemma and the classical theory of FDE, we can decompose the space C as the sum $C = P_\alpha \bigoplus Q_\alpha$, where P_α, Q_α are subspaces, invariant under the flow defined by the linear equation

$$\dot{x}(t) = L(\alpha)x_t, \tag{10}$$

and $P_\alpha = \mathrm{sp}[\Phi_\alpha], \Phi_\alpha = (\varphi_1, \varphi_2)$, where φ_1, φ_2 are real solutions of the linear equation (10) corresponding to the eigenvalues $\lambda(\alpha), \bar{\lambda}(\alpha)$. The function Φ_α has an explicit representation as

$$\begin{aligned}
\Phi_\alpha(\theta) &= \Phi_\alpha(0)e^{B(\alpha)\theta}, \theta \in [-r, 0], \\
B(\alpha) &= \nu_0 B_0 + B_1(\alpha) \\
B_0 &= \begin{pmatrix} 0 & 1 \\ -1 & 0 \end{pmatrix} \\
B_1(\alpha) &= \begin{pmatrix} (\alpha-\alpha_0)\cdot\zeta(\alpha) & (\alpha-\alpha_0)\cdot\gamma(\alpha) \\ -(\alpha-\alpha_0)\cdot\gamma(\alpha) & (\alpha-\alpha_0)\cdot\zeta(\alpha) \end{pmatrix}.
\end{aligned}$$

The functions ζ and γ are continuously differentiable.

Even though the parameter α may contain the delays in the equation, the smoothness in α in hypothesis (H_1) will be satisfied. On the other hand, not all solutions of (9) are differentiable with respect to α. Therefore, it is not reasonable to assume that the functions $L(\alpha)$ and $f(\alpha,\cdot)$ are differentiable in α. As a consequence, we must be careful in the statement of the Hopf bifurcation theorem. We do know that we will have differentiability in α along periodic solutions of (9), which are the solutions of interest. This is the motivation behind the hypotheses in the next result.

Theorem 3.3 (Hopf bifurcation) *In addition to* (H_1), *suppose that*

(H_2) *For any $K > 0, \varphi \in C$ with $\dot{\varphi} \in C, \|\dot{\varphi}\| \leq K$, the functions $L(\alpha)\varphi, f(\alpha,\varphi)$ are C^1 in α.*

(H_3) $\zeta(\alpha_0) \neq 0$.

Then there is an $\epsilon > 0$ such that, for any $a \in \mathbb{R}, |a| < \epsilon$, there is a C^1-manifold $\Gamma_a \subset \mathbb{R}^k$ of codimension one, Γ_a is C^1 in a,

$$\Gamma_0 = \{\alpha \in \mathbb{R}^k : \mathrm{Re}\lambda(\alpha) = 0, |\alpha - \alpha_0| < \epsilon\}$$

and, for each $\alpha \in \Gamma_a$, there are C^1-functions $\omega(\alpha, a), x^*(\alpha, a)$ with $x^*(\alpha, a)$ being an $\omega(\alpha, a)$-periodic solution of equation (9). Furthermore, $\omega(\alpha_0, 0) = \omega_0 = \frac{2\pi}{\nu_0}, x^*(\alpha_0, 0) = 0$.

Theorem 3.3 states that there is a Hopf bifurcation across the manifold Γ_0. The proof of this theorem is in Hale (1979). As remarked earlier, a scalar bifurcation function $G(\alpha, a)$ is constructed with the property that the periodic orbits of (9) near the origin are in one-to-one correspondance with the zeros of G. Furthermore, if there are only two eigenvalues of (10) for α_0 on the imaginary axis, then the stability properties of the periodic orbit are determined from the stability properties of the corresponding zero a_0 of the scalar differential equation

$$\dot{a} = G(\alpha, a). \tag{11}$$

See, for example, de Oliveira and Hale (1980).

Remark 3.3. If we strenghten the differentiable requirements in hypotheses (H_1) and (H_2), then it is possible to show that the function $G(\alpha, a)$ as well as the functions $\omega(\alpha, a), x^*(\alpha, a)$ have additional differentiability properties with respect to α, a.

Remark 3.4. Hypothesis (H_2) may appear to be difficult to satisfy. However, this in not the case. In fact, if we suppose that $\alpha = (r_1, \ldots, r_p)$ with each $r_j \geq 0$ and

$$f(\alpha, \varphi) = F(\varphi(r_1), \ldots, \varphi(r_p))$$

and the function $F : \mathbb{R}^p \to \mathbb{R}^n$ is C^1, then (H_2) is satisfied.

If F is C^k, then each periodic orbit of (9) has initial data which is C^k. We then can show that the bifurcation function is C^k.

4 Discrete versus distributed delays

In the modeling of physical phenomena where the past history is important, we often assume that the influence of the past occurs at discrete points. If the model is an FDE, this implies that the function f in (1) is given by

$$f(\varphi) = g(\varphi(-r_1), \varphi(-r_2), \ldots, \varphi(-r_p)). \tag{12}$$

On the other hand, this is probably only an approximation to the true situation. The influence of the past should be destributed over an interval $[-r, 0]$ where $r > \max\{r_1, \ldots, r_p\}$. One possible way in which to incorporate this more general situation into the model is to assume that f is given by the relation

$$f(\varphi) = g\left(\int_{-r}^{0} d\eta_1(\theta)\varphi(\theta), \ldots, \int_{-r}^{0} d\eta_p(\theta)\varphi(\theta)\right), \tag{13}$$

where the functions $\eta_1, \ldots, \eta_p$ are functions of bounded variation continuous from the right. If each of the functions η_j is a step function with jump one at r_j, then the functions f in (12) and (13) are the same.

One important problem is to determine when a property that has been observed in an equation with discrete delays persists when the delays are distributed and each distribution function η_j is close to the above mentioned step function. Of course, we must be precise about the concept of closeness. In this situation, we say $\eta^{(k)}$ converges to η as $k \to \infty$ if, for each $\varphi \in C$,

$$\int_{-r}^{0} d(\eta^{(k)} - \eta)\varphi \to 0 \text{ as } k \to \infty.$$

This is equivalent to saying that the total variation of $\eta^{(k)} - \eta$ approaches zero as $k \to \infty$.

In this general setting, the Hopf bifurcation theorem of the previous section applies with the parameter α lying in the dual space of C. For a parital discussion, see Hale (1979).

Due to the complications that arise in the study of distributed delays, we encounter very often the following situtation in the literature. It is first assumed that the past history in infinite. The next step is to assume that some or all of the distribution functions are simple exponential functions or at least of the type that the derivatives are very simple. When appropriate assumptions are made, it is possible to reduce the number of delays by increasing the order of the differential equation. In some cases, this approach leads to a much simpler problem, but it remains to discuss the sensitivity of the obtained results to small perturbations of the distribution function. Very little attention has been devoted to this latter problem although it is obviously important. For the Hopf bifurcation, see Hale (1979). For a stability problem, see Lenhart and Travis (1985). See also Busenberg and Travis (1982), Busenberg and Hill (1988).

5 Hopf bifurcation surface

In Section 3, we have seen that it is possible to consider the Hopf bifurcation with the delays chosen as the bifurcation parameters. Therefore, it is important to know the structure of the curves in the delay space that correspond to the situation where there are eigenvalues on the imaginary axis of the linear variation equation about an equilibrium point. In this section, we reproduce some recent results of Huang (1990) and Bélair and Mackey (1989) for the first Hopf bifurcation curve; that is, the curve for which all eigenvalues have nonpositive real parts. The verification of the results requires many nontrivial computations and we therefore present only the diagrams with a few remarks.

Huang (1990) has considered the characteristic equation

$$\lambda + ae^{-\lambda r_1} + be^{-\lambda r_2} = 0, \tag{14}$$

which corresponds to the linear equation

$$\dot{x}(t) = -ax(t - r_1) - bx(t - r_2). \tag{15}$$

For fixed values of a, b, the first Hopf bifurcation curve takes one of the types shown in Figures 5.1-5.6. In each of these situations except when $b = 0$, there are $r_1^0 < r_1^1$ such that, for fixed $r_1 \in (r_1^0, r_1^1)$, there is a possibility of a Hopf bifurcation as r_2 is varied. For any case and fixed $r_1 \in (r_1^0, r_1^1)$, there can be only a finite number of Hopf bifurcations with respect to the parameter r_2. For the situation that $-b > a > 0 > b$, the origin is unstable for all delays $r_1 \geq 0, r_2 \geq 0$. In each case, except when $a = b > 0$ or $b = 0$, the first Hopf bifurcation curve is not a smooth curve. At the points of discontinuity in the tangent vector to this curve, there are two points of eigenvalues on the imaginary axis. For most values of the constants a, b, these are distinct pairs. Therefore, a complete analysis of the dynamics near the equilibrium of a nonlinear perturbation of (15) cannot be accomplished using only the Hopf bifurcation theorem. In general, we expect very complicated behavior near these points. As $b \to a$, the points of discontinuity of the tangent vector approaches infinity and one arrives at Figure 5.1.

We remark that the dotted curves in these figures also are Hopf bifurcation curves. However, they are not the first Hopf bifurcation curves. There are purely imaginary eigenvalues on these dotted curves, but also there are eigenvalues with positive real parts.

Bélair and Mackey (1989) (see also an der Heiden (1979)) consider the characteristic equation

$$\lambda + \frac{Q}{1 + R\lambda} + e^{-\lambda r} = 0, \tag{16}$$

which corresponds to the characteristic equation of the second order equation with one delay

$$R\ddot{x}(t) + \dot{x}(t) + R\dot{x}(t - r) + Qx(t) + x(t - r) = 0. \tag{17}$$

Equation (16) also is the characteristic equation for the linear equation with one distributed delay and one discrete delay:

$$\dot{x}(t) + \frac{Q}{R} \int_{-\infty}^{0} e^{\theta/R} x(t + \theta) d\theta + x(t - r) = 0. \tag{18}$$

This is a situation where the special form of the distributed delay leads to an equation of higher order with one delay (see Section 4).

The parameter Q can be used as a measure of the strength of the distributed delay. For this reason, the plots of the first Hopf bifurcation curve are made in the (Q, r)-plane with R fixed. The different situations are depicted in Figures 5.6-5.8. As in the situation for equation (14) with two discrete delays, for a fixed value of Q, there are never more than a finite number of Hopf bifurcations. The points of discontinuity in the tangent vectors again correspond to two pairs of eigenvalues on the imaginary axis.

6 Delay differential equations and maps

In recent years, there has been some discussion about the relationship between the dynamics defined by maps and the dynamics defined by delay differential equations with a singular perturbation term (see, for example, Chow and Mallet-Paret 1983, Chow, Diekmann and Mallet-Paret 1985, Mallet-Paret and Nussbaum 1986, Ivanov and Sharkovsky 1990). More specifically, suppose that $\epsilon > 0$ is a small parameter and consider the equation

$$\epsilon\dot{x}(t) + x(t) = f(x(t-\tau)). \tag{19}$$

For $\epsilon = 0$, we have the map $y = f(x)$ on $\mathbb{R}$. Is the dynamics of this map related to equation (19)? For example, if the map has a periodic point of period two, does there exist, for ϵ small, a periodic solution of (19) of period approximately two and is it close to a square wave? Under some conditions on the nonlinear function f, this is known to be true. On the other hand, it is not known to be true in the general case. If we ask the same question for periodic points of f of higher period, then it is known that equation (19) does not behave as the map.

From the above remarks, it seems as if it is not reasonable to expect delay differential equations to behave as maps. However, it is possible that the question is not posed in the proper way. The left hand side of (19) may be considered as an "approximation" to the function $x(t+\epsilon)$:

$$x(t+\epsilon) = x(t) + \epsilon\dot{x}(t) + \cdots. \tag{20}$$

If it is a good approximation, then the appropriate map for comparison should be

$$x(t+\epsilon) = f(x(t-\tau)). \tag{21}$$

Since ϵ is a parameter which we want to vary, the equation (21) no longer can be considered as a map on $\mathbb{R}$, but must be considered in a function space. If we want to compare this map to a delay differential equation, we could consider the function space to be $C = C([-r,0],\mathbb{R})$ with r chosen to be greater than $\tau + \epsilon_0$ and $0 < \epsilon \leq \epsilon_0$. Of course, to have the solutions of (21) remain in C, we must impose restrictions on the initial data. Even if a solution remains in C, its limit as $t \to \infty$ may not be in C. For example, suppose that f has three fixed points $a < b < c$ with a, c asymptotically unstable and b unstable and every orbit of f has its limit set as one of these points. If $\phi \in C$ has values which lie in both of the intervals (a,b) and (b,c), then the limit of the solution x_t through φ will be a step function. The same remark holds for the orbit of functions $\varphi \in C$ that approach a period two orbit of the map f. As a consequence, we must exercise care in the selection of the space in which (21) and (22) will be compared. In our discussion to follow, we attempt to make these remarks somewhat more precise.

Let us proceed first by taking a better approximation to $x(t+\epsilon)$. More precisely, for a given integer N, we consider the equation

$$\left(\frac{\epsilon}{N}\frac{d}{dt} + 1\right)^N x(t) = f(x(t-\tau)). \tag{22}$$

For N sufficiently large, the operator

$$A_N^{-1} = \left(\frac{\epsilon}{N}\frac{d}{dt} + 1\right)^{-N} \tag{23}$$

should be a good approximation to the translation semigroup $e^{-\epsilon d\backslash dt}$ evaluated at $-\epsilon$ acting on uniformly continuous functions on $\mathbb{R}$ and taking $x(t)$ to $x(t+\epsilon)$.

From this remark, it is therefore reasonable to expect that we should be able to compare the solutions of (22) which are defined and bounded on $\mathbb{R}$ with the solutions of (21) which are defined and bounded on $\mathbb{R}$. If we suppose that the corresponding semigroups are dissipative (that is, orbits of bounded sets are bounded and each orbit evetually enters and remains in a fixed bounded set), then these globally defined and bounded solutions are precisely the ones that belong to the global attractors. We recall that the global attractor is a compact invariant set with the property that the ω-limit set of any bounded set belongs to it.

Let $L^p = L^p((-r,0),\mathbb{R}^n)$ for some $1 \leq p \leq \infty$. We choose the space of initial data for (21) to be L^p and the space of initial data for (22) to be $L^p \times \mathbb{R}^n$. Let P_{L^p} be the projection of $L^p \times \mathbb{R}^n$ onto L^p. For a fixed $\epsilon, 0 < \epsilon \leq \epsilon_0$, and a fixed integer N, let $\mathcal{A}_{\epsilon,N}$ be the global attractor for (22) in $L^p \times \mathbb{R}^N$. Let us suppose that (21) has a maximal bounded invariant set $\mathcal{A}_\epsilon$ in L^p. An important problem is to compare the flow of (21) on $\mathcal{A}_\epsilon$ with the flow of (22) on $\mathcal{A}_{\epsilon,N}$. The first interesting thing to do would be to show that

$$\text{dist}_{L^p}(P_{L^p}\mathcal{A}_{\epsilon,N}, \mathcal{A}_\epsilon) \to 0 \quad \text{as} \quad N \to \infty. \tag{24}$$

At first sight, it appears that it would be impossible to consider relation (24) because the space of initial data for (22) is $L^p \times \mathbb{R}^N$. However, this is not the case. Let us suppose that the function f is analytic. Then the elements of $\mathcal{A}_{\epsilon,N}$ are analytic functions (see Nussbaum (1973)). Therefore, the elements of $\mathcal{A}_{\epsilon,N}$ are determined uniquely by their projections onto L^p, and to attempt to verify (24) is meaningful. Relation (24) is a first step in the comparison of the size of attractors of (22) with the maximal bounded invariant set of (21). If we show that $||\mathcal{A}_{\epsilon,N}||_{L^p}$ is uniformly bounded in N, then we should be able to use the stability properties of the attractor $\mathcal{A}_\epsilon$ and the appropriate variation of constants formula to prove that (24) is true. We remark that this in not the usual type of problem that is encountered when semigroups are compared on infinite time intervals since the semigroup for (22) is compact for $t \geq 1$ and the semigroup for the limit equation (21) is not compact.

Of course, the next step is to discuss the corresponding flows in more detail. For example, if we suppose that $\mathcal{A}_\epsilon$ contains a periodic orbit of period $2(\tau + \epsilon)$ which is stable, is there a periodic solution of (22) for N large which is periodic with a period close to $2(\tau + \epsilon)$? Other more complicated compact invariant sets in $\mathcal{A}_\epsilon$ also should be considered.

We remark that, with the emphasis being placed upon the parameter N (the order of the differential delay equation), the parameter ϵ will play a minor role in much of the analysis. Of course, many interesting problems will arise when $N \to \infty$ and $\epsilon \to 0$.

We also should attempt the same analysis for equations with several delays: for example, for a nonlinear function

$$F(\varphi) = f(\varphi(-r_1), \varphi(-r_2)).$$

We are in the process of making these remarks more rigorous.

In recent papers of Vallée, Dubois, Côte and Delisle (1987) and Vallée and Marriott (1989), equation (22) has been used as models of hybrid bistable optical devices when N components contribute to the total reponse time. The parameter ϵ^{-1} is a measure of the response time. The models are tested against experimental results and there is particular interest in the comparison in the chaotic regions. Also, are these models reasonable for explaining the onset of chaos? In the chaotic region, the ratio of the delay to the response time does not seem to be large and this had led to discrepencies between the model (19) ($N = 1$) and the experimental observations. In the several component model, the above authors have investigated (22) numerically and found better agreement with the experiments when $N > 1$.

They also analyzed precisely the linear variational equation near the equilibrium and have shown that the curve for the first Hopf bifurcation becomes insensitive to the delay for large values of N. We reproduce these results since they are relatively simple and serve as a good motivation for the comparison of (21) and (22) as outlined above. In the physical problem, the function f depends upon a parameter μ which can vary. The characteristic equation for the linear variational equation has the form

$$\left(1 + \frac{\epsilon\lambda}{N}\right)^N + B_\mu e^{-\lambda r} = 0, \tag{25}$$

where B_μ is the derivative of f evaluated at the equilibrium. For N large this equation is almost the same as the equation

$$e^{(\epsilon + r)\lambda} + B_\mu = 0. \tag{26}$$

From this equation, for N large, we see that the threshold for the existence of a Hopf bifurcation (that is, purely imaginary solutions of equation (25)) is essentially independent of N since it requires that the parameter μ be chosen so that B_μ is approximately -1. Equation (26) is the linear variational equation for (21) and thus the expected correspondance of dynamics.

We remark that other authors also have considered approximating the translation semigroup $e^{\epsilon d \backslash dt}$ to higher order in an attempt to obtain better corespondence between the delay differential equation and the map (see Zhao, Wang and Huo 1988, Longtin 1988).

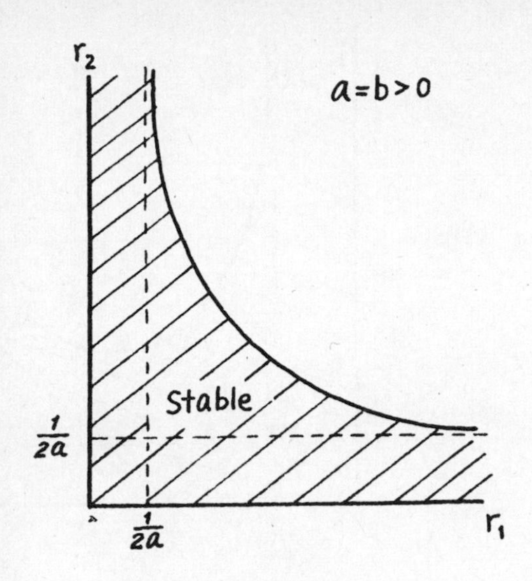

Fig. 5.1. $a = b > 0$

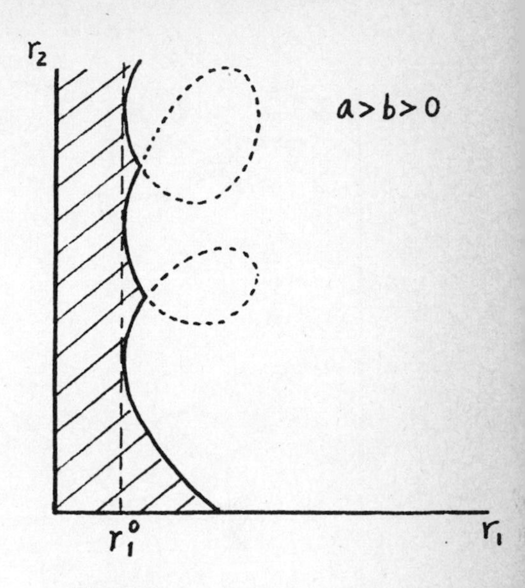

Fig. 5.2. $a > b > 0$

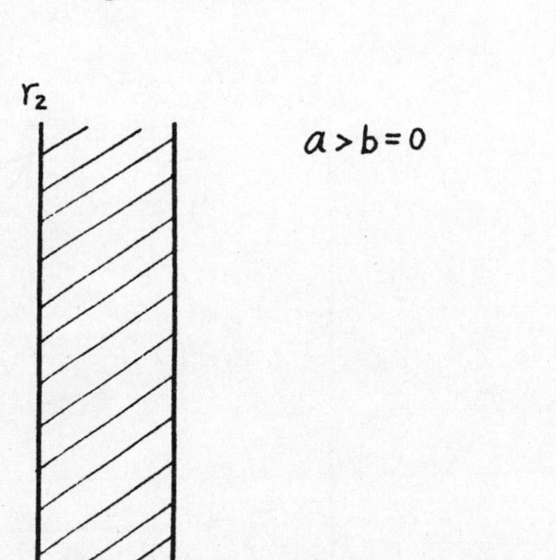

Fig. 5.3. $a > b = 0$

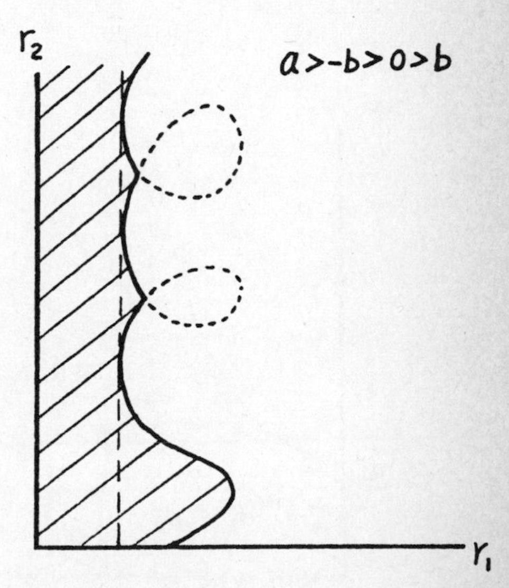

Fig. 5.4. $a > -b > 0 > b$

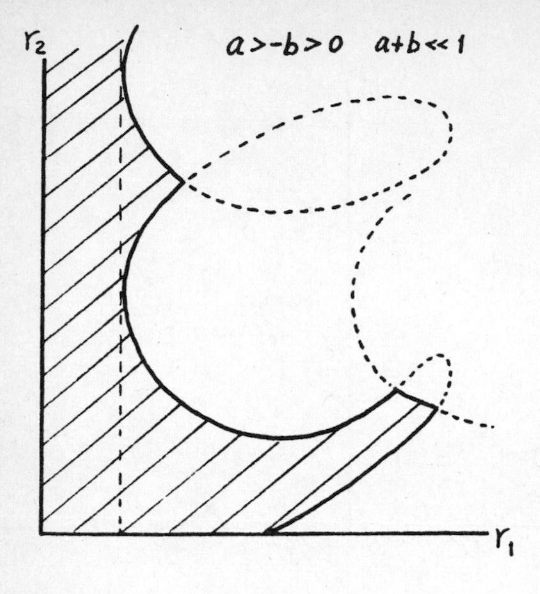

Fig. 5.5. $a > -b > 0 \quad a + b \ll 1$

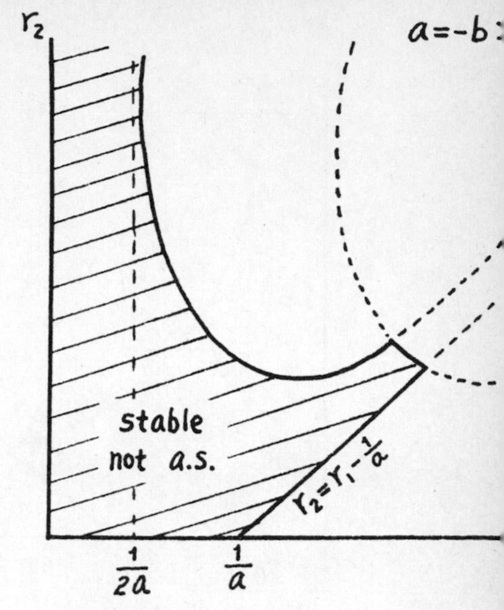

Fig. 5.6. $a = -b > 0$

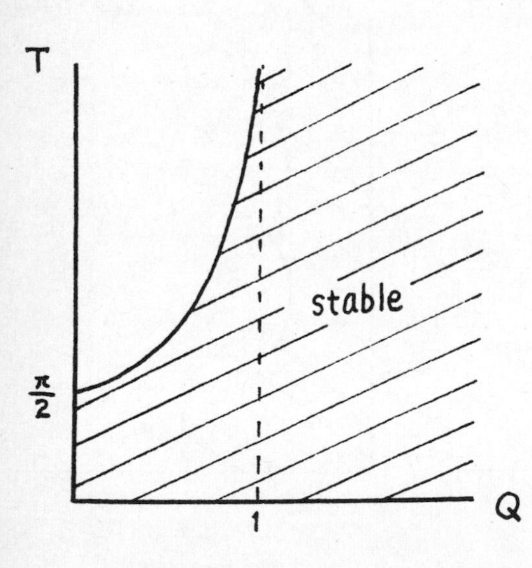

Fig. 5.7. $0 < R < \sqrt{2} - 1$

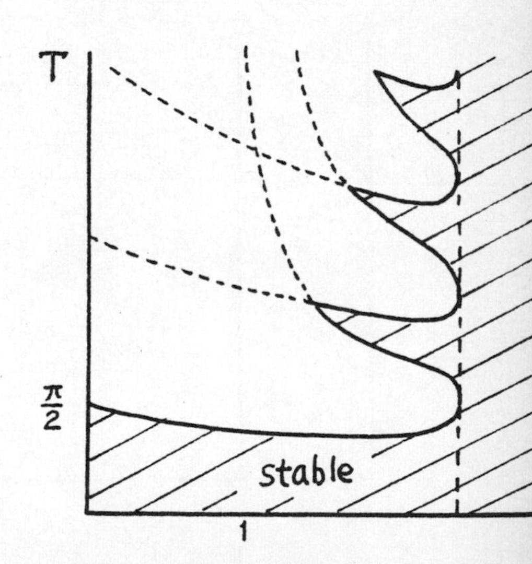

Fig. 5.8. $\sqrt{2} - 1 < R < 1$

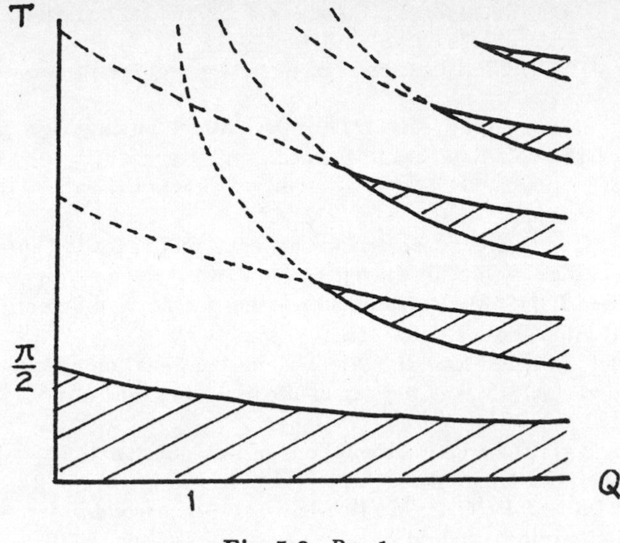

Fig. 5.9. $R > 1$

References

1. an der Heiden, U. (1979): Periodic solutions of second order differential equations with delay. J. Math. Anal. Appl. **70**, 599-609
2. Bélair, J., Mackey, M.C. (1989): Consumer memory and price fluctuations in commodity markets: an integrodifferential model. J. Dyn. Diff. Equations **1**, 299-325
3. Busenberg, S., Travis, C.C. (1982): On the use of reducible functional differential equations. J. Math. Anal. Appl. **89**, 46-66
4. Busenberg, S., Hill, T. (1988): Construction of differential equation approximations to delay differential equations. Appl. Anal. **31**, 35-56
5. Cao, Y. (1989): The discrete Lyapunov function for scalar differential delay equations. J. Differential Equations. To appear
6. Chow, S.-N., Diekmann, O., Mallet-Paret, J. (1985): Multiplicity of symmetric periodic solutions of a nonlinear Volterra integral equation. Japan J. Appl. Math. **2**, 433-469
7. Chow, S.-N., Hale, J.K. (1982): Methods of Bifurcation Theory. Springer-Verlag
8. Chow, S.-N., Lin, X.-L., Mallet-Paret, J. (1989): Transition layers for singularly perturbed delay differential equations with monotone nonlinearities. J. Dynamics Diff. Eqns. **1**, 3-43
9. Chow, S.-N., Mallet-Paret, J. (1983): Singularly perturbed delay differential equations. In Coupled Oscillators (Eds. J. Chandra and A. Scott), North-Holland, 7-12
10. de Oliveira, J.C., Hale, J.K. (1980): Dynamic behavior from the bifurcation function. Tôhoku Math. J. **32**, 577-592
11. Hale, J.K. (1979): Nonlinear oscillations in equations with delays. Lect. Appl. Math. **17**, 157-185. Am. Math. Soc.
12. Hale, J.K. (1985): Flows on centre manifolds for scalar functional differential equations. Proc. Royal Soc. Edinburgh **101A**, 193-201

13. Hale, J.K. (1986): Local flows for functional differential equations. Contemporary Mathematics **56**, 185-192. Am. Math. Soc.
14. Huang, W. (1990): Global geometry of the stable regions for two delay differential equations. To appear
15. Ivanov, A.F., Sharkovsky, A.N. (1990): Oscillations in singularly perturbed delay equations. Dynamics Reported. To appear
16. Lenhart, S.N., Travis, C.C. (1985): Stability of functional partial differential equations. J. Differential Equations **58**, 212-227
17. Longtin, A. (1988): Nonlinear oscillations, noise and chaos in neural delayed feedback. Ph. D. Thesis, McGill University, Montreal
18. Mallet-Paret, J. (1988): Morse decomposition for delay differential equations. J. Differential Equations. **72**, 270-315
19. Mallet-Paret, J., Nussbaum, R. (1986): Global continuation and asymptotic behavior for periodic solutions of a delay differential equation. Ann. Math. Pura Appl. **145**, 33-128
20. Nussbaum, R. (1973): Periodic solutions of analytic functional differential equations are analytic. Mich. Math. J. **20**, 249-255
21. Vallée, R., Dubois, P., Côté, M., Delisle, C. (1987): Second-order differential delay equation to describe a hybrid bistable device. Phys. Rev. A **36**, 1327-1332
22. Vallée, R., Marriott, C. (1989): Analysis of an Nth-order nonlinear differential delay equation. Phys. Rev. A **39**, 197-205
23. Zhao, Y., Wang, H., Huo, Y. (1988): Anomalous behavior of bifurcation in a system with delayed feedback. Phys. Letters A **133**, 353-356

A Brief Survey of Persistence in Dynamical Systems

Paul Waltman

Department of Mathematics and Computer Science, Emory University, Atlanta, Georgia 30322

1 Introduction

The equations governing interacting populations in a closed environment often take the form

$$x_i' = x_i f_i(x_1, x_2, \ldots, x_n) \tag{1.1}$$
$$x_i(0) = x_{i0} \geq 0, \quad i = 1, 2, ..., n$$

To avoid technical conditions assume that f is such that solutions of initial value problems are unique and extend to all of R. The form of the equations makes the positive cone invariant and the coordinate axes and the bounding faces invariant (and represent lower order dynamical systems).

The notion of persistence attempts to capture the idea that if the above differential equation represents a model ecosystem, all components of the ecosystem survive. In this survey we attempt to show how the idea has led to an interesting class of abstract mathematical problems which have applications in biology. While the emphasis is on the mathematical problems, the references give an introduction to the applications.

The system (1.1) is said to be *persistent* if

$$\liminf_{t \to \infty} x_i(t) > 0, \quad i = 1, 2, ..., n$$

for every trajectory with positive initial conditions.

The system (1.1) is said to be *uniformly persistent* if there exists a positive number α such that

$$\liminf_{t \to \infty} x_i(t) \geq \alpha, \quad i = 1, 2, ..., n$$

for every trajectory with positive initial conditions. The term persistent was first (?) used in this context by Freedman and Waltman [12], with lim sup instead of lim inf. Other definitions are relevant, see Freedman and Moson [8] for a discussion. Earlier use of the term persistence corresponded to a stability condition. See Harrison [22] and the references cited therein. The work of [12] was continued in [13] and [14] where three level food chains and competition problems were investigated. Other work in this spirit includes Hallam and Ma [23], Hutson and Law [31], Freedman and So [10] [11], Freedman and Rai [9], Li and Hallam [38], and So [42]. A similar notion appears in Hofbauer [25], and Schuster, Sigmund, and Wolf [40], where the term cooperative is used. (This later became permanence.)

One can distinguish two approaches:

i) Analysis of the flow on the boundary

ii) Use of a Liapunov-like function

Each approach has lead to interesting mathematical problems. This survey will try to point out examples of theorems in each approach. The interest has been focused on three-population models since this is the simplest case of interest but Kirlinger [36] [37] and So [42] have higher dimensional examples. The survey article of Hutson and Schmitt [33] and the book of Hofbauer and Sigmund [28] contain more complete references.

A key tool in some of the applications using the first method was the following lemma:

Lemma Butler-McGehee *Let P be an isolated hyperbolic rest point in the omega limit set,* $\omega(x)$*, of a point x. Then either* $\omega(x)=\{P\}$ *or there exists points* q_1 *and* q_2 *in* $\omega(x)$ *with* q_1 *and* q_2 *different from P but with* $q_1 \in M^+(P)$ *and* $q_2 \in M^-(P)$ *where* M^+ *and* M^- *are the stable and unstable manifolds of P.*

The central difficulty in extending to the general situation is to develop the counterpart of this lemma in the setting of dynamical systems. Those using the Liapunov-like technique have to construct the appropriate function. The key to extending this idea to more general situations was to develop a theory with as few hypotheses as possible on the Liapunov function. Each of these techniques is explored in an abstract setting in the following sections.

2 An abstract persistence theorem

The general setting is that of topological dynamics in a metric space. We review the basic definitions and set up the dynamical system appropriate for systems of the form (1.1). Let X be a locally compact metric space with metric d and let E be a closed subset of X with boundary ∂E and interior E^0. Let π be a dynamical system defined on E which leaves its boundary invariant. (A set B in X is said to be invariant if $\pi(B,t)=B$.) Denote the flow on the boundary by π_∂. The flow is said to be *dissipative* if for each $x \in E, \omega(x)$ is not empty and there exists a bounded set G such that the invariant set $\Omega = \cup_{x\in E}\omega(x)$ lies in G. A nonempty invariant subset M of X is called an *isolated invariant set* if it is the maximal invariant set of a neighborhood of itself. Such a neighborhood is called an *isolating neighborhood.*

The *stable (or attracting)* set of a compact invariant set A is denoted by W^+ and is defined as

$$W^+(A) = \{x | x \in X,\ \omega(x) \neq \phi,\ \omega(x) \subset A\}.$$

The *unstable set*, W^- is defined by

$$W^-(A) = \{x | x \in X,\ \alpha(x) \neq \phi,\ \alpha(x) \subset A\}$$

where $\alpha(x)$ is the alpha limit set.

The weakly stable and unstable sets are defined as

$$W_w^+(A) = \{x | x \in X,\ \omega(x) \neq \phi,\ \omega(x) \cap A \neq \phi\}$$

and

$$W_w^-(A) = \{x | x \in X,\ \alpha(x) \neq \phi,\ \alpha(x) \cap A \neq \phi\}.$$

Lemma. *(Butler and Waltman [4], Dunbar, Rybakowski, and Schmitt [6]) Let M be a compact isolated invariant set for π defined on a locally compact metric space. Then for any $x \in W_w^+(M) \backslash W^+(M)$ it follows that $\omega(x) \cap W^+(M) \backslash M \neq \phi$ and $\omega(x) \cap W^-(M) \backslash M \neq \phi$. A similar statement holds for $\alpha(x)$.*

The following set of definitions are motivated by the technique used in the proof in [FW2],[FW3]. Unfortuately some of the terms are the same as those in the work of Conley [5] in dynamical systems where they have a different meaning. This overlap is regrettable but the definitions are now well established in both places. Let M,N be isolated invariant sets (not necessarily distinct). M is said to be *chained* to N, written $M \to N$, if there exists an element x, $x \notin M \cup N$ such that $x \in W^-(M) \cap W^+(N)$. A finite sequence $M_1, M_2, \ldots, M_k$ of isolated invariant sets will be called a *chain* if $M_1 \to M_2 \to \ldots \to M_k$ ($M_1 \to M_1$, if k=1.) The chain will be called a *cycle* if $M_k = M_1$.

π will be said to be *persistent* if for all $x \in E^0$, $\liminf_{t\to\infty} d(\pi(x,t), \partial E) > 0$ and π will be said to be *uniformly persistent* if there exists an ϵ_0 such that for all $x \in E^0$, $\liminf_{t\to\infty} d(\pi(x,t), \partial E) \geq \epsilon_0 > 0$. π_∂ is said to be isolated if there exists a covering M=$\cup_{i=1}^k M_i$ of $\Omega(\pi_\partial)$ by pairwise disjoint, compact, isolated invariant sets $M_1, M_2, \ldots, M_k$ for π_∂ such that each M_i is also an isolated invariant set for π. (This is a sort of "hyperbolicity" assumption; for example, it prevents interior rest points, or other invariant sets, from accumulating on the boundary.) M is called an *isolated covering*. π_∂ will be called *acyclic* if there exists some isolated covering M=$\cup_{i=1}^k M_i$ of π_∂ such that no subset of the M_i's forms a cycle. An isolated covering satisfying this condition will be called acyclic.

The following theorem provides a criterion for uniform persistence in terms of the flow on the boundary.

Theorem. *[4],[3] Let π be a continuous dynamical system on a locally compact metric space E with invariant boundary. Assume that π is dissipative and the boundary flow π_∂ is isolated and is acyclic with acyclic covering M. Then π is uniformly persistent if and only if*

$$(H) \quad \textit{for each } M_i \in M,\ W^+(M_i) \cap E^0 = \phi.$$

The acyclic condition in the abstract persistence theorem is suggestive of a Morse decomposition. This idea appears in the work of Freedman and So [FS2], who were interested in persistence for maps and in the work of Hofbauer [Ho2] and Garay [15]. We sketch a portion of Garay's paper.

Let S denote the maximal compact invariant set in E and consider π restricted to $S \cap \partial E$, denoted by $\pi_{S\cap\partial E}$. The collection M=$\{M_1, M_2, \ldots, M_n\}$ is a *Morse decomposition* if the M_i's are pairwise disjoint, compact, isolated invariant sets for $\pi_{S\cap\partial E}$ with the property that for each $x \in S \cap \partial E$ there are integers $i(x)$ and $j(x)$ with $i \leq j$ and $\omega(x) \subset M_i$ and $\alpha(x) \subset M_j$ and if i=j, the $x \in M_i = M_j$. Note that the above definition makes a requirement on all x while the chained definition makes a requirement for some x. Garay generalized the above persistence theorem in the spirit of Conley. We state the following result (which is equivalent to the above theorem) to show the nature of the setting. It is a corollary of Garay's main theorem.

Theorem. *Let X be a locally compact metric space and let E be a closed subset of X. Suppose we are given a dissipative dynamical system π on E for which ∂E is invariant. Let M=$\{M_1, M_2, \ldots, M_n\}$ be a Morse decomposition for $\pi_{S\cap\partial E}$. Further assume that for each i*

i) There exists a $\gamma > 0$ such that the set $\{x \in int(E) | d(x, M) < \gamma\}$ contains no entire trajectory

ii)$Int(E) \cap W^+(M_i) = \phi$

Then π is uniformly persistent.

To show that this is equivalent to the above theorem, Garay shows that "Any acyclic covering of $\Omega(\pi_{\partial E})$ is a Morse decomposition for $\pi_{S\cap\partial E}$ and conversely." The first condition is a type of "hyperbolicity" requirement. Garay's main result requires the above two conditions only on a more restricted set.

3 An example

Consider three logistic-like competing populations

$$x_i' = x_i f_i(x_1, x_2, x_3) \quad i = 1, 2, 3.$$

By competitive, one means that $\partial f_i / \partial x_j < 0$ for $i \neq j$. By logistic-like, one intends that there exist numbers K_i, i=1,2,3, such that whenever $x_i = K_i$ for some i and all of the x_j's, $j \neq i$, are zero, then f_i is zero and $\partial f_i / \partial x_i < 0$. Thus there are three rest points, called *axial rest points*, denoted by E_i, i.e., $E_1 = (K_1, 0, 0)$ and the origin, denoted by E_0. Suppose that the boundary has only one additional rest point (x_1^*, x_2^*) which we will label E^* and that $f_3(x_1^*, x_2^*, 0) > 0$. We are supposing that x_1 outcompetes x_3 and that x_3 outcompetes x_2. (Other rest points could be allowed. We are making the example as simple as possible.) We suppose also that all rest points are hyperbolic.

One must check the acyclic condition and (H). Because of the invariance of the dynamical systems on the boundary, the stable and unstable manifolds of the axial rest points all lie in the boundary. Moreover, planar competitive systems do not have limit cycles, so E^* is a global attractor in the $x_1 - x_2$ plane. In particular, the stable manifold of E^* connects to the origin and to infinity. Moreover, the origin is unstable. Hence if the covering is taken to be the five rest points, there are no cycles.

The linearization of E^* takes the form

$$\begin{bmatrix} x_1\frac{\partial f_1}{\partial x_1} & x_1\frac{\partial f_1}{\partial x_2} & x_1\frac{\partial f_1}{\partial x_3} \\ x_2\frac{\partial f_2}{\partial x_1} & x_2\frac{\partial f_2}{\partial x_2} & x_2\frac{\partial f_2}{\partial x_3} \\ 0 & 0 & f_3 \end{bmatrix}$$

where everything is evaluated at the rest point. The matrix decomposes and the positive eigenvalue given by $f_3 > 0$ has an eigenvector pointing into the positive cone – hence (H) is satisfied. The example is pivotal – if one removes the requirement for the existence of E^*, an acyclic covering may not exist (for example, if x_2 outcompetes x_1). The system may or may not be uniformly persistent. The following example was given by May and Leonard [39]. (See also Schuster, Sigmund, and Wolf [41].)

$$\begin{aligned} x_1' &= x_1(1 - x_1 - \alpha x_2 - \beta x_3) \\ x_2' &= x_2(1 - \beta x_1 - x_2 - \alpha x_3) \\ x_3' &= x_3(1 - \alpha x_1 - \beta x_2 - x_3). \end{aligned} \tag{3.1}$$

If $0 < \beta < 1 < \alpha$ and $\alpha + \beta > 2$, then (H) is satisfied but there is no acyclic covering. The abstract persistence theorem does not apply and, indeed, $\limsup_{t\to\infty} x_i(t) = 1$ and $\liminf_{t\to\infty} x_i(t) = 0$. The Liapunov approach, however, can yield information.

4 The Liapunov approach

The Liapunov approach has appeared in various forms, for example as "persistence functions", "average Liapunov functions", etc. The principal contributors include Gard and Hallam [18], Gard, Hallam, Svoboda [19], Gard [16][17], Hutson and Vickers [34], Hutson[30], Jansen [35], Hofbauer, Hutson, and Jansen [27], Kirlinger [36][37], Hofbauer and So [29], and Schuster, Sigmund, and Wolf [41].

The nicest statement may be one due to Fonda, [7], who stated his result in terms of repellers. (See also [29].) Let X and π be as above. A subset S of X is said to be a *uniform repeller* if there exists an $\eta > 0$ such that for all $x \in X\backslash S, \liminf_{t\to\infty} d(\pi(x,t), S) > \eta$. In terms of the original definition for the ordinary differential equations, the system is uniformly persistent if the boundary is a uniform repeller.

Theorem. *Let S be a compact subset of X such that $X\backslash S$ is positively invariant. A necessary and sufficient condition for S to be a uniform repeller is that there exists a neighborhood U of S and a continuous function $P:X \to R^+$ satisfying:*

i) $P(x)=0$ if and only if x is in S

ii) For all $x \in U \backslash S$ there is a $T_x > 0$ such that $P(\pi(x, T_x)) > P(x)$.

Although nicely stated the condition appears difficult to verify. The following corollary is easier (and is essentially the work of Hutson and Hofbauer).

Corollary. *Let S be as above and $P \in C(X, R^+) \cap C^1(X \backslash S, R)$ be such that $P(x)=0$ if and only if x is in S and let there exist a lower semicontinuous function $\Psi : X \to R$, bounded below, and an $\alpha \in [0,1]$, such that*
i) $P'(x) \geq [P(x)]^\alpha \, \Psi(x)$, $x \in X \backslash S$
ii) For all $x \in \Sigma$, $\sup_{T>0} \int_0^T \Psi(\pi(x,s)ds > 0$
where Σ denotes S or, whenever S is positively invariant, $\bar{\Omega}(S)$. Then S is a uniform repeller.

If one takes $P(x)=x_1x_2x_3$ and $\Psi(x) = \sum_{i=1}^{3} f_i(x)$, then a short computation shows that if $\alpha + \beta < 2$, then the system of May and Leonard is uniformly persistent.

5 Persistence for dynamical systems on non-locally compact space

The preceding result would apply to dynamical systems generated by autonomous ordinary differential equations models. However, the models of population ecology encompass much broader systems, so it is of interest to have an even more general theory. Results in this direction may be found in Dunbar, Rybakowski, and Schmitt [6], Hutson and Moran [32], and Burton and Hutson [2], who consider persistence either for delay equations or reaction-diffusion equations. The previously cited paper [33] gives an account of work in this direction.

Hale and Waltman [24] attempt to recover the abstract persistence theorem in a setting appropriate for delay and reaction-diffusion equations. The general idea is easy; replace the dynamical system by a semi-dynamical system and remove the local compactness hypothesis. However, certain technical difficulties arise, in particular, the lack of backward orbits or the nonuniqueness of backward orbits. The general tool is the theory of dissipative systems, as found, for example, in the book of Hale [21]. The key idea is to work on the "global attractor". Several of the definitions need to be modified to take advantage of this setting.

As before, a complete metric space, denoted by X, (with metric d) is the basic setting. The dynamical system is replaced by a C^0-semigroup on X. Let T(t):$X \to X$, $t \geq 0$, satisfy
i) T(0)=I,
ii) T(t+s)=T(t)T(s) for t,s$\geq$ 0,
iii) T(t)x is continuous in t,x.
The semigroup T(t) is said to be *asymptotically smooth* if for any bounded subset B of X, for which T(t)B$\subset$ B for t$\geq$ 0, there exists a compact set K such that

$d(T(t)B, K) \to 0$ as $t \to \infty$. The semigroup T(t) is said to be *point dissipative in* X if there is a bounded nonempty set B in X such that, for any $x \in X$, there is a $t_0 = t_0(x, B)$ such that T(t)x∈B for t≥ t_0.

A basic result on the existence of global attractors is contained in the following statement, [21].

Theorem. *If*

i) T(t) is asymptotically smooth,

ii) T(t) is point dissipative in X,

iii) $\gamma^+(U)$ is bounded in X if U is bounded in X,

then there is a nonempty global attractor A in X.

Assume that the metric space X is the closure of an open set X^0; that is, $X = X^0 \cup \partial X^0$, where ∂X^0 (assumed to be nonempty) is the boundary of X^0. Suppose that the C^0-semigroup T(t) on X satisfies

$$T(t) : X^0 \to X^0$$

$$T(t) : \partial X^0 \to \partial X^0$$

and let $T_0(t) = T(t)|_{X^0}$, $T_\partial = T(t)|_{\partial X^0}$. The set ∂X^0 is a complete metric space (with metric d). If T(t) satisfies the above conditions then T_∂ will satisfy the same conditions in ∂X^0 and there will be a global attractor A_∂ in ∂X^0.

The difficulty with backward orbits requires some modification of the original definitions. A set B in X is said to be *invariant* if T(t)B=B for t≥ 0; that is, the mapping T(t) takes B onto B for each t≥ 0. This implies that there is a negative orbit through each point of an invariant set. To denote the alpha limit set of a specific orbit through the point x, we use $\alpha_\gamma(x)$.

The *stable set* of a compact invariant set A is as defined before

$$W^+(A) = \{x | x \in X,\ \omega(x) \neq \phi,\ \omega(x) \subset A\}.$$

The *unstable set*, W^-, requires modification and is defined by

$$W^-(A) = \{x | x \in X, \text{ there exists a backward orbit } \gamma^-(x) \text{ such that}$$

$$\alpha_\gamma(x) \neq \phi,\ \alpha_\gamma(x) \subset A\}.$$

The other definitions including weakly stable and unstable sets require no modification. Let $\tilde{A}_\partial$ denote the set of omega limit points of the flow of the boundary.

Theorem. *Suppose T(t) satisfies the basic assumptions and*

i) T(t) is asymptotically smooth,

ii) T(t) is point dissipative in X,

iii) $\gamma^+(U)$ is bounded in X if U is bounded in X,

iv) $\tilde{A}_\partial$ is isolated and has an acyclic covering $M = \{M_1, M_2, \ldots, M_k\}$.

Then T(t) is uniformly persistent if and only if for each $M_i \in M$

$$W^+(M_i) \cap X^0 = \phi.$$

The above theorem may be applied to a system of delay equations which model competing populations. Consider the system

$$x'(t) = r_1 x(t)[1 - x(t-1) - \mu_1 y(t)]$$

$$y'(t) = r_2 y(t)[1 - y(t-1) - \mu_2 x(t)]$$

where r_i, i=1,2, are sufficiently small. If, in addition, μ_1 and μ_2 are less than one, the system is persistent. The appropriate space X is the positive cone of $C[-1,0] \times C[-1,0]$. For any pair of initial functions $(\phi, \psi) \in X$, let $x(t, \phi, \psi)$ be a solution and define $T(t)(\phi, \psi) \in X$, $t \geq 0$, by

$$T(t)(\phi,\psi)(\theta) = (x(t+\theta,\phi,\psi), y(t+\theta,\phi,\psi)), \quad -1 \leq \theta \leq 0.$$

Straightforward computations [24] show that the conditions of the theorem are satisfied.

A more significant example may be found in Thieme and Castillo-Chavez [43] who used the result to show that the infectives persist in a model of HIV/AIDS. The model is that of a structured population and invloves both ordinary and partial differential equations. The idea of persistence although not as a direct application of the theorems presented here, appears in Busenberg, Cooke, and Thieme [1] who also were considering an AIDS model.

Acknowledgment

Several colleagues–S. Ellermeyer, H.I. Freedman, T. Gard, V. Hutson, and J.W.-H. So–read a preliminary version of this survey and made valuable comments which are very much appreciated.

This research was supported by NSF grant DMS-8901992.

References

1. Busenberg, S., Cooke, K., Thieme, H.: Demographic change and persistence of HIV/AIDS in a heterogeneous population. To appear, SIAM J. Appl. Math
2. Burton, T.A., Hutson, V. (1989): Repellers with infinite delay. J. Math. Anal. Appl. **137**, 240-263
3. Butler, G., Freedman, H.I., Waltman, P. (1986): Uniformly persistent systems. Proc. AMS **96**, 425-430
4. Butler, G., Waltman, P. (1986): Persistence in dynamical systems. JDE **63**, 255-263
5. Conley, C. (1978): Isolated Invariant Sets and the Morse Index. CBMS vol. 38, Amer. Math. Soc., Providence
6. Dunbar, S.R., Rybakowski, K.P., Schmitt, K. (1986): Persistence in models of predator-prey populations with diffusion. JDE **65**, 117-138
7. Fonda, A. (1988): Uniformly persistent semi-dynamical systems. Proc. AMS **104**, 111-116
8. Freedman, H.I., Moson, P. (1990): Persistence definitions and their connections. Proc. AMS, **109**, 1025-1033

9. Freedman, H.I., Rai, B. (1987): Persistence in a predator-prey-competitor-mutualist model. Janos Bolyai Math. Soc., 73-79
10. Freedman, H.I., So, J.W.-H. (1987): Persistence in discrete models of a population which may not be subject to harvesting. Nat. Res. Modeling **1**, 135-145
11. Freedman, H.I., So, J.W.-H. (1989): Persistence in discrete semidynamical systems. SIAM J. Math. Anal. **20**, 930-938
12. Freedman, H.I., Waltman, P. (1977): Mathematical analysis of some three-species food chain models. Math. Biosc. **33**, 257-276
13. Freedman, H.I., Waltman, P. (1984): Persistence in a model of three interscting predator-prey populations. Math. Biosc. **68**, 213-231
14. Freedman, H.I., Waltman, P. (1985): Persistence in a model of three competitive populations. Math Biosc. **73**, 89-101
15. Garay, B.M. (1989): Uniform persistence and chain recurrence. J. Math. Anal. Appl. **139**, 372-381
16. Gard, T.C. (1980): Persistence in food chains with general interactions. Math. Biosc. **51**, 165-174
17. Gard, T.C. (1987): Uniform persistence in multispecies population models. Math. Biosc. **85**, 93-104
18. Gard, T.C., Hallam, T.G. (1979): Persistence in food webs I: Lotka-Volterra food chains. Bull. Math. Biol. **41**, 477-491
19. Gard, T.C., Hallam, T.G., Svoboda, L.J. (1979): Persistence and extinction in three species Lotka-Volterra competitive systems. Math. Biosc. **46**, 117-124
20. Gatica, J.A., So, J.W.-H. (1988): Predator-prey models with periodic coefficients. Applic. Anal. **27**, 143-152
21. Hale, J.K. (1988): Asymptotic Behavior of Dissipative Systems. Amer. Math. Soc., Providence
22. Harrison, G. (1979): Response of a population to stress: resistance and other stability concepts. Am. Nat. **113**, 659-669
23. Hallam, T.G., Ma, Z. (1986): Persistence in population models with demographic fluctuations. J. Math. Bio. **24**, 327-340
24. Hale, J.K., Waltman, P. (1989): Persistence in infinite dimensional systems. SIAM J. Math. Anal. **20**, 388-395
25. Hofbauer, J. (1980): A general cooperation theorem for hypercycles. Monat. Math. **91**, 233-240
26. Hofbauer, J.: A unified approach to persistence. Preprint
27. Hofbauer, J., Hutson, V., Jansen, W. (1987): Coexistence for systems governed by difference equations of Lotka-Volterra type. J. Math Bio. **25**, 553-570
28. Hofbauer, J., Sigmund, K. (1988): Dynamical Systems and the Theory of Evolution. Cambridge University Press
29. Hofbauer, J., So, J.W.-H. (1989): Uniform persistence and repellors for maps. Proc. AMS **107**, 1137-1142
30. Hutson, V. (1984): A theorem on average Liapunov functions. Monats. Math. **98**, 267-275
31. Hutson, V., Law, R. (1985): Permanent coexistence in general models of three interacting species. J. Math. Biol. **21**, 285-298
32. Hutson, V., Moran, W. (1982): Persistence of species obeying difference equations. J. Math. Bio. **15**, 203-213
33. Hutson, V., Schmitt, K.: Permanence in dynamical systems. Preprint
34. Hutson, V., Vickers, G.T. (1983): A criterion for permanent coexistence of species with an apllication to a two-prey one-predator system. Math. Biosc. **63**, 253-269

35. Jansen, W. (1987): A permanence theorem for replicator and Lotka-Volterra systems. J. Math. Biol. **25**, 411-422
36. Kirlinger, G. (1986): Permanence in Lotka-Volterra equations: linked predator-prey systems. Math. Biosc. **82**, 165-191
37. Kirlinger, G. (1988): Permanence of some ecological systems with several predator and one prey species. J. Math. Biol. **26**, 217-232
38. Li, J., Hallam, T.G. (1988): Survival in continuous structured population models. J. Math. Bio. **26**, 421-433
39. May, R.M., Leonard, W.J. (1975): Nonlinear aspects of competition between three species. SIAM J. Appl. Math. **29**, 243-253
40. Schuster, P. Sigmund, K. Wolf, R. (1979): Dynamical systems under constant organization III. Cooperative and competitve behaviour of hypercycles. JDE **32**, 357-386
41. Schuster, P. Sigmund, K. Wolf, R. (1979): On ω-limits for competition between three species. SIAM J. Appl. Math. **37**, 49-54
42. So, J.W.-H. (1990): Persistence and extinction in a predator-prey model consisting of nine prey genotypes. J. Australian Math. Soc. Ser.B **31**, 347-365
43. Thieme, H.R., Castillo-Chavez, C. (1989): On the role of variable infectivity in the dynamics of the human immunodeficiency virus epidemic. In Mathematical and Statistical Approaches to AIDS Epidemiology (C. Castillo-Chavez, ed.), Lecture Notes in Biomathematics **83**, Springer Verlag, Berlin-Heidelberg-New York

Periodic Orbits of Planar Polynomial Liénard Systems with a Small Parameter

Felix Albrecht[1] *and Gabriele Villari*[2]

[1] Department of Mathematics, University of Illinois, Urbana IL 61801
[2] Dipartimento di Matematica, Università degli Studi di Firenze, Italy

1 Introduction

Consider the Liénard equation

$$\ddot{x} + p(x)\dot{x} + x = 0,$$

where p is a real polynomial of degree n, and the equivalent planar vector field $(y, -x - yp(x))$, i.e. the first-order system on $\mathbb{R}^2$

$$\begin{aligned} \dot{x} &= y \\ \dot{y} &= -x - yp(x). \end{aligned}$$

An isolated nontrivial periodic orbit of this vector field is called a limit cycle. In [2] Lins, de Melo and Pugh proposed the following conjecture: the number of limit cycles of the Liénard vector field defined by p does not exceed $[\frac{n}{2}]$ (here $[\beta]$ denotes the integer part of β). In the same paper it is shown that this bound is the best possible and that the conjecture is true for $n \leq 2$.

The more general question about the existence of a bound (depending only on n) for the number of limit cycles of planar polynomial vector fields of degree $\leq n$ was stated as part of Hilbert's 16th problem (see e.g. [3] for a history of the problem and an extensive bibliography).

The main result in this paper is the following.

Theorem 1. *For every polynomial p of degree n there exists an $\epsilon_0 > 0$ such that for all ϵ satisfying $|\epsilon| \leq \epsilon_0$ the vector field $(y, -x - \epsilon yp(x))$ has at most $[\frac{n}{2}]$ limit cycles.*

2 Background

This section recalls certain known facts from perturbation theory applied to Liénard systems and establishes the notation. Throughout the paper p will denote a polynomial of degree $\leq n$ and P will be defined by $P(x) = \int_0^x p(\xi)d\xi$.

Consider the Liénard equation

$$\ddot{x} + \epsilon p(x)\dot{x} + x = 0, \tag{1}$$

where ϵ is a real parameter, and the equivalent first- order systems

$$\begin{aligned} \dot{x} &= y \\ \dot{y} &= -x - \epsilon y p(x) \end{aligned} \tag{2}$$

and

$$\begin{aligned} \dot{x} &= y - \epsilon P(x) \\ \dot{y} &= -x. \end{aligned} \tag{3}$$

Observe that the vector field defined by (2) is transversal to the positive y-axis and denote by $S \subset \mathbb{R} \times \mathbb{R}$ the open set of pairs (s, ϵ), $s > 0$, with the property that the positive semi-orbit of (2) through the point $(0, s)$ intersects the positive y-axis again at a first point $(0, \lambda(s, \epsilon))$. Thus $s \longmapsto \lambda(s, \epsilon)$ is the Poincaré mapping defined by (2) and the function λ is analytic.

Set $V(x, y) = \frac{1}{2}(x^2 + y^2)$ and, for $(s, \epsilon) \in S$,

$$E(s, \epsilon) = V(0, \lambda(s, \epsilon)) - V(0, s).$$

By a standard argument (see e.g. [1], Section 12.10)

$$E(s, \epsilon) = \epsilon \oint_{C_s} y p dx + \psi(s, \epsilon), \tag{4}$$

where $C_s = \{(x, y) | x^2 + y^2 = s^2\}$, $\psi(s, 0) = 0$ and $\lim_{\epsilon \to 0} \frac{\psi(s,\epsilon)}{\epsilon} = 0$ for all $s > 0$. The function $G : (0, \infty) \to \mathbb{R}$ defined by the above line integral,

$$G(s) = \oint_{C_s} y p dx,$$

is analytic, hence so is the function ψ. It is immediate that

$$\frac{\partial \psi}{\partial \epsilon}(\sigma, 0) = \frac{\partial^i \psi}{\partial s^i}(\sigma, 0) = 0$$

for all $\sigma > 0$ and all $i \geq 0$. Therefore, by a simple induction argument,

$$\begin{aligned} \lim_{\epsilon \to 0} \frac{1}{\epsilon} \frac{\partial^i \psi}{\partial s^i} &= \frac{\partial^{i+1} \psi}{\partial s^i \partial \epsilon}(\sigma, 0) = \frac{\partial}{\partial s}\left(\frac{\partial^i \psi}{\partial s^{i-1} \partial \epsilon}\right)(\sigma, 0) \\ &= \frac{\partial}{\partial s}\left(\lim_{\epsilon \to 0} \frac{1}{\epsilon} \frac{\partial^{i-1} \psi}{\partial s^{i-1}}\right)(\sigma, \epsilon), \end{aligned}$$

hence

$$\lim_{\epsilon \to 0} \frac{1}{\epsilon} \frac{\partial^i \psi}{\partial s^i}(\sigma, \epsilon) = 0 \tag{5}$$

for all $\sigma > 0$ and all $i \geq 0$.

In order to compute $G(s)$ let $p(x) = \sum_{k=0}^{n} a_k x^k$. Then

$$G(s) = -s^2 \int_0^{2\pi} \sin^2 t \quad p(s \cos t) dt = -\sum_{k=0}^{n} a_k s^{k+2} \int_0^{2\pi} \sin^2 t \cos^k t dt. \tag{6}$$

Recall the recursion formula for the indefinite integral

$$J_{\mu,\nu} = \int \sin^\mu t \cos^\nu t dt = \frac{\sin^{\mu+1} t \cos^{\nu-1} t}{\mu + \nu} + \frac{\mu - 1}{\mu + \nu} J_{\mu,\nu-2}, \; \mu + \nu \neq 0.$$

It follows that in (6) the coefficients of the odd powers of s are zero and thus G is an even polynomial in s of the form

$$G(s) = s^2 \sum_{i=0}^{[\frac{n}{2}]} A_{2i} s^{2i}. \tag{7}$$

Remark 1. Since the zeros of $s \longmapsto E(s, \epsilon)$ yield all the nontrivial periodic orbits of (2) it follows by analyticity that either each or else none of these orbits is a limit cycle. Notice also that the function G is completely determined by the even part of the polynomial p; it is independent of the odd part. If p is an odd polynomial it is known that E is identically zero on S, i.e. (2) has no limit cycles (see also Section 4 below). Otherwise the periodic orbits of (2) for $\epsilon \neq 0$ lie in a bounded region of $\mathbb{R}^2$ and are hence finite in number.

Another known result which will be needed later is the following.

Lemma 1. (see, for instance, [4]) *Let P be a non-even polynomial and P_0 be its odd part. If P_0 does not change sign at any $x \neq 0$ then (3) has no nontrivial periodic orbits (and hence Theorem 1 is trivially true). Otherwise every nontrivial periodic orbit of (2) intersects the x-axis outside the segment $\{(x, 0) \| x| \leq a\}$, where $a > 0$ is the smallest positive zero of P_0 at which P_0 changes sign.*

3 Main Result

Equality (4) and the preceding considerations imply that a nontrivial periodic orbit of (2) intersects the positive y–axis at a point $(0, s)$ with the property that s lies near a zero of G, provided $|\epsilon|$ is sufficiently small. In order to obtain a bound for $|\epsilon|$ which is valid for *every* such orbit it will be necessary to establish that there exists a compact set in $\mathbb{R}^2$ containing all the periodic orbits of (2), or of the equivalent system (3), for every sufficiently small $|\epsilon|$, $\epsilon \neq 0$. This result is stated below; its proof will be given in Section 4.

Theorem 2. *Let P be a non–even polynomial of degree ≥ 1, with $P(0) = 0$. There exist γ_0 and K_0, $0 < \gamma_0 < K_0$, depending on P, such that the set $\{(x,y)|\gamma_0^2 \leq x^2 + y^2 \leq K_0^2\}$ contains all the nontrivial periodic orbits of (3) for $|\epsilon| \leq 1$, $\epsilon \neq 0$.*

It is now possible to prove the main result.

Proof of Theorem 1. In view of Remark 1 assume that the polynomial P is not even. Theorem 2 implies that the zeros of the function $s \longmapsto E_\epsilon(s) = E(s,\epsilon)$ defined in Section 2 lie in the segment $[\gamma_0, K_0]$ for all $|\epsilon| \leq 1$, $\epsilon \neq 0$. Equality (4) shows that for every $\delta > 0$ one can choose an $\epsilon_0 > 0$ so that for $|\epsilon| \leq \epsilon_0$, $\epsilon \neq 0$ and for each zero σ_ϵ of E_ϵ there is a (unique) positive zero $\overline{\sigma}$ of G such that $|\sigma_\epsilon - \overline{\sigma}| < \delta$. Assume that $\overline{\sigma}$ is a zero of order ℓ of G, $\ell \geq 1$, i.e. $G^{(j)}(\overline{\sigma}) = 0$ for $0 \leq j \leq \ell-1$, $G^\ell(\overline{\sigma}) \neq 0$. Then for $\delta > 0$ sufficiently small and the corresponding ϵ_0 the function E_ϵ, $|\epsilon| \leq \epsilon_0$, $\epsilon \neq 0$, has at most ℓ zeros in the δ–neighborhood of $\overline{\sigma}$. Indeed, otherwise repeated use of Rolle's theorem would imply that $\frac{\partial^\ell E_\epsilon}{\partial s^\ell}$ has a zero in this neighborhood which is impossible since by (4) and (5)

$$\frac{1}{\epsilon}\frac{\partial^\ell E_\epsilon}{\partial s^\ell}(\sigma) = G^\ell(\sigma) + \frac{1}{\epsilon}\frac{\partial^i \psi}{\partial s^i}(\sigma,\epsilon) \neq 0$$

for $|\sigma - \overline{\sigma}| < \delta$. It is now clear that for an appropriate $\epsilon_0 > 0$ and all $|\epsilon| \leq \epsilon_0$, $\epsilon \neq 0$, the number of distinct zeros of E_ϵ in the compact interval $[\gamma_0, K_0]$ does not exceed the number of positive zeros of G, i.e. of the even part of p, counting their multiplicities.

Finally, the even polynomial G is of the form

$$G(s) = s^2\varphi(s)\prod_{h=1}^{m}(s^2 - \overline{\sigma}_h^2)^{d_h},$$

where each $\overline{\sigma}_h \geq 0$, $\sum_{h=1}^{m} d_h \leq [\frac{n}{2}]$ and the function φ is nowhere zero. In particular $\overline{\sigma}_h$ is a zero of G of order d_h and hence E_ϵ has at most $[\frac{n}{2}]$ zeros. By Remark 1 the proof of Theorem 1 is complete. □

4 Proof of Theorem 2

This section is devoted to the proof of Theorem 2. Under the additional assumption that the polynomial P has odd degree this theorem is proved in [2]. For completeness a slightly modified proof of this result is given below.

Lemma 2. *If P is a polynomial of odd degree and $P(0) = 0$ then the conclusion of Theorem 2 holds.*

Proof . Let $c > 0$ be such that the segment $[-c, c]$ contains all the zeros of P and p in its interior. Choose $s_0 > 0$ so that for $(s,\epsilon) \in S$, with $s \geq s_0$,

$|\epsilon| \leq 1$, the positive semi- orbit of (3) through the point $(0,s)$ intersects the lines $x = c$ and $x = -c$ for the first time at $(c, \lambda_1(s,\epsilon))$, $(c, \lambda_2(s,\epsilon))$, $(-c, \lambda_3(s,\epsilon))$ and $(-c, \lambda_4(s,\epsilon))$, with $\lambda_1(s,\epsilon) > \epsilon P(c)$, $\lambda_2(s,\epsilon) < \epsilon P(c)$, $\lambda_3(s,\epsilon) < \epsilon P(-c)$, $\lambda_4(s,\epsilon) > \epsilon P(-c)$ and $\lambda_1(s,\epsilon) - \lambda_2(s,\epsilon) > 1$, $\lambda_4(s,\epsilon) - \lambda_3(s,\epsilon) > 1$.

Fix $s \geq s_0$ and $\epsilon \neq 0$, $|\epsilon| \leq 1$ and denote by $t \longmapsto (x(t), y(t))$ the integral curve of (3) with initial condition $(0,s)$. The derivative of the function $t \longmapsto \frac{1}{2}(x^2(t) + y^2(t)) = V(x(t), y(t))$ is $-\epsilon x(t) P(x(t))$, therefore

$$\begin{aligned} V(c, \lambda_2(s,\epsilon)) - V(c, \lambda_1(s,\epsilon)) &= \epsilon \int_{\lambda_1(s,\epsilon)}^{\lambda_2(s,\epsilon)} P(x_1(y))dy, \\ V(-c, \lambda_4(s,\epsilon)) - V(-c, \lambda_3(s,\epsilon)) &= \epsilon \int_{\lambda_3(s,\epsilon)}^{\lambda_4(s,\epsilon)} P(x_2(y))dy, \end{aligned} \tag{8}$$

where $y \longmapsto x_1(y)$ and $y \longmapsto x_2(y)$ are the parametrizations by y of the appropriate arcs of the orbits of (3) through $(0,s)$ lying in $\{(x,y) | x \geq c\}$ and $\{(x,y) | x \leq -c\}$ respectively. Since P has odd degree and is strictly monotone for $|x| \geq c$ the right–hand sides in the above equalities have the same sign and

$$|V(c, \lambda_2(s,\epsilon)) - V(c, \lambda_1(s,\epsilon))| > |\epsilon P(c)|.$$

Now

$$|V(-c, \lambda_3(s,\epsilon)) - V(c, \lambda_2(s,\epsilon))| \leq |\epsilon| \int_{-c}^{c} \left| \frac{xP(x)}{y_\epsilon(x) - \epsilon P(x)} \right| dx,$$

where $x \longmapsto y_\epsilon(x)$ is the parametrization by x of the appropriate arc of the orbit of (3) through $(0,s)$. The integrand can be made arbitrarily small for s_0 sufficiently large and $|\epsilon| \leq 1$; in particular, for $s \geq s_0$ and $|\epsilon| \leq 1$, $\epsilon \neq 0$,

$$|V(-c, \lambda_3(s,\epsilon)) - V(c, \lambda_2(s,\epsilon))| < \frac{1}{3} |\epsilon P(c)|.$$

It is obvious that one can obtain the same bound for $V(0, \lambda(s,\epsilon)) - V(-c, \lambda_4(s,\epsilon))$ and $V(c, \lambda_1(s,\epsilon)) - V(0,s)$ and hence $V(0, \lambda(s,\epsilon)) \neq V(0,s)$. Thus it has been shown that the periodic orbits of (3) lie in some closed disc about the origin. In order to complete the proof of Lemma 2 observe that by Lemma 1 this disc can be replaced by an annulus. □

The proof for Lemma 2 fails if the polynomial P in (3) has even degree since in this case the right–hand sides in (8) have opposite signs. The analysis of this situation requires a preliminary investigation.

Let P_e be an even polynomial of degree ≥ 2, with $P_e(0) = 0$. Consider the system

$$\begin{aligned} \dot{x} &= y - \epsilon P_e(x), \quad \epsilon > 0 \\ \dot{y} &= -x. \end{aligned} \tag{9}$$

Recall the phase portrait of (9) (see e.g. [2]). Every orbit of (9) is symmetric with respect to the y–axis and there is a nonperiodic orbit Γ_ϵ separating $\mathbb{R}^2$ into two unbounded components, one of which is the union of all the periodic orbits of (9). Furthermore Γ_ϵ intersects the y-axis in a unique point lying below the x-axis if $\lim_{x\to\infty} P_e(x) = +\infty$ and above the x-axis if $\lim_{x\to\infty} P_e(x) = -\infty$.

Denote by D_ϵ the union of all the nontrivial periodic orbits of (9). Clearly D_ϵ is open in $\mathbb{R}^2$ and contains the positive, respectively the negative, y-axis. It is easy to see that $0 < \epsilon_2 < \epsilon_1$ implies $D_{\epsilon_1} \subset D_{\epsilon_2}$. Define a first integral $W_\epsilon : D_\epsilon \to \mathbb{R}$ of (9) by

$$W_\epsilon(z) = \begin{cases} \frac{1}{2}\rho^2(z) & \text{if } \lim_{x\to\infty} P_e(x) = +\infty \\ \frac{1}{2}\sigma^2(z) & \text{if } \lim_{x\to\infty} P_e(x) = -\infty \end{cases},$$

where $(0, \rho(z))$, respectively $(0, \sigma(z))$, is the (unique) intersection point of the orbit of (9) through $z = (x, y) \in D_\epsilon$ with the positive, respectively the negative, y-axis. Set $W(x, y, \epsilon) = W_\epsilon(x, y)$ and observe that $W : D \to \mathbb{R}$ is analytic, where $D = \{(x, y, \epsilon) | (x, y) \in D_\epsilon\}$ is open in $\mathbb{R}^3$. Notice that for $\epsilon \to 0$ one has $W_\epsilon(x, y) \to V(x, y) = \frac{1}{2}(x^2 + y^2)$.

Lemma 3. *There exists a nowhere vanishing analytic function $k : D \to \mathbb{R}$ such that*

$$\begin{aligned} &\frac{\partial W}{\partial x}(x, y, \epsilon) = k(x, y, \epsilon)x, \quad \frac{\partial W}{\partial y}(x, y, \epsilon) = k(x, y, \epsilon)(y - \epsilon P_e(x)), \\ &\frac{\partial k}{\partial y}(x, y, \epsilon)x = \frac{\partial k}{\partial x}(x, y, \epsilon)(y - \epsilon P_e(x)) - \epsilon k(x, y, \epsilon)p_0(x), \end{aligned} \tag{10}$$

where the odd polynomial $p_0 = P_e'$ is the derivative of P_e.

Proof. The third equality (10) follows from the other two and from the analyticity of W. Since W_ϵ is constant along the orbits of (9) in D_ϵ one has, for every $(x, y, \epsilon) \in D$,

$$\frac{\partial W}{\partial x}(x, y, \epsilon)(y - \epsilon P_e(x)) = \frac{\partial W}{\partial y}(x, y, \epsilon)x, \tag{11}$$

with $\frac{\partial W}{\partial x}$ and $\frac{\partial W}{\partial y}$ not vanishing simultaneously. Therefore x and $\frac{\partial W}{\partial x}$ are simultaneously zero on D and so are $y - \epsilon P_e(x)$ and $\frac{\partial W}{\partial y}$. Define $k : D \to \mathbb{R}$ by

$$k(x, y, \epsilon) = \begin{cases} \frac{1}{x}\frac{\partial W}{\partial x}(x, y, \epsilon) & \text{if } x \neq 0 \\ \frac{1}{y - \epsilon P_e(x)}\frac{\partial W}{\partial y}(x, y, \epsilon) & \text{if } x = 0. \end{cases}$$

The principle of analytic continuation and (11) imply that k is analytic and also the first two equalities (10). □

Lemma 4. *The function k has the following properties:*

(i) $k(0, y, \epsilon) = 1$ for all $y > 0$, $\epsilon > 0$ if $\lim_{x\to\infty} P_e(x) = +\infty$, or for all $y < 0$, $\epsilon > 0$ if $\lim_{x\to\infty} P_e(x) = -\infty$.

(ii) If $p_0(x) \neq 0$ for $x \neq 0$ then $k(x, y, \epsilon) \geq 1$ on D. Otherwise for every $0 < \alpha < 1$ there exists an $R_0 > 0$ such that $k(x, y, \epsilon) > \alpha$ for all $(x, y, \epsilon) \in D$ for which $x^2 + y^2 \geq R_0^2$ and $0 < \epsilon \leq 1$.

Proof. For $(r,s,\epsilon) \in D_\epsilon$ denote by $t \longmapsto (x_e(t,r,s,\epsilon),\ y_e(t,r,s,\epsilon))$ the (periodic) integral curve of (9) with initial condition (r,s). Let

$$\tau_\epsilon(r,s) = \max\{t \le 0 | x_e(t,r,s,\epsilon) = 0\}.$$

The mappings x_e, y_e and τ are analytic and

$$\int_0^{\tau_\epsilon(r,s)} \frac{d}{dt}(V(x_e(t,r,s,\epsilon), y_e(t,r,s,\epsilon)))dt = \frac{1}{2}y_e^2(\tau_\epsilon(r,s),r,s,\epsilon) - \frac{1}{2}(r^2+s^2).$$

Thus

$$W(r,s,\epsilon) = -\epsilon \int_0^{\tau_\epsilon(r,s)} x_e(t,r,s,\epsilon)P_e(x_e(t,r,s,\epsilon))dt + \frac{1}{2}(r^2+s^2). \tag{12}$$

Now assume that $r \ge 0$ if $\lim_{x\to\infty} P_e(x) = +\infty$, or $r \le 0$ if $\lim_{x\to\infty} P_e(x) = -\infty$. Then

$$W(r,s,\epsilon) = \epsilon \int_s^{\sqrt{2W(r,s,\epsilon)}} P_e(\xi(y,r,s,\epsilon))dy + \frac{1}{2}(r^2+s^2),$$

where $y \longmapsto \xi(y,r,s,\epsilon)$ is the parametrization by y of the appropriate arc of the integral curve. Since $\xi(\sqrt{2W(r,s,\epsilon)},r,s,\epsilon) = 0$ and $P_e(0) = p_0(0) = 0$ one obtains

$$\frac{\partial W}{\partial x}(r,s,\epsilon) = \epsilon \int_s^{\sqrt{2W(r,s,\epsilon)}} p_0(\xi(y,r,s,\epsilon))\frac{\partial \xi}{\partial x}(y,r,s,\epsilon)dy + r \tag{13}$$

and

$$\frac{\partial^2 W}{\partial x^2}(r,s,\epsilon) = \epsilon \int_s^{\sqrt{2W(r,s,\epsilon)}} \frac{\partial \varphi}{\partial x}(y,r,s,\epsilon)dy + 1, \tag{14}$$

where $\varphi(y,r,s,\epsilon)$ denotes the integrand in (13). Since $W(0,s,\epsilon) = \frac{1}{2}s^2$ for every $s > 0$ if $\lim_{x\to\infty} P_e(x) = +\infty$ or for every $s < 0$ if $\lim_{x\to\infty} P_e(x) = -\infty$ it follows from (10) and (14) that

$$k(0,s,\epsilon) = \frac{\partial^2 W}{\partial x^2}(0,s,\epsilon) = 1$$

for $s > 0$, respectively for $s > 0$, and hence property (i) has been proved.

Observe now that the symmetry of the phase portrait of (9) implies that $W(x,y,\epsilon) = W(-x,y,\epsilon)$ for all $(x,y,\epsilon) \in D$. Therefore $k(x,y,\epsilon) = k(-x,y,\epsilon)$ and thus it suffices to prove property (ii) for $(x,y,\epsilon) \in D$ with $x \ge 0$ (or $x \le 0$).

Let again $(r,s,\epsilon) \in D$. Compute the variation of k along the arc of the integral curve $t \longmapsto (x_e(t,r,s,\epsilon),\ y_e(t,r,s,\epsilon))$ of (9) between $(0, y_e(\tau_\epsilon(r,s),r,s,\epsilon))$ and (r,s). The third equality (10) yields

$$\frac{\partial}{\partial t}k(x_e(t,r,s,\epsilon),y_e(t,r,s,\epsilon),\epsilon) = \epsilon k(x_e(t,r,s,\epsilon),y_e(t,r,s,\epsilon),\epsilon)p_0(x_e(t,r,s,\epsilon)),$$

hence

$$k(r,s,\epsilon) = k(0, y_e(\tau_\epsilon(r,s), r, s, \epsilon), \epsilon) \exp\left(\epsilon \int_{\tau_\epsilon(r,s)}^{0} p_0(x_e(t,r,s,\epsilon))dt\right)$$

or, by (i),

$$k(r,s,\epsilon) = \exp\left(\epsilon \int_{\tau_\epsilon(r,s)}^{0} p_0(x_e(t,r,s,\epsilon))dt.\right) \tag{15}$$

The first statement in (ii) is now obvious. For the second statement it will be assumed below that $\lim_{x\to\infty} P_e(x) = +\infty$. The other case is treated in the same way by considering the set of points $(x,y,\epsilon) \in D$, $x \leq 0$.

Let $b > 0$ be such that $P_e(b) \geq 0$ and $p_0(x) \geq 0$ for $x \geq b$. If $s > \epsilon P_e(b)$, the orbit of (9) through the point (b,s) is periodic, intersects the line $x = b$ again at (b,u), with $u < \epsilon P_e(b)$, and its two arcs contained in the band $\{(x,y)|0 \leq x \leq b\}$ and lying respectively above and below the curve $y = \epsilon P_e(x)$ admit parametrizations by x:

$$x \longmapsto \eta_1(x,b,s,\epsilon) > \eta_1(b,b,s,\epsilon) = s$$

$$x \longmapsto \eta_2(x,b,u,\epsilon) < \eta_2(b,b,u,\epsilon) = u.$$

Notice that these functions are strictly monotone. Let $L > 0$, $M > 0$ be such that $|p_0(x)| < L$, $|P_e(x)| < M$ on $[0,b]$, and let

$$s_0 > M - 2\frac{Lb}{\log\alpha},$$

where α is the constant in the statement of (ii), $0 < \alpha < 1$. Consider the compact set $\{(x,y)|W(x,y,\epsilon) = W(b,s_0,\epsilon), 0 \leq \epsilon \leq 1\}$, where $W(x,y,0) = V(x,y) = \frac{1}{2}(x^2+y^2)$, and denote by $R_0 > 0$ the radius of a disc containing this set. Henceforth it will be assumed that $0 < \epsilon \leq 1$.

Let $(x_1,y_1) \in D_\epsilon$ be such that $x_1^2 + y_1^2 \geq R_0^2$ and $x_1 > 0$. The closed curve $\{(x,y)|W(x,y,\epsilon) = W(x_1,y_1,\epsilon)\}$ intersects the line $x = b$ at $(b, s_1(\epsilon))$ and $(b, u_1(\epsilon))$, with $s_1(\epsilon) > s_0 > 0$, $u_1(\epsilon) < -s_0 < 0$ (obviously $s_1(\epsilon)$ and $u_1(\epsilon)$ depend also on x_1 and y_1). In order to compute $k(x_1,y_1,\epsilon)$ by means of (15) one distinguishes the following three cases.

(I) $x_1 \leq b$, $y_1 > 0$. One has, for $0 \leq x \leq x_1$,

$$\eta_1(x,b,s_1(\epsilon),\epsilon) - \epsilon P_e(x) > \eta_1(x_1,b,s_1(\epsilon),\epsilon) - M = y_1 - M > s_1(\epsilon) - M.$$

Since $s_1(\epsilon) - M > s_0 - M > -2\frac{Lb}{\log\alpha}$ one has

$$\epsilon\left|\int_0^{x_1} \frac{p_0(x)dx}{\eta_1(x,b,s_1(\epsilon),\epsilon) - \epsilon P_e(x)}\right| < \frac{\epsilon x_1}{2b}\log\frac{1}{\alpha} < \frac{1}{2}\log\frac{1}{\alpha}. \tag{16}$$

It follows that $k(x_1,y_1,\epsilon) > \exp(-\frac{1}{2}\log\frac{1}{\alpha}) = \sqrt{\alpha} > \alpha$.

(II) $x_1 \geq b$. In this case

$$\epsilon \int_{\tau_\epsilon(x_1,y_1)}^{0} p_0(x_e(t,x_1,y_1,\epsilon))dt = \epsilon \int_0^b \frac{p_0(x)dx}{\eta_1(x,b,s_1(\epsilon),\epsilon) - \epsilon P_e(x)} + \epsilon I(x_1,y_1,\epsilon),$$

where $I(x_1, y_1, \epsilon) \geq 0$ since $p_0(x) \geq 0$ for $x \geq b$. Also,

$$\epsilon \left| \int_0^b \frac{p_0(x)dx}{\eta_1(x,b,s_1(\epsilon),\epsilon) - \epsilon P_e(x)} \right| < \frac{1}{2} \log \frac{1}{\alpha} \tag{17}$$

by the same reasoning as that used for proving (16). Thus, by (15), $k(x_1, y_1, \epsilon) > \exp(-\frac{1}{2} \log \frac{1}{\alpha}) > \alpha$.

(III) $x_1 \leq b$, $y_1 < 0$. Proceeding as in case (I) one gets, for $0 \leq x \leq x_1$,

$$\epsilon P_e(x) - \eta_2(x,b,u_1(\epsilon),\epsilon) > -M - \eta_2(x_1,b,u_1(\epsilon),\epsilon) = -M - y_1 > -M - u_1(\epsilon).$$

Since $u_1(\epsilon) < -s_0$ one concludes that $-M - u_1(\epsilon) > \frac{2Lb}{\log \frac{1}{\alpha}}$, hence

$$\epsilon \left| \int_{x_1}^b \frac{p_0(x)dx}{\eta_2(x,b,u_1(\epsilon),\epsilon) - \epsilon P_e(x)} \right| < \frac{1}{2} \log \frac{1}{\alpha}$$

and

$$\epsilon \int_{\tau_\epsilon(x_1,y_1)}^{0} p_0(x_e(t,x_1,y_1,\epsilon))dt = \epsilon \int_0^b \frac{p_0(x)dx}{\eta_1(x,b,s_1(\epsilon),\epsilon) - \epsilon P_e(x)} + \epsilon I(b,u_1(\epsilon),\epsilon)$$
$$+ \epsilon \int_{x_1}^b \frac{p_0(x)dx}{\epsilon P_e(x) - \eta_2(x,b,u_1(\epsilon),\epsilon)} > -\log \frac{1}{\alpha}.$$

Thus, again, $k(x_1, y_1, \epsilon) > \alpha$. This completes the proof of (ii). □

Remark 2. Lemmas 3 and 4 hold, with obvious modifications, for system (9) with $\epsilon < 0$. Indeed, it suffices to replace the polynomial P_e by $-P_e$ and to define the first integral W_ϵ accordingly. Notice also that the second assertion of property (ii) in Lemma 4 admits the following generalization: for every $0 < \alpha < 1$ and every $\epsilon_0 > 0$ there exists an $R_0 > 0$ such that $k(x, y, \epsilon) > \alpha$ for all $(x, y, \epsilon) \in D$ for which $x^2 + y^2 \geq R_0$ and $|\epsilon| \leq \epsilon_0$, $\epsilon \neq 0$.

It is now possible to prove Theorem 2.

Proof of Theorem 2. If the polynomial P has odd degree the theorem is true by Lemma 2. Assume that P has even degree ≥ 2 and write $P = P_e + P_0$, where P_e and P_0 are respectively the even and the odd parts of P. Consider system (9) with this P_e (respectively with $-P_e$ if $\epsilon < 0$ in system (3)) and the corresponding first integral $W_\epsilon : D_\epsilon \to \mathbb{R}$. For definiteness, let $\lim_{x \to \infty} \epsilon P_e(x) = +\infty$, and $c > 0$ be such that the segment $[-c, c]$ contains all the zeros of P_0 and of its derivative in its interior. It will be shown that there exists a $q_0 > \epsilon|P(-c)|$ with the following property: if $q \geq q_0$ and $\epsilon \neq 0$, $|\epsilon| \leq 1$, are such that the positive semi-orbit of (3) through the point $(-c, q)$ intersects the ray $x = -c$, $y > \epsilon|P(-c)|$ again at a first point $(-c, h(q, \epsilon))$ then $W_\epsilon(-c, q) \neq W_\epsilon(-c, h(q, \epsilon))$. The assertion of the theorem is then immediate by Lemma 1.

Consider $q > 0$ and $\epsilon \neq 0$ such that $q > \epsilon|P(-c)|$ and if $h(q,\epsilon)$ exists then the positive semi-orbit of (3) through $(-c,q)$ intersects the lines $x = c$ and $x = -c$ for the first time at $(c, h_1(q,\epsilon))$, $(c, h_2(q,\epsilon))$ and $(-c, h_3(q,\epsilon))$, with $h_1(q,\epsilon) > \epsilon P(c)$, $h_2(q,\epsilon) < \epsilon P(c)$, $h_3(q,\epsilon) < \epsilon P(-c)$, $h_1(q,\epsilon) - h_2(q,\epsilon) > 1$, $h(q,\epsilon) - h_3(q,\epsilon) > 1$. Notice that there exists a $q_0 > 0$ such that this holds for all $q \geq q_0$ and all $\epsilon \neq 0$, $|\epsilon| \leq 1$.

Let (q,ϵ) be such a pair and $t \longmapsto (x(t,-c,q,\epsilon), y(t,-c,q,\epsilon))$ be the integral curve of (3) with initial condition $(-c,q)$. The derivative of W_ϵ along this curve, written for simplicity $t \longmapsto (x(t), y(t))$, is

$$\dot{W}_\epsilon(x(t),y(t)) = \frac{\partial W_\epsilon}{\partial x}(x(t),y(t))(y(t) - \epsilon P(x(t))) - \frac{\partial W_\epsilon}{\partial y}(x(t),y(t))x(t).$$

By (11) and (10)

$$\begin{aligned}\dot{W}_\epsilon(x(t),y(t)) &= -\epsilon P_0(x(t))\frac{\partial W_\epsilon}{\partial x}(x(t),y(t)) \\ &= -\epsilon x(t) P_0(x(t)) k(x(t),y(t),\epsilon).\end{aligned} \tag{18}$$

It follows from (3) that

$$W_\epsilon(c,h_2(q,\epsilon)) - W_\epsilon(c,h_1(q,\epsilon)) = \epsilon \int_{h_1(q,\epsilon)}^{h_2(q,\epsilon)} P_0(x_1(y))k(x_1(y),y,\epsilon)dy,$$

$$W_\epsilon(-c,h(q,\epsilon)) - W_\epsilon(c,h_3(q,\epsilon)) = \epsilon \int_{h_3(q,\epsilon)}^{h(q,\epsilon)} P_0(x_2(y))k(x_2(y),y,\epsilon)dy,$$

where $y \longmapsto x_1(y)$ and $y \longmapsto x_2(y)$ are the parametrizations by y of the appropriate arcs of the orbit through $(-c,q)$ lying in $\{(x,y)|x \geq c\}$ and $\{(x,y)|x \leq -c\}$ respectively. Since the odd function P_0 is nonzero and strictly monotone for $|x| \geq c$ one gets by Lemma 4 for $\alpha = \frac{1}{2}$ and a suitable choice of q_0:

$$\begin{aligned}W_\epsilon(c,h_2(q,\epsilon)) - W_\epsilon(c,h_1(q,\epsilon)) &< -\frac{\epsilon}{2}P_0(c)(h_1(q,\epsilon) - h_2(q,\epsilon)), \\ W_\epsilon(-c,h(q,\epsilon)) - W_\epsilon(-c,h_3(q,\epsilon)) &< -\frac{\epsilon}{2}P_0(c)(h(q,\epsilon) - h_3(q,\epsilon)),\end{aligned} \tag{19}$$

if $\epsilon P_0(c) > 0$ and the reverse inequalities if $\epsilon P_0(c) < 0$. In order to obtain the desired inequality $W_\epsilon(-c,h(q,\epsilon)) - W_\epsilon(-c,q) \neq 0$ it remains to be shown that

$$\begin{aligned}|W_\epsilon(c,h_1(q,\epsilon)) - W_\epsilon(-c,q)| &< \frac{1}{2}|\epsilon P_0(c)|(h_1(q,\epsilon) - h_2(q,\epsilon)), \\ |W_\epsilon(-c,h_3(q,\epsilon)) - W_\epsilon(c,h_2(q,\epsilon))| &< \frac{1}{2}|\epsilon P_0(c)|(h(q,\epsilon) - h_3(q,\epsilon)).\end{aligned} \tag{20}$$

These inequalities cannot be proved in the same way as (19) since even for fixed $\epsilon \neq 0$ the function k is not bounded on $\{(x,y) \in D_\epsilon||x| \leq c\}$. It is therefore necessary to use a different approach.

Recall that $W_\epsilon(x,y) = W_\epsilon(-x,y)$ and that W_ϵ is defined by (12). With the notation introduced earlier

$$
\begin{aligned}
W_\epsilon(c, h_1(q,\epsilon)) - W_\epsilon(-c,q) &= W_\epsilon(c,h_1(q,\epsilon)) - W_\epsilon(c,q) \\
&= \epsilon \int_0^{T_\epsilon(c,q)} x_e(t,c,q,\epsilon)P_e(x_e(t,c,q,\epsilon))dt - \frac{1}{2}(c^2+q^2) \\
&- \epsilon \int_0^{T_\epsilon(c,h_1(q,\epsilon))} x_e(t,c,h_1(q,\epsilon),\epsilon)P_e(x_e(t,c,h_1(q,\epsilon),\epsilon))dt \\
&+ \frac{1}{2}(c^2 + h_1^2(q,\epsilon)).
\end{aligned}
$$

Reparametrize by x the portions of the two integral curves of (9) with initial conditions (c,q) and $(c,h_1(q,\epsilon))$ determined by the limits of the corresponding integrals above:

$$
x \longmapsto \eta_1(x,c,q,\epsilon), \quad x \longmapsto \eta_1(x,c,h_1(q,\epsilon),\epsilon), \quad 0 \le x \le c.
$$

One obtains

$$
\begin{aligned}
&W_\epsilon(c,h_1(q,\epsilon) - W_\epsilon(-c,q) \\
&= \epsilon \int_0^c xP_e(x) \frac{\eta_1(x,c,q,\epsilon) - \eta_1(x,c,h_1(q,\epsilon),\epsilon)}{(\eta_1(x,c,q,\epsilon) - \epsilon P_e(x))(\eta_1(x,c,h_1(q,\epsilon),\epsilon) - \epsilon P_e(x))} dx \qquad (21) \\
&+ \frac{1}{2}(h_1^2(q,\epsilon) - q^2).
\end{aligned}
$$

In order to estimate $h_1(q,\epsilon) - q$ and $h_1^2(q,\epsilon) - q^2$ notice that

$$
h_1(q,\epsilon) - q = -\int_{-c}^{c} \frac{x dx}{y_1(x,-c,q,\epsilon) - \epsilon P(x)},
$$

where $x \longmapsto y_1(x,-c,q,\epsilon)$ is the parametrization by x of the appropriate portion of the integral curve of (3) with initial condition $(-c,q)$. Differentiation of $(x,y) \longmapsto \frac{1}{2}(x^2+y^2)$ along this arc and subsequent integration yield

$$
h_1^2(q,\epsilon) - q^2 = -\epsilon \int_{-c}^{c} \frac{xP(x)dx}{y_1(x,-c,q,\epsilon) - \epsilon P(x)}.
$$

It is now obvious that for q_0 sufficiently large $|h_1(q,\epsilon) - q|$ and $\frac{|h_1^2(q,\epsilon)-q^2|}{|\epsilon|}$ are arbitrarily small for $q \ge q_0$ and $|\epsilon| \le 1$, $\epsilon \ne 0$.

Next set $\varphi(x) = \eta_1(x,c,q,\epsilon) - \eta_1(x,c,h_1(q,\epsilon),\epsilon)$ and observe that for $0 \le x \le c$

$$
\varphi(x) = q - h_1(q,\epsilon) + \int_c^x \left[\frac{\xi}{\eta_1(\xi,c,h_1(q,\epsilon),\epsilon) - \epsilon P_e(x)} - \frac{\xi}{\eta_1(\xi,c,q,\epsilon) - \epsilon P_e(x)} \right] d\xi.
$$

Differentiation yields

$$
\varphi'(x) = \frac{x\varphi(x)}{(\eta_1(x,c,q,\epsilon) - \epsilon P_e(x))(\eta_1(x,c,h_1(q,\epsilon),\epsilon) - \epsilon P_e(x)}
$$

and hence

$$\varphi(x) = (q - h_1(q,\epsilon)) \exp \int_c^x \frac{\xi d\xi}{(\eta_1(\xi, c, q, \epsilon) - \epsilon P_e(x))(\eta_1(\xi, c, h_1(q,\epsilon), \epsilon) - \epsilon P_e(x))}.$$

It follows that $|\varphi(x)| \leq |h_1(q,\epsilon) - q|$ for $0 \leq x \leq c$ and therefore (21) implies that $q_0 > 0$ can be chosen so that

$$\frac{|W_\epsilon(c, h_1(q,\epsilon) - W_\epsilon(-c,q)|}{|\epsilon|} < \frac{1}{2}|P_0(c)|$$

for $|\epsilon| \leq 1$, $\epsilon \neq 0$. The other inequality (20) is proved in a similar manner observing that $|h_3(q,\epsilon) - h_2(q,\epsilon)|$ is bounded for $q \geq q_0$. This completes the proof of Theorem 2. □

Remark 3. If P_0 does not change sign at any $x \neq 0$ then the right–hand side of (18) has constant sign along every nontrivial orbit of (3). Lemma 1 is a straightforward consequence of this observation.

Notice that Theorem 2 can be replaced by a more general statement: for every $\epsilon_0 > 0$ the constants γ_0 and K_0 can be chosen so that the conclusion of the theorem holds for all $|\epsilon| \leq \epsilon_0$, $\epsilon \neq 0$ (see also Remark 2). Notice also that the preceding proof requires only that $P_e \not\equiv 0$, $P_0 \not\equiv 0$ and remains valid if the degree of P is odd. Finally, for every polynomial p and $\epsilon = 1$ system (2) (or the equivalent system (3)) has at most finitely many limit cycles.

Acknowledgment

This paper was written while the second author held a C.N.R. fellowship sponsored by NATO and was visiting the University of Illinois at Urbana-Champaign. He expresses his gratitude to the Department of Mathematics for its warm hospitality.

References

1. Arnold, V.I. (1973): Ordinary differential equations. MIT Press
2. Lins, A., de Melo, W., Pugh, C.C. (1976): On Liénard's equation, geometry and topology. Rio de Janeiro, Lect. Notes in Math. **597**, 335–357, Springer-Verlag
3. Lloyd, N.G. (1988): Limit cycles of polynomial systems–some recent developments. New Directions in Dynamical Systems, London Math. Soc. Lecture Notes Series 127, Cambridge Univ. Press, 192–234
4. Villari, G. (1987): A new system for Liénard's equation. Bollettino Un. Mat. Ital. (7), 1-A, 375–382

Hopf Bifurcation in Quasilinear Reaction–Diffusion Systems

Herbert Amann

Mathematical Institute, University of Zurich, Switzerland

1 Introduction and main results

During the last two decades, reaction–diffusion systems have been widely studied, usually in the form

$$\partial_t u_r - a_r \Delta u_r = f_r(u) \quad \text{in } \Omega \times (0,\infty) \quad r = 1,\ldots,N\ , \tag{1}$$

where Ω is a bounded domain in R^n such that $\overline{\Omega}$ is an n-dimensional smooth (i.e., C^∞-) submanifold of R^n, the 'diffusion coefficients', a_r, are positive, and the 'reaction terms', f_r, are smooth functions of $u := (u_1,\ldots,u_N)$. Of course, a_r and f_r can also depend smoothly upon $x \in \overline{\Omega}$ and $-a_r \Delta u_r$ can be replaced by $\mathcal{A}_r u_r$, where $\mathcal{A}_r$ is a strongly uniformly elliptic second order differential operator.

The system (1) has to be complemented by boundary conditions, which are usually Dirichlet boundary conditions:

$$u_r | \partial\Omega = 0\ , \quad 1 \le r \le N\ , \tag{2}$$

or Neumann, that is, 'no flux' conditions:

$$\partial_\nu u_r = 0\ , \quad 1 \le r \le N\ , \tag{3}$$

where $\nu := (\nu^1,\ldots,\nu^n)$ is the outer unit normal on $\partial\Omega$ and ∂_ν the normal derivative along $\partial\Omega$, or a combination of (2) and (3).

The basic existence, uniqueness, and continuity questions for problem (1)–(3) are by now well understood and a great deal is known concerning the qualitative behavior of the semiflow generated by (1)–(3) (cf. [9] for pioneering work in this field).

If we introduce the 'flux vectors'

$$j_r(u) := -a_r \nabla u_r\ , \quad r = 1,\ldots,N\ , \tag{4}$$

we can rewrite (1) as

$$\partial_t u_r + \mathrm{div} j_r(u) = f_r(u) \quad \text{in } \Omega \times (0,\infty)\ , \quad r = 1,\ldots,N\ , \tag{5}$$

and the no flux conditions (3) are equivalent to

$$(j_r(u) \mid \nu) = 0 \quad \text{on } \partial\Omega \times (0,\infty)\,, \quad r = 1,\dots,N\,, \tag{6}$$

where $(\cdot \mid \cdot)$ is the inner product in R^n. This formulation of the reaction–diffusion system (1), (3) reveals the importance of the flux vectors $j_r(u)$, since (5) is nothing else but a basic conservation law which governs a great variety of physical, physico-chemical, biological etc. processes (e.g., [6, 8]). However, in a general N-component system the flux vector $j_r(u)$, belonging to the r-th quantity u_r, will not be of the simple form (4), but will depend on the other quantities u_s, $s \neq r$, too. General physical principles (e.g., [6]) imply that in a great many cases $j_r(u)$ is given by

$$j_r(u) := -\sum_{s=1}^{N} [a_{rs}(u)\nabla u_s + \alpha_{rs}(u)u_s]\,, \tag{7}$$

where a_{rs} and α_{rs} depend smoothly on u and $x \in \overline{\Omega}$. (For simplicity we suppress here the x-dependence in our notation.) Moreover, instead of 'no flux' boundary conditions one often finds 'prescribed flux' boundary conditions for some components of u, which may be nonlinear, and Dirichlet boundary conditions for the remaining components. This means that the boundary conditions are of the form

$$\delta_r(j_r(u) \mid \nu) + (1-\delta_r)u_r = \delta_r g_r(u) \quad \text{on } \partial\Omega \times (0,\infty)\,, \tag{8}$$

where $\delta_r \in C(\partial\Omega, \{0,1\})$, so that δ_r is constant on each component of $\partial\Omega$, and g_r is a smooth function of u (and of $x \in \partial\Omega$). Observe that (8) reduces to the Dirichlet boundary condition (2) if $\delta_r = 0$, whereas it is a nonlinear Neumann condition

$$(j_r(u) \mid \nu) = g_r(u)$$

if $\delta_r = 1$.

By inserting (7) in (5) and (8) we obtain a strongly coupled system of quasilinear evolution equations subject to nonlinear boundary conditions. Using matrix notation, we can write this system in the form

$$\begin{aligned} \partial_t u - \partial_j(a(u)\partial_j u) + a_j(u)\partial_j u + a_0(u)u &= f(u) \;\text{ in }\; \Omega \times (0,\infty)\,, \\ \delta(a(u)\partial_\nu u + b(u)u) + (1-\delta)u &= \delta g(u) \text{ on }\; \partial\Omega \times (0,\infty)\,, \end{aligned} \tag{9}$$

where $a(u) := [a_{rs}(u)]_{1 \le r,s \le N}$, $a_j(u)$, $a_0(u)$, and $b(u)$ are suitable $N \times N$-matrices depending smoothly on u, where $\delta := \mathrm{diag}[\delta_1, \dots, \delta_N]$, and where the summation convention is being used, j running from 1 to n.

In fact, we assume that the system (9) depends smoothly upon a real parameter λ. To be more precise, we assume that G is a domain in R^N containing 0 and being starshaped with respect to 0, and that

$$a, a_j, a_0 \in C^\infty(\overline{\Omega} \times G \times R, L(R^N))\,, \quad j = 1,\dots,n\,, \tag{10}$$

where $L(E, F)$ is the Banach space of all bounded linear operators mapping the Banach space E into the Banach space F, where $L(E) := L(E, E)$, and where $L(R^N)$ is identified with the space of $N \times N$-matrices. We also assume that

$$b \in C^\infty(\partial\Omega \times G \times R, L(R^N)) \tag{11}$$

and that

$$f \in C^\infty(\overline{\Omega} \times G \times R, R^N)\,, \quad g \in C^\infty(\partial\Omega \times G \times R, R^N)\ . \tag{12}$$

Finally, we suppose that

$$\sigma(a(x, \eta, \lambda)) \subset \{z \in C; \operatorname{Re} z > 0\}\,, \quad (x, \eta, \lambda) \in \overline{\Omega} \times G \times R\ , \tag{13}$$

where $\sigma(\cdot)$ denotes the spectrum, and — for simplicity — that

$$\delta_1 = \delta_2 = \ldots = \delta_N\ . \tag{14}$$

We fix $p > n + 1$ arbitrarily and put

$$H^1_{p,B} := \{v \in H^1_p(\Omega, R^N); (1 - \delta)\gamma_\partial v = 0\}\ ,$$

where γ_∂ is the trace operator on $\partial\Omega$. Moreover,

$$V := \{v \in H^1_{p,B}; v(\overline{\Omega}) \subset G\}\ ,$$

so that, thanks to Sobolev's embedding theorem, V is an open subset of $H^1_{p,B}$, where the latter is a closed subspace of the Sobolev space $H^1_p(\Omega, R^N)$, whence a Banach space. Given $(v, \lambda) \in V \times R$, we put

$$\mathcal{A}(v, \lambda)u := -\partial_j(a(\cdot, v, \lambda)\partial_j u) + a_j(\cdot, v, \lambda)\partial_j u + a_0(\cdot, v, \lambda)u \tag{15}$$

and

$$\mathcal{B}(v, \lambda)u := \delta(a(\cdot, v, \lambda)\partial_\nu u + b(\cdot, v, \lambda)\gamma_\partial u) + (1 - \delta)\gamma_\partial u \tag{16}$$

for $u \in H^2_p(\Omega, R^N)$. Then (9) can be rewritten as

$$\begin{aligned} \partial_t u + \mathcal{A}(u, \lambda)u &= f(\cdot, u, \lambda) \qquad \text{in } \Omega \times (0, \infty)\ , \\ \mathcal{B}(u, \lambda)u &= \delta g(\cdot, u, \lambda) \qquad \text{on } \partial\Omega \times (0, \infty)\ . \end{aligned} \tag{17)$_\lambda$$

Finally, we assume that

$$(f(\cdot, 0, \cdot), g(\cdot, 0, \cdot)) = (0, 0)\ . \tag{18}$$

Given $(v, \lambda) \in V \times R$, it follows from (13) and (14), thanks to [3, Theorem 4.4], that $(\mathcal{A}(v, \lambda), \mathcal{B}(v, \lambda))$ is normally elliptic in the sense of [3, Section 1]. Hence we deduce from [3, Corollary 9.4], (10)–(12), and (18) that, given any

$$u_0 \in V\ , \tag{19}$$

there exists a unique maximal classical solution

$$u(\cdot, u_0, \lambda) \in C([0, t^+(u_0, \lambda)), V) \cap C^\infty(\overline{\Omega} \times (0, t^+(u_0, \lambda)), R^N) \tag{20}$$

of $(17)_\lambda$ satisfying $u(0,u_0,\lambda)=u_0$ (cf. also formulas (3) and (4) of [3]). Moreover, [3, Corollary 7.4 and Theorem 10.5] guarantee that

$$(t,u_0)\mapsto u(t,u_0,\lambda) \tag{21}$$

is a smooth local semiflow on V depending smoothly on $\lambda\in R$. Observe that (18) implies that 0 is a restpoint for this semiflow, independently of $\lambda\in R$.

In this paper we address the question whether $0\in V$ is a bifurcation point for periodic orbits of this semiflow. This is known to be the case for semilinear reaction–diffusion systems of the form (1), where f depends on $\lambda\in R$, provided, for example, the linearized problem (at $u=0$) possesses a pair of simple eigenvalues $\pm i\omega_0$ crossing the imaginary axis as λ crosses 0, say, and there are no other eigenvalues at $\lambda=0$, which are integer multiples of $i\omega_0$ (e.g., [5,9]). It is clear from this eigenvalue condition that Hopf bifurcation, that is, bifurcation of periodic orbits from the rest point 0, is caused by the coupling in the reaction terms in the case of semilinear reaction–diffusion systems with symmetric diagonal principal part, whose prototype is given by (1)–(3). In the case of strongly coupled quasilinear reaction–diffusion systems of the form (17), it is to be expected that Hopf bifurcation can be caused by the coupling in the principal part, even if the reaction terms are decoupled.

To see this, we consider a simple two component **model problem**, given by (5) and (8), where $\delta_1=\delta_2=1$, and

$$\begin{aligned} j_1(u) &:= -a_{11}(u,\lambda)\nabla u_1 - a_{12}(u,\lambda)\nabla u_2\ , \\ j_2(u) &:= a_{21}(u,\lambda)\nabla u_1\quad , \end{aligned} \tag{22}$$

where we assume that $a_{11},a_{12},a_{21}\in C^\infty(R^2\times R,R)$ and

$$a_{11}(\eta,\lambda)>0\,,\quad a_{12}(\eta,\lambda)a_{21}(\eta,\lambda)>0\,,\quad (\eta,\lambda)\in R^2\times R\ . \tag{23}$$

We also assume that

$$f(u,\lambda):=(\varphi(u_1,\lambda)u_1,0) \tag{24}$$

and that

$$\begin{aligned} g_1(u,\lambda) &:= -\,\alpha_{11}(u,\lambda)u_1 - \alpha_{12}(u,\lambda)u_2\ , \\ g_2(u,\lambda) &:= \alpha_{21}(u,\lambda)u_1\ , \end{aligned} \tag{25}$$

where $\varphi\in C^\infty(R\times R,R)$ and $\alpha_{11},\alpha_{12},\alpha_{21}\in C^\infty(R^2\times R,R)$. Letting

$$a(u,\lambda):=\begin{bmatrix} a_{11}(u,\lambda) & a_{12}(u,\lambda) \\ -a_{21}(u,\lambda) & 0 \end{bmatrix}$$

our system takes the form

$$\begin{aligned} \partial_t u - \partial_j(a(u,\lambda)\partial_j u) &= f(u,\lambda) \quad \text{in } \Omega\times(0,\infty)\ , \\ a(u,\lambda)\partial_\nu u &= g(u,\lambda) \quad \text{on } \partial\Omega\times(0,\infty)\ , \end{aligned} \tag{26}$$

and (23) guarantees that condition (13) is satisfied, where now $G:=R^2$. Setting

$$\alpha(u,\lambda):=\begin{bmatrix} \alpha_{11}(u,\lambda) & \alpha_{12}(u,\lambda) \\ -\alpha_{21}(u,\lambda) & 0 \end{bmatrix}$$

we assume — to simplify some computations below — that there exists a positive constant β so that

$$\alpha(0,\lambda) = \beta a(0,\lambda), \quad \lambda \in R . \tag{27}$$

It is useful to interpret (22)–(27) as a **two populations model**. Then — assuming that $a_{12}(u,\lambda) > 0$ — the first quantity in (22) means that 'the flux of the first population goes in the direction where the density of its own species decreases and also to the places where the density of the second population ('the enemies') is low'. Thus $a_{11}(u)\nabla u_1$ can be interpreted as a 'social friction' term, which prevents 'overcrowdedness'.The second quantity simply means that the second population 'runs in the direction where there is a higher density of the first species'. Similarly, the boundary condition

$$(j_r(u,\lambda) \mid \nu) = g_r(u,\lambda) \quad \text{on } \partial\Omega, \quad r = 1,2 \ ,$$

thanks to (25), means that 'the first population tends to stay away from those parts of the boundary, where there is already a lot of its own species or a lot of enemies', whereas the second population 'wants to go to those places on the boundary, where there is a lot of the first species'. This is true for small populations, thanks to (27).

Thus the above model essentially says that the 'second species chases the first one', that the first one 'runs away from the second species' and that it 'diffuses' too, that is, 'runs away from places of high density of its own species'. Consequently, it seems reasonable to expect that there exists periodic behavior if no diffusive behavior of the first population occurs. Thus, if one wants to have a periodic behavior in the presence of diffusion, one will have to 'produce' the first species at an appropriate rate in order to compensate for the 'loss' caused by diffusion. It will be a simple consequence of our general results that this is indeed the case. This shows that periodic behavior is caused by the strong coupling of the highest order term — by 'cross diffusion terms' — and not by the reaction terms. It is now obvious how (22)–(27) can be interpreted if $a_{12}(u,\lambda) < 0$.

We associate with $(17)_\lambda$ the linear elliptic eigenvalue problem at $u = 0$:

$$\begin{aligned} [-\mathcal{A}(0,\lambda) + \partial_2 f(\cdot,0,\lambda)]\, v &= \mu(\lambda) v \quad \text{in } \Omega , \\ [-\mathcal{B}(0,\lambda) + \delta\partial_2 g(\cdot,0,\lambda)]\, v &= 0 \qquad\quad \text{on } \partial\Omega \end{aligned} \tag{28$_\lambda$}$$

Since the L_p-realization of this problem has a compact resolvent, this eigenvalue problem is well posed.

We can now formulate the main existence theorem for Hopf bifurcation.

Theorem 1. *Suppose that*

1. $\{\pm i\omega_0\}$ *are simple eigenvalues of* $(28)_0$, *where* $\omega_0 > 0$.
2. $(28)_0$ *has no eigenvalues of the form* $ik\omega_0$ *for* $k \in Z \setminus \{\pm 1\}$.
3. $\partial_\lambda Re\mu(0) \neq 0$, *where* $\mu(\lambda)$ *is the unique eigenvalue of* $(28)_\lambda$ *for* λ *in a neighborhood of 0 satisfying* $\mu(0) = i\omega_0$.

Then the system $(17)_\lambda$ *has in a neighborhood of* $(0,0) \in V \times R$ *a unique one-parameter family* $\{\gamma(s); 0 < s < \varepsilon\}$ *of noncritical periodic orbits. More precisely: there exist* $\varepsilon > 0$ *and*

$$(u(\cdot), T(\cdot), \lambda(\cdot)) \in C^\infty((-\varepsilon, \varepsilon), V \times R \times R)$$

satisfying

$$(u(0), T(0), \lambda(0)) = (0, \frac{2\pi}{\omega_0}, 0)$$

such that

$$\gamma(s) := \gamma(u(s))$$

is for $0 < s < |\varepsilon|$ *a noncritical periodic orbit of* $(17)_{\lambda(s)}$ *of period* $T(s)$ *passing through* $u(s) \in V$*. If* $0 < s_1 < s_2 < \varepsilon$*, then* $\gamma(s_1) \neq \gamma(s_2)$*. The family* $\{\gamma(s); 0 < s < \varepsilon\}$ *contains every noncritical periodic orbit of (17) lying in a suitable neighborhood of* $(0, T(0), 0) \in V \times R \times R$*.*

It should be remarked that the existence of a unique smooth continuation $\mu(\lambda)$ of $i\omega_0$ for λ near 0 is part of the Theorem.

Observe that condition (3) is the standard 'Hopf condition' guaranteeing that a pair of simple eigenvalues crosses the imaginary axis at $\lambda = 0$ with nonzero speed ([5, 9, 11]).

In order to apply the Theorem to our **example** (22)–(27), we have to study the linear elliptic eigenvalue problem

$$\begin{aligned} a(0,\lambda)\Delta v + \partial_1 f(0,\lambda)v &= \mu(\lambda)v \quad \text{in } \Omega\,, \\ a(0,\lambda)\partial_\nu v + \alpha(0,\lambda)v &= 0 \quad \text{on } \partial\Omega\,, \end{aligned} \tag{29}$$

which, thanks to assumption (27), is equivalent to the system:

$$\begin{aligned} a(0,\lambda)\Delta v + \partial_1 f(0,\lambda)v &= \mu(\lambda)v \quad \text{in } \Omega\,, \\ \partial_\nu v + \beta v &= 0 \quad \text{on } \partial\Omega\,. \end{aligned} \tag{30}$$

Consider the scalar eigenvalue problem

$$\begin{aligned} \Delta\varphi &= \kappa\varphi \quad \text{in } \Omega\,, \\ \partial_\nu\varphi + \beta\varphi &= 0 \quad \text{on } \partial\Omega\,, \end{aligned} \tag{31}$$

where $\varphi : \Omega \to C$, and denote by

$$0 > \kappa_0 > \kappa_1 \geq \kappa_2 \geq \ldots$$

the sequence of eigenvalues of (31), each one counted according to its multiplicity. By using the fact that the orthonormalized sequence (φ_k) of corresponding eigenfunctions forms an orthonormal basis in $L_2(\Omega, C)$ and that the spectrum and the eigenfunctions of (31) are independent of $q \in (1, \infty)$, if (31) is being considered as an eigenvalue problem in $L_q(\Omega)$, it is not difficult to see that the eigenvalues of (30) are given by $\{\mu_{k,j}(\lambda);\ k \in N,\ j = 1,2\}$, where $\mu_{k,1}(\lambda)$ and $\mu_{k,2}(\lambda)$ are the eigenvalues of the matrices

$$a(0,\lambda)\kappa_k + \partial_1 f(0,\lambda) \in L(R^2)\,, \quad k \in N\ .$$

Thanks to the special form of f in (24), we see that

$$a(0,\lambda)\kappa_k + \partial_1 f(0,\lambda) = \begin{bmatrix} a_{11}(0,\lambda)\kappa_k + \varphi(0,\lambda) & a_{12}(0,\lambda)\kappa_k \\ -a_{21}(0,\lambda)\kappa_k & 0 \end{bmatrix}, \quad k \in N.$$

Denoting by $\sigma(\lambda)$ the spectrum of (29), that is, the set of eigenvalues of (29), we easily deduce from the above that

$$\sigma(\lambda) \subset [\mathrm{Re}\, z < 0] \quad \text{if} \quad \varphi(0,\lambda) < -a_{11}(0,\lambda)\kappa_0 \tag{32}$$

and that

$$\sigma(\lambda) \subset [\mathrm{Re}\, z \leq 0]\,, \quad \sigma(\lambda) \cap iR = \{\pm i\omega_0\}\ ,$$

where $\omega_0 := \kappa_0\sqrt{a_{11}(0,\lambda)a_{21}(0)}$, provided $\varphi(0,\lambda) = -a_{11}(0,\lambda)\kappa_0$.

We now assume that

$$\varphi(0,\lambda) < -a_{11}(0,\lambda)\kappa_0 \quad \text{if} \quad \lambda < 0\,,\ \varphi(0,0) = -a_{11}(0,0)\kappa_0\ . \tag{33}$$

Then it follows that

$$\partial_\lambda \mathrm{Re}\mu(0) = 1/2(\partial_2 a_{11}(0,0)\kappa_0 + \partial_2\varphi(0,0))\ .$$

We deduce from (32), (33) and [7], for example, that zero is an asymptotically stable critical point of the semiflow induced by (22)–(27), provided $\lambda < 0$, which loses its stability at $\lambda = 0$. If

$$\partial_2 a_{11}(0,0)\kappa_0 + \partial_2\varphi(0,0) \neq 0\ , \tag{34}$$

it follows from the Theorem that a ‘branch’ of periodic orbits of period close to

$$\frac{2\pi}{\kappa_0\sqrt{a_{12}(0,0)a_{21}(0,0)}}$$

bifurcates off the restpoint as λ crosses 0. Observe that conditions (33) and (34) are satisfied, for example, if a_{11} is independent of λ, that is, $a_{11}(u,\lambda) = a_{11}(u)$, and $\varphi(u,\lambda) = \lambda - a_{11}(u)\kappa_0$. This shows that our heuristic considerations based upon the population model interpretation were correct.

2 Proof of the Theorem

Given $q \in (1,\infty)$, we put

$$H^s_{q,B} := \begin{cases} \{v \in H^s_q;\quad (1-\delta)\gamma_\partial v = 0\}, & \frac{1}{q} < s < 1 + \frac{1}{q}\,, \\ H^s_q\,, & -1 + \frac{1}{q} < s < \frac{1}{q} \\ (H^{-s}_{q',B})'\,, & -2 + \frac{1}{q} < s < -1 + \frac{1}{q}\,, \end{cases} \tag{35}$$

$H^s_q := H^s_q(\Omega, R^N)$ being the Bessel potential spaces and the duality pairing being induced by $\langle u, v\rangle := \int_\Omega \langle u(x), v(x)\rangle\, dx$, where $\langle u(x), v(x)\rangle$ is the standard duality pairing in C^N. It follows from [3, Proposition 5.4] that (except for equivalent norms)

$$H^s_{q,B} = [H^{s_0}_{q,B}, H^{s_1}_{q,B}]_{\frac{s-s_0}{s_1-s_0}}\,, \tag{36}$$

for $s_0 < s < s_1$ and $s, s_0, s_1 \in (-2+1/q, 1+1/q) \setminus Z + 1/q$, where $[\cdot,\cdot]_\theta$, $0 < \theta < 1$, denotes the complex interpolation functor.

Given $u \in C(\overline{\Omega}, G)$ and $\lambda \in R$, we put

$$\begin{aligned} a(u,\lambda)(v,w) := {} & \langle \partial_j v, a(\cdot,u,\lambda)\partial_j w\rangle + \langle v, a_j(\cdot,u,\lambda)\partial_j w + a_0(\cdot,u,\lambda)w\rangle \\ & + \langle \gamma_\partial v, b(\cdot,u,\lambda)\gamma_\partial w\rangle_\partial\,, \quad (v,w) \in H^1_{p',B} \times H^1_{p,B}\,, \end{aligned}$$

where $\langle v, w\rangle_\partial := \int_{\partial\Omega} \langle v(x), w(x)\rangle\, d\sigma$. It follows from [3, Theorem 6.1] that

$$[(u,\lambda) \mapsto a(u,\lambda)] \in C^\infty(V \times R, L^2(H^{2-s}_{p',B}, H^s_{p,B}; R)) \tag{37}$$

for $1 \le s < 1+1/p$, where $L^2(E, F; R)$ denotes the Banach space of all continuous bilinear forms on $E \times F$.

We fix $s \in (1, 1+1/p)$ and put

$$E_1 := H^s_{p,B}\,, \quad E_0 := H^{s-2}_{p,B}\,.$$

Then it follows from (37) that there exists

$$A(\cdot,\cdot) \in C^\infty(V \times R, L(E_1, E_0))$$

satisfying

$$\langle v, A(u,\lambda)w\rangle = a(u,\lambda)(v,w)\,, \quad (u,\lambda) \in V \times R\,, \quad (v,w) \in E_0' \times E_1\,.$$

We also put

$$F(u,\lambda) := f(\cdot,u,\lambda) + \gamma_\partial' g(\cdot,u,\lambda)\,, \quad (u,\lambda) \in V \times R\,,$$

where γ_∂' is the dual of the trace operator $\gamma_\partial \in L(H^\sigma_{p',B}, B^{\sigma-1/p'}_{p'})$ for $1 - 1/p < \sigma < 2-s$, where $B^{\sigma-1/p'}_{p'} := B^{\sigma-1/p'}_{p',p'}(\partial\Omega, R^N)$ are Besov spaces. Then it is easily verified that

$$F \in C^\infty(V \times R, E_\gamma)\,,$$

where $E_\gamma := H^{-\sigma}_{p,B}$.

We associate with $(17)_\lambda$ the abstract Cauchy problem

$$\dot{u} + A(u,\lambda)u = F(u,\lambda)\,, \quad 0 < t < \infty\,, \quad u(0) = u_0 \tag{$(38)_\lambda$}$$

in E_0. Since $(\mathcal{A}(u,\lambda), \mathcal{B}(u,\lambda))$ is normally elliptic for $(u,\lambda) \in V \times R$, it follows from [3, Theorem 5.6] that $-A(u,\lambda)$ is for each $(u,\lambda) \in V \times R$ the infinitesimal generator of a strongly continuous analytic semigroup on E_0. Now the results of [3] imply that $(38)_\lambda$ has for each $u_0 \in V$ a unique maximal solution

$$\begin{aligned} u(\cdot, u_0, \lambda) \in C([0, t^+(u_0,\lambda)), V) &\cap C((0, t^+(u_0,\lambda)), E_1) \\ &\cap C^1((0, t^+(u_0,\lambda)), E_0) \end{aligned}$$

such that

$$(t, u_0) \mapsto u(t, u_0, \lambda)$$

is a local semiflow on V, which depends smoothly upon (t, u_0, λ) for $t > 0$ (cf. also [2]). Moreover, the derivatives of u with respect to the various variables are the unique solutions of the various linear Cauchy problems which are obtained by linearizing $(38)_\lambda$ at $u(\cdot, u_0, \lambda)$ with respect to the corresponding variables (cf. [2, Section 11] and [3, Theorem 10.5]). Finally, it follows from the considerations in [3, Section 9] that $u(\cdot, u_0, \lambda)$ is the unique maximal classical solution of $(17)_\lambda$ satisfying (20) and $u(0, u_0, \lambda) = u_0$.

At this point there are two distinct possibilities to prove the Theorem. Namely, we can either apply the results of Da Prato and Lunardi [11], or we can modify the finite–dimensional approach given in [4].

In the *first case*, we put $V_1 := V \cap E_1$, equipped with the E_1-topology. We then define

$$\Phi \in C^\infty(V_1 \times R, E_0)$$

by

$$\Phi(u,\lambda) := -A(u,\lambda)u + F(u,\lambda)\ .$$

Since

$$\partial_1\Phi(0,0) = -A(0,0) + \partial_1 F(0,0)$$

and $\partial_1 F(0,0) \in L(E_\alpha, E_0)$, where $E_\alpha := H^1_{p,B}$, a standard perturbation theorem for analytic semigroups (e.g., [10, Corollary 3.2.4]) implies that $\partial_1\Phi(0,0) \in L(E_1, E_0)$ is the infinitesimal generator of a strongly continuous analytic semigroup on E_0. Since E_1 is compactly injected in E_0 (e.g., [3, Proof of Lemma 8.1]), we see that $\partial_1\Phi(0,0)$ has a compact resolvent (considered as a linear operator in E_0). Finally, it is easily verified that the eigenvalue problem

$$\partial_1\Phi(0,\lambda)v = \mu(\lambda)v$$

in E_0 is equivalent to $(28)_\lambda$. Hence the Theorem follows from [11, Theorem 2.2].

As for the *second approach*, we replace the independent variable t in $(38)_\lambda$ by $\frac{\tau_0+\tau}{2\pi}t$, where $\tau_0 := 2\pi/\omega_0$ and $\tau \in R$. Then $(38)_\lambda$ transforms into

$$\dot{v} = h(v,\sigma)\,, \quad \sigma := (\lambda,\tau)\,, \quad h(v,\sigma) := \tfrac{\tau_0+\tau}{2\pi}[-A(v,\lambda)v + F(v,\lambda)]\ . \tag{39}$$

Let $v(\cdot,\xi,\sigma)$ be the unique maximal solution of (39) satisfying $v(0,\xi,\sigma)=\xi\in V$. Since $0\in V$ is a critical point for the semiflow generated by (39), there exists a neighborhood $W\times\Sigma$ of 0 in $V\times R^2$ such that

$$[(\xi,\sigma)\mapsto g(\xi,\sigma):=\xi-v(2\pi,\xi,\sigma)]\in C^\infty(W\times\Sigma,E_\alpha)\ .$$

It is easily verified that

$$\partial_1 g(0,0)=1-e^{2\pi L}$$

in E_α, where $\{e^{tL};t\ \geq\ 0\}$ is the semigroup generated (in E_α) by $L:=\frac{\tau_0}{2\pi}[-A(0,0)+\partial_1 F(0,0)]$. By carrying out obvious modifications, one verifies that the proofs of [4, Theorems (26.21) and (26.25)] remain true. For this it suffices to observe that g is a nonlinear Fredholm operator of index 2 and $\{e^{tL};t\geq 0\}$ restricts to a flow on $\ker(1-e^{2\pi L})$. Thus the Theorem follows by this approach too. □

3 Remarks

• It is clear that our regularity conditions can be considerably weakened. Moreover, it is also clear that (15) and (16) can be replaced by more general systems. It is only necessary that $(\mathcal{A}(v,\lambda),\mathcal{B}(v,\lambda))$ are normally elliptic for $(v,\lambda)\in V\times R$ (cf. [3] for more details).

• It is, of course, of considerable interest to study the stability of the bifurcating periodic orbits. We refrain from doing this here, but we mention that again there are two possible approaches to this problem. Namely we can either use the techniques of Da Prato and Lunardi [12], or we can modify the proofs of [4, Theorems (27.11) and (27.14)].

• In concrete applications it is, of course, essential to determine to some extent the eigenvalues of the linear elliptic system $(28)_\lambda$, at least so far that conditions (i)–(iii) of the Theorem can be verified. Whereas a lot is known about the eigenvalue problem for a single elliptic equation, not very much seems to be known for elliptic systems. For example, we are not able to verify conditions (i)–(iii) of the Theorem for the 'simple' problem (29) if assumption (27) is not satisfied or if the coefficients of (22) depend upon $x\in\overline{\Omega}$ for $u=0$.

• The proof of [2, Proposition 6.1], which is the basis for the demonstration in [3] that $(17)_\lambda$ generates a local semiflow on V, contains a mistake. In fact, the assumption in [2, Q1] that E_α be relatively complete with respect to E_β should be dropped. Moreover, hypothesis (Q2) in [2] can be simplified by dropping the 'local regular boundedness' assumption. Thus — using the notations of [2] — the results of the latter paper remain valid under the following simplified assumptions:

(Q1) $\overline{E}\in\mathcal{B}_2\,,\ 0<\gamma\leq\beta<\alpha<1\,,\ \beta$ and α are well related, and (γ,α) and (γ,β) are stable.

(Q2) Λ is a metric space,

$$A \in C^{\rho,1-,\sigma}([0,\tau] \times V_\beta \times \Lambda, \mathcal{H}(\overline{E}))$$

for some $\rho \in (0,1)$, $\sigma \in [0,1) \cup \{1-\}$, and $T > 0$, and

$$f \in C^{0,1-,\sigma}([0,T] \times V_\beta \times \Lambda, E_\gamma) \ .$$

To see this, it suffices to observe that the proof of [1, Lemma 6.2] gives the desired result, since the 'local regular boundedness', which is needed, is a consequence of the continuity hypotheses in (Q2).

References

1. Amann, H. (1986): Quasilinear Evolution Equations and Parabolic Systems. Trans. Amer. Math. Soc. **29**, 191–227
2. Amann, H. (1988): Dynamic Theory of Quasilinear Parabolic Equations—I. Abstract Evolution Equations. Nonlinear Analysis, TM & A. **12**, 895–919
3. Amann, H. (1990): Dynamic Theory of Quasilinear Parabolic Equations–I. Reaction–Diffusion Systems. Diff.–Integral Eq. **3**, 13–75
4. Amann, H. (1990): Ordinary Differential Equations. An Introduction to Nonlinear Analysis, de Gruyter, Berlin
5. Crandall, M.G. , Rabinowitz, P.H. (1977/8): The Hopf Bifurcation Theorem in Infinite Dimensions. Arch. Rat. Mech. Anal. **67**, 53–72
6. de Groot, S.R. , Mazur, P. (1962): Non–Equilibrium Thermodynamics. North Holland, Amsterdam
7. Drangeid, A.K. (1989): The Principle of Linearized Stability for Quasilinear Parabolic Evolution Equations. Nonlinear Analysis, TM &, A **13**, 1091–1113
8. Fife, P. (1979): Mathematical Aspects of Reacting and Diffusing Systems. Lecture Notes in Biomath. No. 28, Springer Verlag, Berlin-Heidelberg-New York
9. Henry, D. (1981): Geometric Theory of Semilinear Parablic Equations. Lecture Notes in Math. No. 849, Berlin-Heidelberg-New York
10. Pazy, A. (1983): Semigroups of Linear Operators and Applications to Partial Differential Operators. Springer Verlag, Berlin-Heidelberg-New York
11. Da Prato, G. , Lunardi, A. (1986): Hopf Bifurcation for Fully Nonlinear Equations in Banach Spaces. Ann. Inst. Henri Poincaré–Analyse non Linéaire. **3**, 315–329
12. Da Prato, G. , Lunardi, A. (1988): Stability, Instability and Center Manifold Theorem for Fully Nonlinear Autonomous Parabolic Equations in Banach Spaces. Arch. Rat. Mech. Anal. **101**, 115–141

Monotone Semi-Flows Which Have a Monotone First Integral

Ovide Arino

Départment de Mathématiques, Faculté des Sciences, Avenue de l'Université, 64000 Pau, France

Abstract

We present a fairly general study for a class of semi-flows defined on an abstract space, which are monotonically increasing and possess a first integral, also increasing. Examples of that are systems of delay differential equations generated by compartmental models. Under reasonable restrictions, a complete description of the asymptotic behavior can be obtained in situations including almost-periodic dependence. We will not go into these details here. Rather the paper intends to enlighten the main aspects of the theory. Finally, a comparison with related literature is made.

Introduction

My purpose here is to present some results on the asymptotic behavior of solutions of monotone semi-flows which have a monotone first integral. First, I will present the class of equations which motivated my interest in this particular question. It goes back a little more than ten years to some work I did in collaboration with Pierre Séguier. I discussed this work with Ken Cooke during a Conference in Italy (Cortona, July, 1979). Ken encouraged me to improve the results. The paper which stemmed from this effort appeared in the Journal of Mathematical Analysis and Applications (1984); it was my first ever paper in such a journal. This, by the way, made my choice of a subject for this Conference quite obvious. The work with P.Séguier was restricted to a scalar equation and the techniques we used there are not easily extendable to the vector case. In collaboration with E.Haourigui, I developed a more general approach. This allowed us to extend my previous results with P.Séguier to the case of a system of delay differential equations, which had been introduced by Istvan Györi ([7]) as a general compartmental model.

In Sect. 1, I will briefly illustrate on this class of equations the methods we used and the results we obtained. It should be clear, looking at this example, that it is only a particular example of a broader class; indeed, in collaboration with F.Bourad ([5]), I extend the results obtained previously with E.Haourigui to abstract dynamical systems. In Sect. 2, I will give a brief presentation of the appropriate setting, and derive the results in the autonomous case. The general situation will be considered elsewhere. As far as I know, monotonicity was not very popular at the time I started working on delay differential equations of monotone type, not counting, of course, specific areas such as parabolic equations where it led to such concepts as upper, lower solutions. Things changed when M.Hirsch began to produce his fundamental results on monotone dynamical systems ([9]). A number of people have since pursued the idea of applying Hirsch's results to monotone systems generated by delay or functional differential equations. Probably, the most significant contribution in that respect was done by H.Smith ([13]). In the last section of this paper, I will show how some of Hirsch's results can be used in the autonomous case to study the asymptotic behavior of solutions for systems with a first integral. A few remarks on related works will also be found there.

1 A class of delay differential equations with a first integral

The first equation I considered is

$$\frac{dx}{dt} = f(t-1, x(t-1)) - f(t, x(t)), \tag{1}$$

which can also be written in the integral form:

$$x(t) = C - \int_{t-1}^{t} f(s, x(s))ds. \tag{2}$$

here, $f(t,x)$ is a function: $\Re \times \Re \to \Re$, continuous in (t,x), increasing in x, with, in addition, enough regularity to ensure uniqueness of solutions. It is convenient to assume that $f(t,0) = 0$, which means that 0 is a trivial solution of (1). This condition is motivated by biological considerations. Equation (1) may be viewed as the non-autonomous form of a model introduced by K.Cooke and J.Yorke ([6]). A major difference, however, is that these authors did not assume monotonicity in their model or at least did not make any use of it.

Let us define the following nonlinear functional

$$\mathcal{J}(t,\phi) = \phi(0) + \int_{-1}^{0} f(t+s, \phi(s))ds. \tag{3}$$

It is then a matter of straightforward computation to observe that $\mathcal{J}(t, x_t) =$ Constant, if x is a solution of (1). This is the property we refer to when we say that (1) has a first integral. We will now state in a theorem the main properties and asymptotic results we obtained for (1).

Theorem 1.1 **([4])** *Let ϕ, ψ be given in $C([-1,0],\Re)$, $t_0 \in \Re$. Denote by $x(t)$ (resp. $y(t)$) the solution of (1) such that $x_{t_0} = \phi$ (resp. $y_{t_0} = \psi$).*

1) Suppose first that $\phi \geq 0$ (resp. $\phi \leq 0$). Then, $0 \leq x(t) \leq \mathcal{J}(t_0, \phi)$, $t \geq t_0$, (resp. $\mathcal{J}(t_0,\phi) \leq x(t) \leq 0$, $t \geq t_0$).

2) Suppose now $\phi < \psi$. Then, $\mathcal{J}(t_0,\phi) < \mathcal{J}(t_0,\psi)$, and $x(t) \leq y(t)$, for $t \geq t_0$, and $x(t) < y(t)$ for $t \geq t_0 + 1$ (monotonicity of the first integral and the solution operator).

3) $\mathcal{J}(t_0, -|\phi|) \leq x(t) \leq \mathcal{J}(t_0, |\phi|)$, for $t \geq t_0$, so, in particular, every solution of (1) is bounded.

4) Suppose $\mathcal{J}(t_0,\phi) = \mathcal{J}(t_0,\psi)$. Then, $\lim_{t\to\infty} x(t) - y(t) = 0$.

5) Suppose $f(t,x)$ is periodic in t, of period T. Let $\alpha \in \Re$. Then, (1) has exactly one periodic solution p, of period T, such that $\mathcal{J}(t_0, p_{t_0}) = \alpha$; any other periodic solution q of (1) such that $\mathcal{J}(t_0, q_{t_0}) = \alpha$ is just a translate of p. Finally, each solution of (1) is asymptotically periodic.

The results obtained in ([4]) go far beyond the periodic case. However, for a better understanding of the present paper, we will restrict ourselves to this case. The main technique involved in the proof of these results is the comparison of solutions and, through that, an estimate of successive minima and maxima of a solution.

The method seems to depend a lot on the fact that solutions are scalar. At the time these results appeared, I.Györi ([7]) introduced a general model which in fact looks very much like the vector formulation of equation (1); namely,

$$\frac{dx_i}{dt} = -\sum_{j=1}^{n} g_{j,i}(t, x_i(t)) + \sum_{j=1}^{n} g_{i,j}(t - \tau_{i,j}, x_j(t - \tau_{i,j})), \quad i = 1, \ldots, n. \tag{4}$$

If n=1, this reduces to (1). We may observe that (4) is slightly different from Györi's model in which $g_{i,j}(t,u) = g_{i,j}(u)$, but a forcing term, a sort of source term, is allowed. The difference can be interpreted by saying that the system modelled by (4) is closed with no input or output while Györi's model assumes the system may be open, however asymptotically closed.

The main feature of (4), shared with Györi's model, is that the functions $g_{i,j}(t,u)$ are increasing in u. This, together with some reasonable regularity assumption, yields the same monotonicity property of the solution operator as for (1). It is also easily verified that the function

$$\mathcal{J}(y,\phi) = \sum_{i=1}^{n} \phi_i(0) + \sum_{i,j=1}^{n} \int_{-\tau_{i,j}}^{0} g_{i,j}(t+s, \phi_j(s))\, ds, \tag{5}$$

is a constant along the solutions of (4).

The first three properties stated in Theorem 1.1 can be easily extended to (4) with appropriate care in defining the state space. For each $j \in [1,n]$, we denote by $\bar{\tau}_j = max\{\tau_{i,j} : 1 \leq j \leq n\}$. The optimal state space in the sense of a remark made by H.Smith ([13]) is then

$$\prod_{i=1}^{n} C([-\bar{\tau}_i, 0], \Re) \equiv X. \tag{6}$$

Let us now recall the following definitions which are more or less standard in the literature of monotone flows.

We first need to specify the orders we will consider. On $\Re^n$, it is the "usual" order:

$$x \leq y \quad \text{iff} \quad x_i \leq y_i, \quad \text{for} \quad 1 \leq i \leq n. \tag{7}$$

On X,

$$\phi \leq \psi \text{ iff } \phi_i \leq \psi_i \text{ [that is, } \phi_i(\theta) \leq \psi_i(\theta), \text{ for all } \theta] \text{ for } 1 \leq i \leq n. \tag{8}$$

We write $x < y$ to mean that $x \leq y$ and $x \neq y$; this convention holds for any order. Both the orders we introduced are associated to a cone with a non-empty interior. This gives the possibility of considering a relation stronger than $<$. We write

$$x \ll y \text{ iff } y - x \text{ lies in the interior of the positive cone.} \tag{9}$$

We denote by $|.|$ the l^1 norm in $\Re^n$, $|x| = \sum_{i=1}^{n} |x_i|$; likewise, we denote by $|.|$ the norm on X, $|\phi| = \sum_{i=1}^{n} max\{|\phi_i(\theta)| : -1 \leq \theta \leq 0\}$.

Theorem 1.2 ([3]) *Let ϕ, ψ be given in X, t_o in $\Re$. Denote by $x(t)$ (resp. $y(t)$) the solution of (4) such that $x_{t_0} = \phi$ (resp. $y_{t_0} = \psi$).*

1) $\mathcal{J}(t_0, -|\phi|) \leq |x(t)| \leq \mathcal{J}(t_0, |\phi|)$, for $t \geq t_0$. So, in particular, every solution of (4) is bounded.

2) Suppose $\phi < \psi$. Then, $\mathcal{J}(t_0, \phi) < \mathcal{J}(t_0, \psi)$, and $x(t) \leq y(t)$ for $t \geq t_0$; moreover, $x(t) \ll y(t)$ for $t \geq t_0 + \bar{\tau}$, where $\bar{\tau} = max\{\bar{\tau}_j : 1 \leq j \leq n\}$.

3) Suppose $\mathcal{J}(t_0, \phi) = \mathcal{J}(t_0, \psi)$. Then $\lim_{t\to\infty} x(t) - y(t) = 0$.

Finally, if the functions $g_{i,j}$ are periodic in time, with the same period, then, as in 4) of Theorem 1.1, we can conclude that there is a periodic solution on each "level set", [that is, for each first value of the first integral], and the solutions are asymptotically periodic. The main tool in the derivation of these results is a Lyapunov functional on pairs of solutions, which is defined as follows:

$$V(t, \phi, \psi) = \mathcal{J}(t, [\phi - \psi]^+ + \psi) - \mathcal{J}(t, \psi). \tag{10}$$

Here, $[]^+$ denotes the positive part of an element in X, $[x]^+ = sup\{x, 0\}$. Similarly, if we denote $[]^-$ the negative part of an element in X, that is, $[x]^- = inf\{x, 0\}$, we can define another Lyapunov functional

$$W(t, \phi, \psi) = \mathcal{J}(t, [\phi - \psi]^- + \psi) - \mathcal{J}(t, \psi). \tag{11}$$

It is not difficult to see that all these results can be extended to a much more general situation, using the framework of dynamical systems. This has been done in detail in F.Bourad's thesis ([5]). I will now discuss a few aspects of this theory.

2 Monotone semi-dynamical systems with a monotone first integral

We first consider the case of autonomous systems: the essential ingredients of the proof of asymptotic behavior are more apparent in this simplified situation. The extension to non-autonomous systems in the abstract setting seems to pose some specific problems, which require additional assumptions. We will briefly discuss these questions.

The "abstract" setting

Given a Banach lattice X ([11]); we use the notations $\leq$, $<$, and $\ll$ in the sense defined in Sect. 1. The lattice structure allows us to define $[]^+$ $[]^-$ and these applications are continuous. Moreover, we assume that the positive cone X^+ has a non-empty interior. On X, we suppose a semi-flow is defined, denoted π or $\pi(t, .) : X \to X$, for $t \geq 0$, monotone in the sense that

$$x \leq y \Rightarrow \pi(t,x) \leq \pi(t,y), \text{ for } t \geq 0. \tag{MS}$$

We will need in fact a stronger monotonicity, denoted (SMS), that is, if $x < y$, there exists $t_0 \geq 0$, such that $t \geq t_0 \Rightarrow \pi(t,x) \ll \pi(t,y)$. Note that t_0 may depend on x and y.

Our next two assumptions do not involve the order:

$$\pi(t,0) = 0, \text{ and each positive orbit is precompact.} \tag{12}$$

We now assume the existence of a functional $\mathcal{J}$, $\mathcal{J} : X \to \Re$, continuous, $\mathcal{J}(0) = 0$, and $\mathcal{J}$ is strictly monotone in the sense that

$$x \ll y \Rightarrow \mathcal{J}(x) < \mathcal{J}(y). \tag{SMI}$$

Finally, $\mathcal{J}$ is a first integral of π, that is, for each $x \in X$,

$$\mathcal{J}(\pi(t,x)) = Constant.$$

Proposition 2.1 ([2]) *Denote $V(x) = \mathcal{J}(x^+)$ (resp. $W(x) = \mathcal{J}(x^-)$). Then, $V \geq 0$, V is non-increasing along the solutions of π (resp. $W \leq 0$, W is non-decreasing...).*

The proof of this is not difficult. We refer to ([2],[5]) for it. However, it is the essential ingredient when deriving the asymptotic behavior of the solutions.

Theorem 2.2 ([2]) *Let X be a Banach lattice, such that X^+(the positive cone) has a non-empty interior. Let π be a semi-dynamical system on X, such that:*
i) $\pi(t,0) = 0$.
ii) The positive orbits of π are precompact.
iii) π is both (MS) and (SMS).

iv) π has a first integral $\mathcal{J}$, $\mathcal{J} : X \to \Re$, $\mathcal{J}(0) = 0$, $\mathcal{J}$ is continuous and (SMI). Then, for every $x \in X$, such that $\mathcal{J}(x) = 0$, the solution of π through x, $\pi(t, x)$, $t \geq 0$, approaches zero at infinity.

Proof. Let x be given in X, such that $\mathcal{J}(x) = 0$. From assumption ii), we know that the omega limit set of any solution, [$\omega(x)$ will denote as usual the omega limit set of the solution through x], is non-empty and compact. All we have to do is show that $\omega(x) = \{0\}$.

Let $\bar{x}$ be an element of $\omega(x)$. From the property of V (resp. W) stated in Prop. 2.1, we have $V(\pi(t, \bar{x}))$ =Constant, so $\mathcal{J}([\pi(t, \bar{x})]^+)$ =Constant. More precisely, from the invariance of $\mathcal{J}$ along the solutions of π, we can conclude that

$$\mathcal{J}([\pi(t, \bar{x})]^+) = \mathcal{J}(\pi(t, \bar{x}^+)), \text{ for all } t \geq 0. \tag{13}$$

On the other hand, from $\bar{x} \leq \bar{x}^+$, and the monotonicity of π, we can deduce that $\pi(t, \bar{x}^+) \geq \pi(t, \bar{x})$, and, from $\bar{x}^+ \geq 0$, and the condition i), we also have $\pi(t, \bar{x}^+) \geq 0$. Therefore, it follows from the definition of $[]^+$ as the least upper bound of $[]$ and 0 that $\pi(t, \bar{x}^+) \geq [\pi(t, \bar{x})]^+$. In view of (13) and the condition (SMI) verified by $\mathcal{J}$, we conclude that

$$\pi(t, \bar{x}^+) = [\pi(t, \bar{x})]^+ \text{ for all } t \geq 0. \tag{14}$$

We are not yet done. We want to conclude that $\bar{x} = \bar{x}^+$. Suppose this has been done. By the same argument, we will also conclude that $\bar{x} = \bar{x}^-$, which in particular implies that $\bar{x}$ is both ≥ 0 and ≤ 0. Therefore $\bar{x} = 0$, which is the desired result. In fact, the proof does not work exactly this way. We proceed by contradiction. Suppose $\bar{x}^+$ is neither $= \bar{x}$, nor $= 0$. From (SMS), we conclude that $\pi(t, \bar{x}^+) \gg \pi(t, \bar{x})$, and $\pi(t, \bar{x}^+) \gg 0$, for $t \geq t_0$, for some t_0. This in turn implies that $\pi(t, \bar{x}^+) \gg [\pi(t, \bar{x})]^+$, in contradiction with the equality in (14). So, we have either $\bar{x}^+ = \bar{x}$ or $\bar{x}^+ = 0$. But, we also have $\mathcal{J}(\bar{x}) = 0$, because by continuity of $\mathcal{J}$ and invariance along the solutions, the omega limit set of any element lies in the same level set as the element. So, from the first part of the alternative, we conclude that $\mathcal{J}(\bar{x}^+) = 0$, which in view of (SMI) implies that $\bar{x}^+ = 0$. The proof is complete. □

Corollary 2.3 *We assume the same as in Theorem 2.2. Suppose that for some $\alpha \in \Re$, there exists an element $a \in X$, such that $\pi(t, a) = a$, for all t, and $\mathcal{J}(a) = \alpha$. Then, for every $x \in X$, such that $\mathcal{J}(x) = \alpha$, the solution $\pi(t, x)$ approaches a at infinity.*

The proof consists of just changing the variable x into $x - a$, for $x \in X$, and changing X, π, and $\mathcal{J}$ accordingly so that we can apply Theorem 2.2. This observation motivated the search for a more general result about the asymptotic behavior, valid even in the non-autonomous setting, which ideally would be that "the omega limit set of a solution is the same for all solutions lying in the same level set", or even more precisely, "if $x(t)$ and $y(t)$ are solutions of π, with $\mathcal{J}(x(t)) = \mathcal{J}(y(t))$, then $\lim_{t\to\infty}(x(t) - y(t)) = 0$", whatever the asymptotic behavior of $x(t)$ may be.

The idea is to make a change of variable which in fact is nothing other than a change of the origin of state space using an arbitrary solution $x(t)$ as the new origin. One difficulty is that this leads automatically to a non-autonomous system; another problem is that this system is not necessarily a semi-flow. We will now briefly explain how these difficulties can be overcome.

Regarding the non-autonomous character, we considered in fact the extension of monotone semi-flows from the autonomous to a non-autonomous setting. Using the analog of skew-products ([12]), we can interpret such systems as autonomous semi-systems on a larger space. In addition to the space X, enjoying the same properties as stated in Theorem 2.2, we introduce a space Y, the space of equations defined on X. Y is a metric space, with a group of translations defined on it: for each $y \in Y$, $t \in \Re$, we denote y_t the t-translate of y. There are some properties, describing the asymptotic behavior of the equations. We will not elaborate on them here.

Denote $Z = X \times Y$. By a non-autonomous semi-flow on X, we mean a mapping $\pi : \Re^+ \times Z \to Z$, which is a semi-flow on Z, such that the second component of π is just the translation $y \to y_t$, while the first component, denoted π_1, corresponds more or less to a solution operator associated with an equation. We assume that π satisfies the same assumptions as i),..., iv) of Theorem 2.2, with the obvious modifications.

For example, condition i) reads as

$$\pi(t, 0, y) = (0, y_t).$$

We assume that π is continuous in (t, x, y), and the positive orbits are precompact. We also assume that $\pi_1(t, x, y)$ is (MS) and (SMS) in x. We can in fact deduce an order on Z from the order on X:

$$(x_1, y_1) \leq (x_2, y_2) \text{ iff } x_1 \leq x_2 \text{ [in X] and } y_1 = y_2.$$

We can accordingly define $<$, $\ll$; thus, it is immediate to see that π is (MS) (resp. SMS)) if π_1 is (MS) (resp. (SMS)). Finally, we assume that, associated with π, there is a first integral $\mathcal{J}$, $\mathcal{J} : \Re \times X \times Y \to \Re$, continuous in (t, x, y), increasing in x, $\mathcal{J}(t, 0, y) = 0$, for all $y \in Y$. With these assumptions, we can prove the analog of Prop. 2.1 and Theorem 2.2, by the same method.

In order to extend the validity of Corollary 2.3, we introduce the notion of origin.

Definition 2.4 ([2],[5]) Let $z_0 = (x_0, y_0) \in Z$. We will call z_0 an origin for π if there exists $\bar{y}_0 \in Y$, such that for any $x \in X$,
i) $\pi_1(t, x, y_0) - \pi_1(t, x_0, y_0) = \pi_1(t, x - x_0, \bar{y}_0)$.
ii) $\mathcal{J}(x, y_0) - \mathcal{J}(x_0, y_0) = \mathcal{J}(x - x_0, \bar{y}_0)$.

In the case of systems generated by evolution equations, each y in Y is just the right hand side of the equation: any element z_0 in Z is an origin. The notion of origin merely corresponds to centering the solutions of a given equation y_0 around a given solution $\pi_1(t, x_0, y_0)$. Of course, the properties of the centered equation are not necessarily the same as those of the original equation: for example, if the original equation is periodic, and the solution used as an origin is not periodic, then the centered equation will not be periodic either. However, the main features of the equations under consideration are preserved, that is, (SM) and (SMS). While the notion of origin is trivial in the case of evolution equations, it does not seem to follow necessarily in the case of general semi-dynamical systems, even though we do not have an example indicating that it may fail. Note that if z_0 is an origin, then for $t \geq 0$, $\pi(t, z_0)$ is an origin.

Let z_0 be an origin for π. We may then define the following functionals on pairs of solutions of π associated with the equation y_0: for $z \in Z$, such that $z = (x, y_0)$,

$$V(z, z_0) = \mathcal{J}([x - x_0]^+ + x_0, y_0) - \mathcal{J}(x_0, y_0),$$

$$W(z, z_0) = \mathcal{J}([x - x_0]^- + x_0, y_0) - \mathcal{J}(x_0, y_0).$$

Using these functionals and the same arguments as in the proof of Theorem 2.2, we can prove that:

$$\text{if } \mathcal{J}(z) = \mathcal{J}(z_0), \text{ then } \pi_1(t, z) - \pi_1(t, z_0) \to 0, \text{ as } t \to +\infty. \tag{15}$$

This theory applies in particular to delay differential equation (4) of Sect. 1. In this case, the space Y is a set of equations. Each equation (4) is equivalent to a matrix function $G(t, u) = (g_{i,j}(t, u))_{1 \leq i,j \leq n}$. So, Y is a set of G's, restricted by a few reasonable assumptions on these functions. As indicated above, each element (ϕ, G) can be used as an origin. There is an obvious correspondence between V and W defined by (10) and (11), and the functions $V(z, z_0)$ and $W(z, z_0)$ defined above. The relation (15) yields, in this situation, the asymptotic result stated in part 3 of Theorem 1.2, that is: let G verify the assumptions of Theorem 1.2. Let ϕ, ψ be given in X, such that $\mathcal{J}(\phi, G) = \mathcal{J}(\psi, G)$. If we denote by $x(t)$ (resp. $y(t)$) the solution of (4) associated with G, such that $x_0 = \phi$ (resp. $y_0 = \psi$), then $(x(t) - y(t)) \to 0$ as $t \to +\infty$.

3 Comparison with works by Hirsch, Mierczynski, Smith, Takác and others

First of all, I would like to mention a paper by J.Mierczynski ([10]) on a subject very closely related to the one treated here. In ([10]), this author considers an ordinary differential system of equations in $\Re^n_+$ (the positive orthant in $\Re^n$) of the cooperative type (see below and [13])

$$x'(t) = F(x(t)), \tag{16}$$

which has a first integral, defined in terms of the function $H : \Re^n_+ \to \Re$, such that $grad\, H(x) > 0$, for $x \neq 0$. It is also assumed that $F(0) = 0$, $H(0) = 0$. The main tool in proving the asymptotic results (convergence or unboundness) is a Lyapunov functional defined in terms of the constant solutions of the equation. This example can be studied using the theory described in the previous section.

We will do it briefly now. For this purpose, we have to be more specific regarding the assumptions. F is continuously differentiable, and such that for every $x, y \in \Re^n_+$, $i \in [1, n]$,

$$x_i < y_i,\ x_j = y_j,\ \text{for } j \neq i, \rightarrow F_j(x) < F_j(y),\ \text{for } j \neq i. \tag{17}$$

This is the property of cooperativity. From (17) and the regularity assumption, it follows that the operator solution associated to (16) is defined from $\Re^n_+$ into $\Re^n_+$, and is (MS) and (SMS).

Let $x(t)$ be a solution of (16). The equation centered around $x(t)$ is

$$\frac{dz}{dt} = F(x(t) + z(t)) - F(x(t)).$$

We denote

$$G(t, z) = F(x(t) + z(t)) - F(x(t)). \tag{18}$$

G verifies the same assumptions as F. If, accordingly, we denote

$$K(t, z) = H(x(t) + z) - H(x(t)), \tag{19}$$

K is the transformed first integral: if $z(t)$ is a solution of

$$\frac{dz}{dt} = G(t, z(t)), \tag{20}$$

we have $K(t, z(t))$ =Constant. We still have$\nabla_z K(t, z) > 0$, for $z \neq 0$.

As in Sect. 2, we define

$$V(t, z) = K(t, z^+)\ [\text{resp. } W(t, z) = K(t, z^-)]. \tag{21}$$

Lemma 3.1 *V is ≥ 0, non-increasing along the solutions of (20) [resp. W is ≤ 0, non-decreasing].*

Proof. The function $V(t, z)$ is differentiable in t, Lipschitz continuous in z. In fact, it is differentiable in z except possibly at points z where one of the coordinates is 0. We can and will restrict our attention to points z with all components $\neq 0$. Given such a point z, the components of z can be split into two disjoint subsets:

$$I = \{i : z_i > 0\};\quad J = \{i : z_i < 0\}.$$

We will use the notation x_I to denote the vector $\bar{x}$ such that $\bar{x}_i = x_i$, for $i \in I$, $\bar{x}_i = 0$, for $i \notin I$, and similarly for x_J.

Let $t_0 \in \Re$. We want to evaluate the derivative of V with respect to (20), at (t_0, z), that is, $V'_{(20)}(t_0, z)$ according to a classical notation, or simply $V'(t_0, z)$. We have

$$\begin{aligned} V'(t_0, z) &= \frac{\partial}{\partial t} V(t_0, z) + \nabla_y V(t_0, z) G(t_0, z), \\ &= [\nabla H(x(t_0) + z_I) - \nabla H(x(t_0))]\, x'(t_0) + \nabla_I H(x(t_0) + z_I) G_I(t_0, z). \end{aligned}$$

Here, ∇_I means the collection of derivatives with respect to the x_i's, $i \in I$. From the fact that H is a first integral of (16), we have $\nabla H(x(t_0))x'(t_0) = 0$. This fact together with the relation (18) leads to the following expression for V':

$$V'(t_0, z) = \nabla_I H(x(t_0) + z_I)F(x(t_0) + z) + \nabla_J H(x(t_0) + z_I)F_J(x(t_0)).$$

We add two final observations:

1) In view of (17), and $z_J < 0$, we have $F_I(x(t_0 + z) < F_I(x(t_0 + z_I)$.

2) From$\nabla H(x)F(x) = 0$, applied to $x = x(t_0) + z_I$, we can deduce that

$$\nabla_I H(x(t_0) + z_I)F_I(x(t_0) + z_I) < -\nabla_J H(x(t_0) + z_I)F_J(x(t_0) + z_I).$$

These two facts combined with the positivity of ∇H yield

$$\begin{aligned} V'(t_0, z) &\leq -\nabla_J H(x(t_0) + z_I)[F_J(x(t_0) + z_I) - F_J(x(t_0))] \\ &\leq 0 \text{ [using once more the relation (17) and } \nabla H > 0.] \end{aligned}$$

The proof of the lemma is complete. □

The above lemma could be stated in the following way: Let $x(t)$, $y(t)$ be solutions of (16). Then, [taking $z(t) = y(t) - x(t)$], the function $H(x(t) + [y(t) - x(t)]^+) - H(x(t))$ is non-increasing. The same function with $[]^-$ instead of $[]^+$ is non-decreasing. In fact, these results are implied by our theory of monotone systems with a monotone first integral. We thought however it might be interesting to see through a simple example how these monotonicity properties connect to each other to yield Lyapunov functionals. We should observe finally that, using the general theory, a number of asymptotic properties could probably be concluded from the study of these functionals. One assumption that we do not make here is the precompactness of the solutions. In fact, this assumption can be weakened by restricting our attention to solutions which have this property. We can also infer it by assuming a little more on the function H, namely, that $H(x) \to +\infty$, as $|x| \to +\infty$.

The last question we will discuss in this paper is: what can the results obtained by Hirsch ([9]) and other investigators (e.g. [15]) do in the case of a monotone system with a monotone first integral? We will concentrate on autonomous systems; this is where Hirsch's results apply. Probably, the most significant achievement in his recent work is the "dichotomy principle", which says that, for a strongly monotone semi-flow π, if $x < y$, either $\omega(x) \ll \omega(y)$, or $\omega(x) = \omega(y) \in E$, where E is the set of equilibria of π. ([9])

Suppose X is a Banach lattice, with $int(X^+) \neq \emptyset$. Assuming a lattice structure is a little more restrictive than in Hirsch's theory. We add the assumption that X is separable. Suppose π is a strongly monotone semi-dynamical system, with a strictly monotone first integral $\mathcal{J}$. In fact, it is probably enough to assume that π is eventually strongly monotone as we did in Sect. 2. Suppose finally that the orbits are precompact, which is also more than in Hirsch's paper ([9]).

Let $x \in X$. We will show, using the dichotomy principle, that $\omega(x)$ is reduced to a single element. This means that, for every $x \in X$, $\pi(t, x)$ converges as $t \to +\infty$, that is to say, π is quasi-convergent.

Choose $e \in X^+\backslash\{0\}$. The dichotomy principle and the separability axiom imply that $\omega(x+\lambda x)$ is reduced to a single element for all λ except at most a denumerable family of λ. Choose a sequence $\lambda_n > 0$, $\lambda_n \to 0$, as $n \to +\infty$, such that for $\lambda \neq \pm\lambda_n$, $\omega(x+\lambda e)$ is a single element. Suppose that $\omega(x) \neq \emptyset$, otherwise there is nothing to prove. From $\omega(x-\lambda_n e) \ll \omega(x) \ll \omega(x+\lambda_n e)$, it follows that

$$\omega(x-\lambda_n e) \leq \inf \omega(x) \leq \sup \omega(x) \leq \omega(x+\lambda_n e).$$

From these inequalities and the invariance of $\mathcal{J}$ along the solutions of π, we obtain

$$\mathcal{J}(x-\lambda_n e) \leq \mathcal{J}(\inf \omega(x)) \leq \mathcal{J}(\sup \omega(x)) \leq \mathcal{J}(x+\lambda_n e).$$

Letting $n \to +\infty$, we obtain $\mathcal{J}(x)$ on both ends of this chain of inequalities. Therefore, we conclude that

$$\mathcal{J}(\inf \omega(x)) = \mathcal{J}(\sup \omega(x)),$$

which, in view of the strict monotonicity of $\mathcal{J}$, implies that

$$\inf \omega(x) = \sup \omega(x),$$

that is, $\omega(x)$ is reduced to a single element.

It is not completely clear form the above considerations that solutions lying in a same level set with respect to $\mathcal{J}$ converge to the same equilibrium. This fact together with the convergence can be deduced at once as an application of a recent theorem due to P.Takác ([15]). This theorem can be used in that case, by just applying it separately to each operator $\pi(t, \;.)$. Takác's theorem states that if an operator T is order-compact, strongly increasing, and every equiilibrium is Lyapunov stable, then: 1) The set of equilibria is linearly ordered, and ii) each solution converges to an equilibrium. Of course, the assumption of order-compactness may be thought of as restrictive in the context of "abstract" semi-flows. However, it is nearly automatic in most systems of interest generated by evolution equations as soon as we know that the solutions are bounded. The fact that the Lyapunov stability has to be checked for the equilibria only is an interesting feature of the theorem; in our situation, it is a straightforward consequence of the existence of the Lyapunov functionals for the system centered around an equilbrium.

As final remarks, we would like to mention some other works connected to ours. H.Smith ([13]) worked on cooperative systems generated by functional differential equations; this is mainly an adaptation of Hirsch's ideas to the setting of F.D.E. The equilibria are supposed to be isolated and hyperbolic, which seems unlikely to occur in the presence of a first integral. Lately, H.Smith and H.Thieme considered strongly monotone systems with a linearly ordered set of equilibria ([14]). J.Haddock, M.Nkashama and J.Wu introduced the notion of pseudo-monotone semi-flows ([8]): the order in their case is restricted to pairs (e, ϕ) where e is an equilibrium. The main motivation for this theory seems to lie in the study of equations similar to (1), with the property that all the constants are solutions. Among them is the neutral type delay differential equation

$$\frac{d}{dt}[x(t) - cx(t-1)] = f(x(t-1)) - f(x(t)).$$

A more general version of this equation, with f(t,x) instead of f(x), has been studied by F.Bourad and myself, as an application of our theory of monotone semi-flows with a monotone first integral ([5],[1]).

References

1. Arino, O., Bourad, F. (1990): On the asymptotic behavior of the solutions of a class of scalar neutral equations generating a monotone semi-flow. J. Diff. Equat. (to appear)
2. Arino, O., Bourad, F., Hassani, N. (1988): Un résultat sur le comportement asymptôtique des solutions de systèmes dynamiques. C. R. Acad. Sci. Paris, t. **307**, Série I, 311-315
3. Arino, O., Haourigui, E. (1987): On the asymptotic behavior of solutions of some delay differential systems which have a first integral. J. Math. Anal. Appl. **122**, 36-47
4. Arino, O., Séguier, P. (1984): On the asymptotic behavior at infinity of solutions of $x'(t) = f(t-1, x(t-1)) - f(t, x(t))$. J. Math. Anal. Appl. **96**, 420-436
5. Bourad, F. (1988): Sur le comportement asymptôtique des systèmes dynamiques monotones. Thesis of the University of Algiers (Algeria)
6. Cooke, K.L., Yorke, J.A. (1973): Some equations modelling growth processes and gonorrhea epidemics. Math. Biosc. **16**, 75-101
7. Győri, I. (1986): Connections between compartmental systems with pipes and integro-differential equations. Math. Modelling **7**, 1215-1238
8. Haddock, J., Nkashama, M., Wu, J.: Asymptotic constancy for pseudo-monotone dynamical systems on function spaces and applications. Communication in this Conference
9. Hirsch, M.W. (1988): Stability and convergence in strongly monotone dynamical systems. J. Reine Angew. Math. **383**, 1-53
10. Mierczynski, J. (1983): Strictly cooperative systems with a first integral. SIAM J.Math. Anal. **18** No. 3, 642-646
11. Schaefer, H. (1974): Banach lattices and positive operators. Springer-Verlag, N.Y.
12. Sell, G.R. (1974): Topological dynamics and ordinary differential equations. Van Nostrand, N.Y.
13. Smith, H. (1987): Monotone semi-flows generated by functional differential equations. J. Diff. Equat. **66**, 420-442
14. Smith, H., Thieme, H.: Quasiconvergence and stability for strongly order preserving semi-flows. SIAM J. Math. Anal. (to appear)
15. Takác, P. (1988): Convergence to equilibrium on invariant d-hypersurfaces for strongly increasing discrete-time semigroups. Preprint

The Coincidence Degree of Some Functional Differential Operators in Spaces of Periodic Functions and Related Continuation Theorems

Anna Capietto[1], *Jean Mawhin*[2], *Fabio Zanolin*[3]

[1] International School for Advanced Studies, Strada Costiera 11, 34014 Trieste, Italy,
[2] Université de Louvain, Institute Mathematique, Chemin du Cyclotron 2, B-1348 Louvain-La-Neuve, Belgium,
[3] Department of Mathematics and Computer Sciences, Via Zanon 6, 33100 Udine, Italy

To Ken Cooke, who witnessed and encouraged the development of earlier continuation theorems.

1 Introduction

In a recent paper [1], we have shown that if Ω is an open bounded set of the space C_ω of continuous and ω-periodic functions with values in $\mathbf{R}^m$ such that the autonomous equation

$$x'(t) - f(x(t)) = 0, \tag{1}$$

with $f : \mathbf{R}^m \to \mathbf{R}^m$ continuous, has no ω-periodic solution on $\partial\Omega$, then the coincidence degree of the operator in C_ω associated to the left-hand member of (1) is equal to $(-1)^m$ times the Brouwer degree of f with respect to $\Omega \cap \mathbf{R}^m$. Of course, we identify here $\mathbf{R}^m$ with the space of constant mappings from $\mathbf{R}$ to $\mathbf{R}^m$.

In other words, this coincidence degree (which is essentially a Leray-Schauder degree of an equivalent fixed point operator, see [9, 10]), is "blind" to the non-trivial ω-periodic solutions and depends only on the equilibria of (1). The positive aspect of this result is that this degree is rather easy to compute and provides then effective continuation theorems for the existence of ω-periodic solutions of nonautonomous differential equations of the form

$$x'(t) - f(x(t)) = e(t, x)$$

with e "small" or having restricted growth, via the homotopy

$$x'(t) - f(x(t)) = \lambda e(t, x), \quad \lambda \in [0, 1].$$

The proof of the mentioned result depends upon the Kupka-Smale approximation theorem [5, 12] for the closed orbits of autonomous systems and on some delicate degree computations.

The aim of this paper consists in stating and proving the corresponding result for the autonomous retarded functional differential equation (RFDE)

$$x'(t) - f(x_t) = 0, \tag{2}$$

where $f : \mathcal{C}_r \to \mathbf{R}^m$ is continuous and takes bounded sets into bounded sets, $\mathcal{C}_r = C([-r,0], \mathbf{R}^m)$ and, for each t, x_t is the element of $\mathcal{C}_r$ defined by $x_t(\theta) = x(t+\theta)$, $\theta \in [-r, 0]$ (see [3] for these notations and a thorough treatment of RFDE). If Ω is like above, it is well known [8, 9] that the coincidence degree of the mapping in C_ω defined by the left-hand member of (2) exists and we shall prove in Theorem 1 that it is again equal, up to a factor $(-1)^m$, to the Brouwer degree of the restriction of f to $\Omega \cap \mathbf{R}^m$.

A basic ingredient in the proof will be an extension to RFDE of the Kupka-Smale theorem due to Mallet-Paret [7] and following earlier generic results for fixed points of a RFDE defined on a compact manifold due to Oliva [11]. See also interesting remarks in [2] and [4].

Although our proof will follow the same main lines as the one given in [1] for the ordinary differential equation case, the different nature of (2) will require at various stages nontrivial variants of the arguments and even completely different ones due in particular to the fact that time-scaling involve modifications of the delay in a RFDE.

We then relate the coincidence degree associated to a RFDE of the form

$$x'(t) - h(t, x_t) = 0, \quad t \in \mathbf{R}$$

with h ω-periodic in t and positively homogeneous of degree $\alpha \neq 1$ in its second variable, to that associated to the autonomous RFDE

$$x'(t) - \bar{h}(x_t) = 0, \quad t \in \mathbf{R}$$

with

$$\bar{h}(\varphi) = \omega^{-1} \int_0^\omega h(s, \varphi) ds$$

and we give related existence theorems.

We finally state and prove a continuation theorem for the ω-periodic solutions of non-autonomous RFDE

$$x'(t) = F(t, x_t), \quad t \in \mathbf{R}$$

based upon the previous degree calculations.

2 A formula for the coincidence degree of an autonomous RFDE in the space of ω-periodic functions

Let $\mathcal{C}_r = C([-r,0],\mathbf{R}^m)$, $(r \geq 0)$ and let $f : \mathcal{C}_r \to \mathbf{R}^m$ be continuous and such that it takes bounded sets into bounded sets. We consider the

$$\omega\text{-}periodic\ solutions \quad (\ \omega > 0 \text{ fixed })$$

of the corresponding RFDE

$$x'(t) = f(x_t), \tag{3}$$

i.e., the functions $x : \mathbf{R} \to \mathbf{R}^m$ of class C^1 such that

$$x(t+\omega) = x(t), \quad t \in \mathbf{R}$$

which satisfy (3) on $\mathbf{R}$. We denote by C_ω the Banach space of continuous ω-periodic functions $x : \mathbf{R} \to \mathbf{R}^m$ with the uniform norm $\| x \| = \max_{t\in\mathbf{R}} |x(t)|$.

If we define $L : D(L) \subset C_\omega \to C_\omega$ by $D(L) = \{x \in C_\omega : x \text{ is of class } C^1\}$ and $Lx = x'$, and $F : C_\omega \to C_\omega$ by $F(x)(t) = f(x_t)$, $t \in \mathbf{R}$ (Nemitzky operator), then it is well known [8, 9] that L is a Fredholm operator of index zero, F is L-completely continuous on C_ω and the existence of ω-periodic solutions of (3) is equivalent to the abstract equation

$$Lx = Fx, \quad x \in D(L).$$

Moreover, if $\Omega \subset C_\omega$ is an open bounded set such that

$$Lx \neq Fx, \quad x \in D(L) \cap \partial\Omega,$$

then the coincidence degree $D_L(L - F, \Omega)$ is defined as the Leray-Schauder degree of an associated fixed point problem.

We denote by d_B the Brouwer degree of a mapping from $\mathbf{R}^m$ into $\mathbf{R}^m$ (see, e.g. [9, 10] for details).

Theorem 1 *Assume that $\Omega \subset C_\omega$ is an open bounded set such that there is no $x \in \partial\Omega$ such that $x'(t) = f(x_t)$, $t \in \mathbf{R}$. Then*

$$D_L(L - F, \Omega) = (-1)^m d_B\left(f|_{\mathbf{R}^m}, \Omega \cap \mathbf{R}^m, 0\right).$$

Proof . First of all, we observe that, as Ω is bounded, there is a constant $R > 0$ such that $\|x\| < R$ for every $x \in cl\Omega$. Furthermore, we point out that the assumption is equivalent to

$$Lx \neq Fx, \tag{4}$$

for all $x \in D(L) \cap \partial\Omega$; therefore, the coincidence degree $D_L(L - F, \Omega)$ is well defined, and as we also have

$$f(c) \neq 0$$

for all $c \in \mathbf{R}^m \cap \partial\Omega$, the Brouwer degree $d_B\left(f|_{\mathbf{R}^m}, \Omega \cap \mathbf{R}^m, 0\right)$ is defined as well.

The proof is performed by means of Mallet-Paret's extension of the Kupka-Smale's theorem [7]; this result ensures the existence of a sequence of C^1-functions (φ_k), $\varphi_k : \mathcal{C}_r \to \mathbf{R}^m$, taking bounded sets into bounded sets, and such that

(a) $(\varphi_k) \to f$ uniformly on closed bounded sets

(b) for every closed bounded subset B of $\mathcal{C}_r$ and for all $k \in \mathbf{N}$, the equation

$$x' = \varphi_k(x_t)$$

has finitely many singular orbits in $\mathcal{C}_r$ (i.e., rest points and closed orbits) with minimal period in $[0, \omega+1]$ which are contained in B and they are all hyperbolic.

Let $N^{k,\mu}$ be the Nemitzky operator induced by the functions $x \longmapsto \mu f(x.) + (1-\mu)\varphi_k(x.)$, $\mu \in [0,1]$. We claim that there is $k_0 > 0$ such that, for all $k \geq k_0$ and for all $\mu \in [0,1]$,

$$Lx \neq N^{k,\mu}x \qquad \text{for all} \quad x \in D(L) \cap \partial\Omega. \tag{5}$$

This fact will imply, in particular, that

$$\varphi_k(z) \neq 0 \qquad \text{for all} \quad z \in \partial\Omega \cap \mathbf{R}^m, \quad k \geq k_0.$$

Then, a classical compactness argument ensures that, for any $k \geq k_0$, there is $\delta_1 = \delta_1(k)$ such that

$$\varphi_k(y) \neq 0 \qquad \text{for all} \quad y \in B(\partial\Omega \cap \mathbf{R}^m, \delta_1).$$

To obtain (5), it is sufficient to observe that the sequence of operators $N^{k,\mu}$ converges, as $k \to +\infty$, to F in C_ω uniformly on $cl\Omega \times [0,1]$ and that, by (4),

$$\inf\{ \| (L-F)x \| : x \in D(L) \cap \partial\Omega \} > 0.$$

Hence, the claim is proved and, using the homotopy property of the coincidence degree (see [9, p. 15]), we can write

$$D_L(L-F, \Omega) = D_L(L - N^{k,1}, \Omega) = D_L(L - N^{k,0}, \Omega),$$

for every $k \geq k_0$ and, in particular,

$$d_B(f|_{\mathbf{R}^m}, \Omega \cap \mathbf{R}^m, 0) = d_B(\varphi_k|_{\mathbf{R}^m}, \Omega \cap \mathbf{R}^m, 0),$$

for every $k \geq k_0$.

Let us fix $k^* \geq k_0$. For brevity, we set

$$\varphi := \varphi_{k^*}, \qquad N_\varphi := N^{k^*,0}, \qquad \delta_1 := \delta_1(k^*).$$

Consider the singular orbits (i.e. rest points and closed orbits) with minimal period in $[0, \omega+1]$ of the equation

$$x' = \varphi(x_t). \tag{6}$$

By the Kupka-Smale's property, there exist finitely many such orbits which are contained in $B(0,R) \subset \mathcal{C}_r$ and they are hyperbolic. Recall that for a rest point, this means that the spectrum of the infinitesimal generator of its linearized equation contains no purely imaginary values and, for a nonconstant periodic solution, this means that the characteristic multiplier $\mu = 1$ of the linearized equation is simple and no other characteristic multiplier satisfies $|\mu| = 1$ (see [3]). We denote these orbits by $S_1, ..., S_n$. They are mutually disjoint (two orbits of (6) may cross in $\mathcal{C}_r$ because uniqueness of the Cauchy problem only holds in the future, but this may not happen to closed orbits). Pick, for each $i = 1, ..., n$ a point (in $\mathcal{C}_r$) $\phi_i \in S_i$. Then, ϕ_i is a periodic point (possibly a rest point). We can assume that ϕ_i is a rest point ζ_i for $1 \leq i \leq p$ ($p \geq 0$ an integer) and a periodic point for $p+1 \leq i \leq n$. We denote its minimal period by T_i ($p+1 \leq i \leq n$). We can also assume that $T_i \leq \omega$ for $p+1 \leq i \leq q$ and $\omega < T_i \leq \omega + 1$ for $q+1 \leq i \leq n$. We denote by k_i the largest integer such that $k_i T_i \leq \omega$ $(p+1 \leq i \leq q)$, so that $(k_i+1)T_i > \omega$ $(p+1 \leq i \leq q)$. We denote by $x^i(\cdot)$ the solution of (6) with $x_0^i = \phi_i$ $(p+1 \leq i \leq n)$.

We claim that for each ω' such that

$$\omega < \omega' < \min\{(k_{p+1}+1)T_{p+1}, ..., (k_q+1)T_q, T_{q+1}, ..., T_n, \omega+1\} := \tau ,$$

the problem

$$x' = \varphi(x_t), \quad x(t+\omega') = x(\omega') \tag{7}$$

has no solution $x(\cdot)$, with $x_t \in B(0,R)$, and hence $|x(t)| < R$, for all t, other than the equilibria $\zeta_1, ..., \zeta_p$.

Indeed, if x satisfies (7) and $x_t \in B(0,R)$, for all t, then $S = \{x_t : t \in \mathbf{R}\}$ is a singular orbit of (6) contained in $B(0,R)$. If it is not a rest point, then $S = S_i$ for some $p+1 \leq i \leq n$ and hence there exists $t_i \in \mathbf{R}$ such that

$$x_t = x^i_{t+t_i}, \qquad t \in \mathbf{R}$$

(indeed t_i is such that $x_{-t_i} = \phi_i = x_0^i$).

In particular,

$$x^i(\omega' + t_i) = x^i(t_i).$$

This is impossible for $q+1 \leq i \leq n$ as then $\omega' < T_i$ and T_i is the smallest period. This is impossible for $p+1 \leq i \leq q$ as in this case $k_i T_i < \omega' < (k_i+1)T_i$. Therefore the claim is proved. □

Now, the solutions of (7) correspond, by the transformation

$$y(t) = x\left(\frac{\omega'}{\omega}t\right), \qquad t \in \mathbf{R},$$

to the solutions of the problem

$$y'(t) = \frac{\omega'}{\omega}\varphi\left(y_t\left(\frac{\omega}{\omega'}(\cdot)\right)\right), \qquad y(t+\omega) = y(t). \tag{8}$$

Thus, problem (8) has, by construction, no nontrivial (i.e. non-equilibrium) solution on $cl\Omega$ and, by assumption, no rest point on $\partial\Omega$ (as its rest points are the same as those of (6) and all its possible solutions in $B(0,R)$ are rest points). Defining

$$\Phi : C_\omega \times [\omega, \tau[\to C_\omega \ ,$$

by

$$\Phi(y,\omega')(t) = \frac{\omega'}{\omega}\varphi\left(y_t\left(\frac{\omega}{\omega'}(\cdot)\right)\right), \qquad t \in \mathbf{R},$$

we shall show that Φ is continuous so that we can apply a homotopy argument.

If $\varepsilon > 0$, $y \in C_\omega$, $\omega' \in [\omega, \tau[$ are given, then as $\left\{ \left(y_t\left(\frac{\omega}{\omega'}(\cdot)\right)\right) : t \in [0,\omega] \right\}$ is compact, there exists $\delta > 0$ such that

$$\left|\varphi(\phi) - \varphi\left(y_t\left(\frac{\omega}{\omega'}(\cdot)\right)\right)\right| \leq \varepsilon$$

whenever $t \in [0,\omega]$ (and hence whenever $t \in \mathbf{R}$) and $\left|\phi - y_t\left(\frac{\omega}{\omega'}(\cdot)\right)\right|_{C_r} \leq \delta.$

Now, y being uniformly continuous, there exists $\delta' > 0$, such that

$$|y(t') - y(t'')| \leq \frac{\delta}{2}$$

when $|t' - t''| \leq \delta'$ and hence if $\omega < \omega'' < \tau$ and $\omega'' \geq \omega'/2$, we shall have

$$\left|y\left(t + \frac{\omega}{\omega''}(\theta)\right) - y\left(t + \frac{\omega}{\omega'}(\theta)\right)\right| \leq \frac{\delta}{2}$$

when

$$\left|\frac{\omega}{\omega''} - \frac{\omega}{\omega'}\right| |\theta| \leq \delta'$$

which will be the case if

$$|\omega'' - \omega'| \leq \frac{\delta'(\omega')^2}{2\omega r}\ .$$

Now, if $z \in C_\omega$ is such that $\| z - y \| \leq \delta/2$, we have

$$\begin{aligned} &\left|z\left(t + \frac{\omega}{\omega''}(\theta)\right) - y\left(t + \frac{\omega}{\omega'}(\theta)\right)\right| \\ &\quad \leq \left|z\left(t + \frac{\omega}{\omega''}(\theta)\right) - y\left(t + \frac{\omega}{\omega''}(\theta)\right)\right| + \left|y\left(t + \frac{\omega}{\omega''}(\theta)\right) - y\left(t + \frac{\omega}{\omega'}(\theta)\right)\right| \\ &\quad \leq \| z - y \| + \frac{\delta}{2} \leq \delta \end{aligned}$$

for all $t \in \mathbf{R}$ and $\theta \in [-r, 0]$ and hence

$$\left|z_t\left(\frac{\omega}{\omega''}(\cdot)\right) - y_t\left(\frac{\omega}{\omega'}(\cdot)\right)\right|_{C_r} \leq \delta.$$

Summarizing, if $z \in C_\omega$ with $\| z - y \| \leq \delta/2$, and if $\omega'' \in [\omega, \tau[$ with

$$|\omega'' - \omega'| \leq \min\left\{ \frac{\omega'}{2}, \frac{\delta'(\omega')^2}{2\omega r} \right\},$$

we shall have

$$\left|\varphi\left(z_t\left(\frac{\omega}{\omega''}(\cdot)\right)\right) - \varphi\left(y_t\left(\frac{\omega}{\omega'}(\cdot)\right)\right)\right| \leq \varepsilon.$$

for all $t \in \mathbf{R}$ and this easily implies the continuity of Φ on $C_\omega \times [\omega, \tau[$.

Now, as (6) has no solution on $\partial\Omega$, the homotopy invariance of the coincidence degree implies that

$$D_L(L - N_\varphi, \Omega) = D_L(L - \Phi(\cdot,\omega'), \Omega) \tag{9}$$

for all $\omega \leq \omega' < \tau$.
Thus, by excision,

$$D_L(L - \Phi(\cdot,\omega'), \Omega) = \sum_{\substack{1\leq j\leq p \\ \zeta_j\in\Omega}} D_L(L - \Phi(\cdot,\omega'), B(\zeta_j,\delta)), \tag{10}$$

where

$$\delta = \min\{\delta_1, \eta/2\}, \qquad \eta = \min\{d(S_i,S_j) : 1 \leq i \neq j \leq n\}.$$

But, by the choice of δ, $L - N_\varphi$ has no solution on $\partial B(\zeta_j,\delta)$ and hence, by homotopy invariance again we have, for all $\omega \leq \omega' < \tau$,

$$D_L(L - \Phi(\cdot,\omega'), B(\zeta_j,\delta)) = D_L(L - N_\varphi, B(\zeta_j,\delta)), \tag{11}$$
$$1 \leq j \leq p,\ \zeta_j \in \Omega.$$

As φ is of class C^1, the same is true for N_φ and hence, by the linearization property of the degree (see e.g. [9, Prop. VIII.3]),

$$D_L(L - N_\varphi, B(\zeta_j,\delta)) = D_L(L - N_\varphi'(\zeta_i), B(0,1)), \tag{12}$$
$$1 \leq j \leq p,\ \zeta_j \in \Omega.$$

where $N_\varphi'(\zeta_i)$ has the form

$$N_\varphi'(\zeta_i)\phi = \varphi'(\zeta_i)\phi = \int_{-r}^{0} \phi(\theta)d\eta^j(\theta)$$

for some function η^j whose elements are of bounded variation ([3]). Moreover, the hyperbolicity of ζ_j implies that the corresponding characteristic equation

$$\det \Delta_j(\mu) = 0,$$

where

$$\det \Delta_j(\mu) = \mu I - \int_{-r}^{0} e^{\mu\theta} d\eta^j(\theta),$$

has all its roots with nonzero real part (see e.g. [3]).

Consequently, the same is true for the characteristic equation of the equations in the family

$$x'(t) = \lambda\varphi'(\zeta_j)x_t, \qquad \lambda \in]0,1]$$

which therefore only admit the trivial ω-periodic solution. A standard argument (see e.g. [9, Th. IV.12]) then shows that the same is true for the family

$$x'(t) = (1-\lambda)\varphi'(\zeta_j)\bar{x} + \lambda\varphi'(\zeta_j)x_t, \qquad \lambda \in [0,1]$$

where

$$\bar{x} = \omega^{-1} \int_0^\omega x(s)ds = Px.$$

Consequently, by the homotopy invariance, and a classical property of coincidence degree (see e.g. [9, Prop. II.12]),

$$\begin{aligned} D_L(L - N'_\varphi(\zeta_i), B(0,1)) &= D_L(L - N'_\varphi(\zeta_i)P, B(0,1)) \\ &= (-1)^m d_B\left(\varphi'(\zeta_i)|_{\mathbf{R}^m}, B(0,1) \cap \mathbf{R}^m, 0\right) \\ &= (-1)^m d_B\left(\varphi|_{\mathbf{R}^m}, B(\zeta_i,\delta) \cap \mathbf{R}^m, 0\right), \qquad (13) \\ & \qquad 1 \le j \le p,\ \zeta_j \in \Omega. \end{aligned}$$

Therefore, combining (9), (10), (11), (12) and (13), we obtain

$$\begin{aligned} D_L(L - N_\varphi, \Omega) &= \sum_{\substack{1\le j\le p \\ \zeta_j \in \Omega}} (-1)^m d_B\left(\varphi|_{\mathbf{R}^m}, B(\zeta_j,\delta) \cap \mathbf{R}^m, 0\right) \\ &= (-1)^m d_B\left(\varphi|_{\mathbf{R}^m}, \Omega \cap \mathbf{R}^m, 0\right), \end{aligned}$$

and the proof is complete.

3 A formula for the coincidence degree of nonlinear homogeneous nonautonomous *RFDE* in the space of ω-periodic functions

We consider the *RFDE*

$$x'(t) = h(t, x_t) + \delta(\alpha)p(t) \qquad (14)$$

where $h: \mathbf{R} \times \mathcal{C}_r \to \mathbf{R}^m$, $(t,\varphi) \mapsto h(t,\varphi)$ is a continuous mapping, taking bounded set into bounded sets, ω-periodic in t and positively homogeneous of order $\alpha \neq 1$ in φ, $\delta(\alpha) = \max(0, (1-\alpha)/|1-\alpha|)$ and $p \in C_\omega$. We define the averaged vector field $\bar{h}: \mathcal{C}_r \to \mathbf{R}^m$ by

$$\bar{h}(\varphi) = (1/\omega) \int_0^\omega h(s,\varphi)ds$$

and the corresponding Nemitzky mappings $H: C_\omega \to C_\omega$ and $\overline{H}: C_\omega \to C_\omega$ by

$$H(x)(t) = h(t, x_t), \overline{H}(x)(t) = \bar{h}(x_t),$$

for all $t \in \mathbf{R}$.

Theorem 2 *Assume that $\bar{h}(z) \neq 0$ for $|z| = 1$ in $\mathbf{R}^m$. Then there exists $r_0 > 0$ such that, if $\alpha < 1$ and $r \ge r_0$ or $\alpha > 1$ and $0 < r \le r_0$, (14) has no ω-periodic solutions x with $\|x\| = r$ and*

$$D_L(L - H - \delta(\alpha)p, B(0,r)) = D_L(L - \overline{H}, B(0,r)).$$

Proof. Let us define $\mathcal{H}: C_\omega \times [0,1] \to C_\omega$ by

$$\mathcal{H}(x,\lambda) = (1-\lambda)\overline{H}(x) + \lambda(H(x) + \delta(\alpha)p).$$

We have only to show that there is some $r_0 > 0$ such that for each $r \geq r_0$ if $\alpha < 1$ or each $0 < r \leq r_0$ if $\alpha > 1$, and for each $\lambda \in [0,1]$, the equation

$$Lx = \mathcal{H}(x;\lambda)$$

has no solution x with $\|x\| = r$.

If this is not the case, there are sequences (r_k) in $\mathbf{R}_+$, (x_k) in C_ω and (λ_k) in $[0,1]$ such that $\|x_k\| = r_k$, $r_k \leq 1/k$ if $\alpha > 1$, $r_k \geq k$ if $\alpha < 1$ and

$$x_k'(t) = (1-\lambda_k)\bar{h}((x_k)_t) + \lambda_k[h(t,(x_k)_t) + \delta(\alpha)p(t)]$$

($k \in \mathbf{N}$). Letting $u_k = x_k/\|x_k\| = r_k^{-1}x_k$, we get

$$u_k'(t) = r_k^{\alpha-1}[(1-\lambda_k)\bar{h}((u_k)_t) + \lambda_k[h(t,(u_k)_t)] + r_k^{-1}\lambda_k\delta(\alpha)p(t), \tag{15}$$

so that

$$|u_k'(t)| \leq r_k^{\alpha-1}\beta + \gamma$$

for some $\beta, \gamma > 0$ and all $t \in \mathbf{R}$. Consequently there are subsequences (λ_{j_k}), (u_{j_k}) and $\lambda^* \in [0,1]$, $v \in C_\omega$, $\|v\| = 1$ such that $(u_{j_k}) \to v$ uniformly on $\mathbf{R}$ and $(\lambda_{j_k}) \to \lambda^*$.

From

$$u_k(t) - u_k(0) = r_k^{\alpha-1}\int_0^t [(1-\lambda_k)\bar{h}(u_k)_s) +$$

$$+\lambda_k h(s,(u_k)_s)]ds + r_k^{-1}\int_0^t \delta(\alpha)p(s)ds,$$

we get

$$v(t) - v(0) = 0, t \in \mathbf{R},$$

so that v is constant and $\|v\| = 1$. From (15) we also get

$$0 = \int_0^\omega [(1-\lambda_k)\bar{h}((u_k)_s) + \lambda_k h(s,(u_k)_s) + r_k^{-\alpha}\delta(\alpha)\lambda_k p(s)]$$

and hence, letting $j_k \to \infty$,

$$0 = \omega\bar{h}(v),$$

a contradiction.

Hence, for $0 < r \leq r_0$ if $\alpha > 1$ and $r \geq r_0$ if $\alpha < 1$, we have

$$D_L(L - H - \delta(\alpha)p, B(0,r)) = D_L(L - \mathcal{H}(\cdot,1), B(0,r)) =$$

$$= D_L(L - \mathcal{H}(\cdot,0), B(0,r)) = D_L(L - \overline{H}, B(0,r)),$$

and the proof is complete. □

By using Theorem 1 and 2, the existence property of degree and its invariance for sufficiently small perturbations of the nonlinear term we immediately deduce the following existence results.

Corollary 1 *Assume that h satisfies the assumptions of Theorem 2 and that*

$$d_B(\bar{h}_{|\mathbf{R}^m}, B(0,1) \cap \mathbf{R}^m, 0) \neq 0.$$

Then, if $\alpha < 1$, the RFDE

$$x'(t) = h(t, x_t) + p(t) \tag{16}$$

has at least one ω-periodic solution for each $p \in C_\omega$ and, if $\alpha > 1$, there exists $\varepsilon_0 > 0$ such that (16) has at least one ω-periodic solution for each $p \in C_\omega$ with $\|p\| \leq \varepsilon_0$.

4 Continuation theorems for ω-periodic solutions of nonautonomous *RFDE*.

Let $F: \mathbf{R} \times C_r \to \mathbf{R}^m$, $(t, \varphi) \mapsto F(t, \varphi)$ be a continuous mapping, taking bounded sets into bounded sets and such that

$$F(t + \omega, \varphi) = F(t, \varphi)$$

for some $\omega > 0$ and all $t \in \mathbf{R}$ and $\varphi \in C_r$. We consider the existence of ω-periodic solutions of the $RFDE$

$$x'(t) = F(t, x_t), t \in \mathbf{R}, \tag{17}$$

i.e. of solutions x such that

$$x(t) = x(t + \omega), t \in \mathbf{R}. \tag{18}$$

As it is the case in any continuation theorem, we introduce a mapping $f: \mathbf{R} \times C_r \times [0,1] \to \mathbf{R}^m$ which is continuous, takes bounded sets into bounded sets and is such that

$$f(t + \omega, \varphi, \lambda) = f(t, \varphi, \lambda)$$

for all $t \in \mathbf{R}$, $\varphi \in C_r$, $\lambda \in [0,1]$,

$$f(t, \varphi, 0) = f_0(\varphi)$$

for all $t \in \mathbf{R}$, $\varphi \in C_r$ (i.e. $f(\cdot, \cdot, 0)$ is autonomous), and

$$f(t, \varphi, 1) = F(t, \varphi)$$

for all $t \in \mathbf{R}$, $\varphi \in C_r$.

Theorem 3 *Let $\Omega \subset C_\omega$ be an open bounded set such that the following conditions are satisfied:*

(p_1) *there is no $x \in \partial\Omega$ such that*

$$x'(t) = f(t, x_t, \lambda), t \in \mathbf{R}, \lambda \in [0,1); \tag{19}$$

(p_2) $$d_B(f_{0|\mathbf{R}^m}, \Omega \cap \mathbf{R}^m, 0) \neq 0.$$

Then (17)-(18) have at least one solution $x \in cl\Omega$.

Proof. We use the framework of coincidence degree as in Theorem 1. The classical Leray-Schauder continuation theorem [6] could be used instead by the equivalence stated at the beginning of Section 2. Besides the spaces and operators considered there, we further define $M := M(x;\lambda): C_\omega \times [0,1] \to C_\omega$:

$$M(x;\lambda)(t) = f(t, x_t; \lambda).$$

Observe that $M(\cdot;0) = M_0$ where

$$M_0: C_\omega \to C_\omega, x \mapsto f_0(x_.).$$

According to [9, Chapter I], M is L-compact on $cl\Omega \times [0,1]$. We remark that x is an ω-periodic solution of $x'(t) = f(t, x_t; \lambda)$, $\lambda \in [0,1]$, if and only if $x \in D(L)$ is a solution of the coincidence equation $Lx = M(x;\lambda)$, $\lambda \in [0,1]$. In particular, (17)-(18) is equivalent to $Lx = M(x;1)$. Without loss of generality, we suppose that (p_1) holds in (19). Otherwise, the result is proved for $x \in \partial\Omega$. Accordingly, by the definition of $M(\cdot,\lambda)$ and using (p_1) we have:

$$Lx \neq M(x;\lambda), \lambda \in [0,1]$$

for all $x \in D(L) \cap \partial\Omega$. Thus we can apply the homotopy property of the coincidence degree and obtain:

$$D_L(L - M_0, \Omega) = D_L(L - M(\cdot;0), \Omega) = D_L(L - M(\cdot;1), \Omega). \tag{20}$$

Assumption (p_1) (for $\lambda = 0$) ensures that Theorem 1 can be applied, so that (20) and (p_2) imply:

$$|D_L(L - M_0, \Omega)| = |(d_B(f_0|_{\mathbf{R}^m}, \Omega \cap \mathbf{R}^m, 0)| \neq 0.$$

Hence, by the existence property of the coincidence degree, there is $\tilde{x} \in D(L) \cap \Omega$ such that $L\tilde{x} = M(\tilde{x};1)$; thus $\tilde{x}(\cdot)$ is a solution to (17)-(18), with $\tilde{x} \in D(L) \cap \Omega$. The proof is complete. □

Theorem 3 is particularly suitable for the study of ω-periodic solutions of perturbed autonomous $RFDE$ of the form

$$x'(t) = f(x_t) + e(t), t \in \mathbf{R},$$

with $e \in C_\omega$, through the homotopy

$$x'(t) = f(x_t) + \lambda e(t), t \in \mathbf{R}, \lambda \in [0,1].$$

References

1. Capietto, A., Mawhin, J., Zanolin, F.: Continuation theorems for periodic perturbations of autonomous systems. Trans. Amer. Math. Soc., to appear
2. Chow, S.N., Mallet-Paret, J. (1978): The Fuller index and global Hopf bifurcation. J. Differential Equations, **29**, 66-85
3. Hale, J.K. (1977): Theory of Functional Differential Equations. 2nd ed., Applied Math. Sci., vol. 3, Springer Verlag, Berlin-Heidelberg-New York
4. Hale, J.K., Magalhaes, L.T., Oliva, W.M. (1984): An Introduction to Infinite Dimensional Dynamical Systems. Geometric Theory. Applied Math. Sci., vol. 47, Springer Verlag, Berlin-Heidelberg-New York
5. Kupka, I. (1963-4): Contribution à la théorie des champs génériques. In Contrib. Differential Equations, **2**, (1963), 457-484; **3**, (1964), 411-420
6. Leray, J., Schauder, J. (1934): Topologie et équations fonctionnelles. Ann. Sci. Ecol. Norm. Sup. **351**, 45-78
7. Mallet-Paret, J. (1977): Generic periodic solutions of functional differential equations. J. Differential Equations, **25**, 163-183
8. Mawhin, J. (1971): Periodic solutions of functional differential equations. J. Differential Equations, **10**, 240-261
9. Mawhin, J. (1979): Topological degree methods in nonlinear boundary value problems. CBMS 40, Amer. Math. Soc., Providence, R.I.
10. Mawhin, J. (1985): Points fixes, points critiques et problèmes aux limites. Séminaire de Mathématiques Supérieures, vol. 92, Les Presses de l'Université de Montréal
11. Oliva, W.M. (1969): Functional differential equations on compact manifolds and an approximation theorem. J. Differential Equations, **5**, 483-496
12. Smale, S. (1963): Stable manifolds for differential equations and diffeomorphisms. Ann. Scuola Norm. Sup. Pisa, **17**, 97-116

On the Stability of Discrete Equations and Ordinary Differential Equations

L.A.V.Carvalho

Institute de Ciências Matemáticas de São Carlos Universidade de São Paulo and Mathematics Department - Pomona College Claremont, California - 91711 - U.S.A.

Dedicated to K.L.Cooke

1 Introduction

In [1] the author has given an extension of Liapunov's direct method to discrete equations. In this paper we present both an improvement of this method as well as an application of it to the study of the asymptotic behavior of ordinary differential equations. As a consequence of the technique used here, a small contribution to the comparative analysis of the asymptotic behavior of the solutions of some o.d.e.'s and of their discretized versions is furnished.

In order to achieve a certain degree of completeness and objectiveness we give below a summary of the results that are of interest to this paper. We restrict ourselves to stability and asymptotic stability properties of autonomous equations.

Let $f : \mathbb{R}^N \to \mathbb{R}^N$ be a given continuous map and consider the discrete equation

$$x_n = f(x_{n-1}),\ n = 1, 2, \dots \tag{1.1}$$

subject to the initial condition

$$x_0 = y\ . \tag{1.2}$$

The solution of (1.1-1.2), which is a sequence $x_0, x_1, x_2, \dots, x_n, \dots$, exists, is unique and depends continuously on y. It will be denoted by $x_n(y)$. We shall suppose that $f(0) = 0$. This makes of the null sequence, $x_n = 0$, a solution of (1.1), which will be, accordingly, denoted by $x_n(0)$ and will be called the *null equilibrium* of (1.1).

Definition 1.1. 1.1 The null equilibrium of (1.1) is said to be *stable* (in the sense of Liapunov) if, given $\varepsilon > 0$, we can find a $\delta > 0$ such that $| y | < \delta$ implies that $| x_n(y) | < \varepsilon$ for all $n \geq 0$.

Here, $| . |$ denotes the Euclidean norm of $\mathbb{R}^N$, i.e.,

$$| x | = \left[\sum_{j=1}^{N} (x^j)^2 \right]^{\frac{1}{2}}$$

where $x = (x^1, x^2, \ldots, x^N)$. Of course, the stability character of $x_n(0)$ does not change if another norm is used in $\mathbb{R}^N$.

Definition 1.2. 1.2 The equilibrium $x_n(0)$ is said to be *asymptotically stable* if it is stable and, moreover, there exists a $\gamma > 0$ such that $| y | < \gamma$ implies that $x_n(y) \to 0$ as $n \to \infty$.

We shall denote the open ball of radius r centered at the origin in $\mathbb{R}^N$ by B_r. Ω, on the other hand, will always denote a certain given neighborhood of the origin in $\mathbb{R}^N$ and, if A is a set, $\overline{A}$ will be its closure. In particular, if $A \subset \Omega$, $\overline{A}$ is its closure relative to Ω, unless otherwise stated.

A map $V : \mathbb{R}^N \to \mathbb{R}$ is said to be *positive semi-definite* (in Ω) if $V(x) \geq 0$ for all $x \in \Omega$. If, moreover, $V(x) > 0$ when $x \neq 0, x \in \Omega$, we say that V is *positive definite* (in Ω). The *variation of V with respect to (1.1)* is the map $\Delta V : \mathbb{R}^N \to \mathbb{R}$ given by $\Delta V(x) = V(f(x)) - V(x)$. This concept is generalized as follows. Given integers p, q, $0 \leq p, q$, we define the (p, q) - *variation of V with respect to (1.1)* as being the map $\Delta_q^p V : \mathbb{R}^N \to \mathbb{R}$ defined by $\Delta_q^p V(x) = V(f^p(x)) - V(f^q(x))$, where f^j is the j-th iterate of f . In particular, we note that $\Delta V = \Delta_0^1 V$ and that, according to our previous notation for a solution of (1.1-2), $\Delta_q^p V(y) = V(x_p(y)) - V(x_q(y))$.

Definition 1.3. 1.3 A continuous map $V : \mathbb{R}^N \to \mathbb{R}$ is said to be a *Liapunov function for (1.1)* whenever there exists Ω such that ΔV is negative semi-definite in Ω.

The following two theorems contain the fundamental results of Liapunov's direct method as applies to the stability of the null equilibrium of equation (1.1)

Theorem 1.4. *Suppose that there exists a positive definite Liapunov function for (1.1). Then, the null equilibrium is stable.*

Theorem 1.5. *Suppose, in addition to the hypothesis of the above theorem, that ΔV is negative definite. Then, $x_n(0)$ is asymptotically stable.*

To extend these results we make the following definitions.

Definition 1.6. A continuous map $V : \mathbb{R}^N \to \mathbb{R}$ is said to be *dichotomic with respect to (1.1)* (in Ω) if there exists an integer $k \geq 2$ such that whenever we have $y \in \Omega$ and $\Delta_{k-1}^k V(y) \geq 0$, we also have $\Delta_0^k V(y) \leq 0$.

Observe that it is not required in the above definition that either $\Delta_{k-1}^k V$ or $\Delta_0^k V$ have a definite sign in Ω, but rather, that they mantain a kind of opposite sign in that neighborhood of the origin, as specified.

Definition 1.7. 1.5 A map V is said to be *strictly dichotomic with respect to (1.1)* if it is dichotomic with respect to this equation and, moreover, it satisfies the condition that whenever $y \in \Omega, y \neq 0$ and $\Delta_{k-1}^k V(y) \geq 0$, then $\Delta_0^k V(y) < 0$.

Note that every Liapunov function is automatically a dichotomic map already for $k = 2$, due to the fact that one such map is non-increasing along the solutions of (1.1).

The refinements of Theorems 1.4, 1.5 given above (see [1]) are as follows:

Theorem 1.8. *Suppose that V is both positive definite and dichotomic with respect to (1.1). Then, the equilibrium $x_n(0)$ is stable.*

Theorem 1.9. *Suppose, in addition to the hypothesis of Theorem 1.8, that V is strictly dichotomic with respect to (1.1). Then, $x_n(0)$ is asymptotically stable.*

Notice the fact that the application of these results takes clear advantage of the fact that $x_k(y)$ can be, so to say, easily computed.

2 An improvement

Given a map $V : \mathbb{R}^N \to \mathbb{R}$ and integers p, q with $p, q \geq 0$, we consider the following sets,

$$\Omega_+(p,q) = \{x \in \Omega : \Delta_q^p V(x) > 0\}$$
$$\Omega_-(p,q) = \{x \in \Omega : \Delta_q^p V(x) < 0\}$$
$$\Omega_0(p,q) = \{x \in \Omega : \Delta_q^p V(x) = 0\} \ ,$$

and put $\Omega_-^0(p,q) = \Omega_-(p,q) \cup \Omega_0(p,q)$. Also, if A is a set which contains the origin 0, we let A^* be the set $\{x \in A : x \neq 0\}$. Then, we reformulate definitions 1.6 and 1.7, respectively, as follows.

Definition 2.1. A continuous map $V : \mathbb{R}^N \to \mathbb{R}$ is said to be *dichotomic with respect to (1.1)* (in Ω) if there exists an integer $k \geq 2$ such that $\overline{\Omega_+}(k, k-1) \subset \Omega_-^0(k, 0)$.

Note that the requirement that $\overline{\Omega_+}(k, k-1) \subset \Omega_-^0(k,0)$ in this definition is weaker than the requirement that $\Delta_0^k V(x) \leq 0$ whenever $\Delta_{k-1}^k V(x) \geq 0$ of Definition 1.6, since it now permits the existence of points x such that $\Delta_{k-1}^k V(x) = 0$ and $\Delta_0^k V(x) > 0$.

Definition 2.2. A map V which is dichotomic with respect to (1.1) is said to be *strictly dichotomic with respect to (1.1)* if it satisfies the further condition that $\overline{\Omega_+^*}(k, k-1) \subset \Omega_-(k,0)$ and $\Omega_0(k,k-1) \cap \Omega_0(k,0) = \{0\}$.

The same observation made just above applies to this definition as well, when compared to its counterpart, namely, Definition 1.7.

From now on, whenever we refer to either dichotomic or strictly dichotomic maps with respect to (1.1), unless explicitly stated on the contrary, we are referring to maps that satisfy either Definition 2.1 or Definition 2.2, respectively.

Given a map $V : \mathbb{R}^N \to \mathbb{R}$, a point $y \in \mathbb{R}^N$ and an integer $k > 0$, we put

$$c_j = \max\{V(x_n(y)) : (j-1)k \leq n \leq jk\}, j = 1, 2, \ldots \tag{2.1}$$

and let

$$j^* = \min\{n : (j-1)k \leq n \leq jk \ \text{ and } \ c_j = V(x_n(y))\} \ , \tag{2.2}$$

and obtain the following useful lemmas.

Lemma 2.3. *If V is positive definite and $c_j = 0$ for some j, then $x_n(y) = 0$ for $n > j^*$ (and, hence, $c_n = 0$ for $n > j$).*

Proof. In fact, in this case we have that $V(x_{j^*}(y)) = 0$ and, since $V(x) = 0$ implies that $x = 0$, the result follows. □

Lemma 2.4. *If V is dichotomic with respect to (1.1), k is as in Definition 2.1 and $j \geq 2$, then $c_j \leq c_{j-1}$.*

Proof. Suppose that $j^* > (j-1)k$. Then, we have $(j-1)k \leq j^* - 1 < j^* \leq jk$ and so, $V(x_{j^*-1}(y)) < V(x_{j^*}(y))$, which means that $x_{j^*-k}(y) \in \Omega_+(k, k-1)$. By hypothesis, we must also have $x_{j^*-k}(y) \in \Omega^0_-(k, 0)$, i.e.,

$$c_j = V(x_{j^*}(y)) \leq V(x_{j^*-k}(y)) \leq c_{j-1}.$$

If, on the other hand, $j^* = (j-1)k$, it immediately follows from (2.1) that $c_{j-1} \geq c_j$, since $(j-1)k$ is also in the range of definition of c_{j-1} . □

Corollary 2.5. *If V is strictly dichotomic with respect to (1.1) and $j^* > (j-1)k$, it follows that $c_{j-1} > c_j$.*

Proof. In fact, this time $x_{j^*-k}(y) \in \Omega_+(k, k-1)$ implies that it also belongs to $\Omega_-(k, 0)$ and so, $c_j = V(x_{j^*}(y)) < V(x_{j^*-k}(y)) \leq c_{j-1}$. □

Lemma 2.6. *If V is strictly dichotomic with respect to (1.1), k is as in Definition 2.2, $j \geq 3$ and $c_{j-1} \neq 0$, then $c_{j-2} > c_j$.*

Proof. Under the above hypothesis, we know from Lemma 2.4 that $c_{j-2} \geq c_{j-1} \geq c_j$. Furthermore, from Corollary 2.5, we know that if $j^* > (j-1)/k$, then $c_{j-1} > c_j$, so that $c_{j-2} > c_j$. In the case that $j^* = (j-1)k$, we have either $c_{j-1} > c_j$ or $c_{j-1} = c_j$. Now, if $c_{j-1} > c_j$, then again $c_{j-2} > c_j$. If $c_{j-1} = c_j$, we do not have $(j-1)^* = (j-2)k$. In fact from $\Omega_0(k, 0) \cap \Omega_0(k, k-1) = \{0\}$ and $c_{j-1} \neq 0$, we obtain $V(x_{j^*-1}) < V(x_{j^*})$ which implies $V(x_{j^*}) < V(x_{j^*-k})$, i.e., $c_j < c_{j-1}$, a contradiction. Therefore $(j-1)^* > (j-2)k$, when $c_{j-1} = c_j$. This means $c_{j-2} > c_{j-1} = c_j$ and the proof is complete. □

We can now prove the following result.

Metatheorem 2.7. *Theorems 1.8 and 1.9 hold also under Definitions 2.1 and 2.2, respectively.*

Proof. Take $R > 0$ with $\Omega \supset \{|y| \leq R\}$, and take any $r > 0$ with $\sup\{|f(y)| : |y| \leq r\}$. Let $\delta = \inf\{V(y)| : r \leq |y| \leq R\}$, and note since V is continuous that $\delta > 0$. If $|y| \leq R$ and $V(y) < \delta$, then $|y| < r$. Noting continuity of f and V, we can take $\mu \in (0, r)$ such that for $n = 0, \ldots, k$ and $|y| < \mu$, one has $|f^n(y)| \leq R$, $V(f^n(y)) \leq \delta/2$. For such n, y, one has $|f^n(y)| < r$, and since $c_1(y) < \delta$, we know that $c_j(y) < \delta$ for all $j \geq 1$. Now a simple inductive argument on the index $n \geq k$ will show that $|f^n(y)| < r$ for all $n \geq 0$. This proves Theorem 1.8.

Now let V be strictly dichotomic, and take y having $|y| < \mu$, with μ as above. Let $c_j = c_j(y)$, and since $\{c_j\}$ is nonincreasing, we let $c_j \downarrow \alpha$ as $j \to \infty$. Now suppose $\alpha > 0$. Then since $\{f^n(y)\}$ lies entirely within a compact set, we have increasing sequences $\{j_i\}$, $\{m_i\}$ of positive integers with $(j_i - 1)k \leq m_i \leq j_i k$, $c_{j_i} = V(f^{m_i}(y))$, and $f^{m_i}(y) \to y_0$. Take any p with $0 \leq p \leq 4k$. As $i \to \infty$, we have $f^{m_i}(y) \to y_0$, so that $f^{p+m_i}(y) \to f^p(y_0)$ and $V(f^{p+m_i}(y)) \to V(f^p(y_0))$. Now note that $c_j \downarrow \alpha$ and $V(f^{p+m_i}(y)) \leq \max(c_{j_i}, \ldots, c_{4+j_i})$, and conclude that $V(f^p(y_0)) \leq \alpha$. For the discrete process $\tilde{x}_n = f^n(y_0)$, we let $\tilde{c}_j = c_j(y_0)$ and see that $\tilde{c}_1 \leq \alpha$, so that $\tilde{c}_j \leq \alpha$ for all $j \geq 1$. In fact, since V is strictly dichotomic, we know that $\alpha \geq \tilde{c}_1 > \tilde{c}_3 \geq \tilde{c}_4$. For any p with $2k \leq p < 4k$, note that $V(f^p(y_0)) \leq \tilde{c}_3$, and set $\gamma = (\alpha - \tilde{c}_3)/2$. Since there is i_p with $V(f^{p+m}(y)) \leq \gamma + V(f^p(y_0))$ for all $i \geq i_p$, we set $i_0 = max\{i_p : 2k \leq p \leq 4k\}$, and see that for $i \geq i_0$ and $2k \leq p \leq 4k$, one has $V(f^{p+m_i}(y)) \leq \gamma + V(f^p(y_0)) \leq \gamma + \tilde{c}_3$. For $j = 3 + j_{i_0}$ this makes $c_j \leq \gamma + \tilde{c}_3$, so that $c_j < \alpha$. This contradicts the fact that $c_j \downarrow \alpha$, and we conclude that $\alpha = 0$. □

The above result constitutes, therefore, a little improvement of the method given in [1].

3 The o.d.e. case

We shall now present a version of the above method which is suitable for ordinary differential equations.

Consider the equation

$$x'(t) = g(x(t)) \tag{3.1}$$

subject to the intitial condition

$$x(0) = y \; . \tag{3.2}$$

Here, we suppose that $g : \mathbb{R}^N \to \mathbb{R}^N$ is continuous, locally Lipschitzian and satisfies $g(0) = 0$. Thus, the solution of (3.1-2) exists, is unique and is defined for all $t \in \mathbb{R}$. It is denoted by $x(t, y)$ and, as a function of y, it is continuous. Observe that the null function $x(t, 0)$ is a solution of (3.1-2) when $y = 0$, called the *null equilibrium* of (3.1).

The definitions of stability and asymptotic stability of the null equilibrium of (3.1) are easily carried out from definitions 1.1 and 1.2 upon the simple substitution of n by t and $x_n(y)$ by $x(t, y)$.

Given a differentiable map $V : \mathbb{R}^N \to \mathbb{R}$, we define the *variation of V with respect to (3.1)* as being the map $V' : \mathbb{R}^N \to \mathbb{R}$ given by $V'(y) = \frac{d}{dt}V(x(t,y)) \mid_{t=0} = GradV(y).g(y)$, where $GradV$ is the gradient of V. We say that V is a *Liapunov function for (3.1)* when there exists a neighborhood Ω of the origin where V' is negative semi-definite. The basic results of Liapunov's direct method for ordinary differential equations are contained in the following analogues of Theorems 1.4, 1.5.

Theorem 3.1. *Suppose that there exists a positive definite Liapunov function for (3.1). Then, the null equilibrium is stable.*

Theorem 3.2. *Suppose, in addition to the hypothesis of the above theorem, that V' is negative definite. Then the null equilibrium is asymptotically stable.*

Following our previous notations, we now define some sets that will play a role similar to those of Section 2. If $T > 0$ is a given constant and Ω is a neighborhood of the origin having $x(t, y)$ defined for $-T \leq t \leq T$, $y \in \Omega$, we put

$$\Omega_-(T) = \{y \in \Omega : V(y) < V(x(-T, y))\}$$

$$\Omega_0(T) = \{y \in \Omega : V(y) = V(x(-T, y))\}$$

$$\Omega'_+ = \{y \in \Omega : V'(y) > 0\}$$

$$\Omega'_0 = \{y \in \Omega : V'(y) = 0\} \ ,$$

and let $\Omega^0_-(T) = \Omega_-(T) \cup \Omega_0(T)$. Also, if A and B are sets, we let $A \setminus B$ denote the set $\{x \in A : x \notin B\}$.

Definition 3.3. We say that a differentiable map $V : \mathbb{R}^N \to \mathbb{R}$ is *dichotomic with respect to (3.1)* if there are a constant $T > 0$ and a neighborhood Ω of the origin in $\mathbb{R}^N$ such that $\overline{\Omega'_+} \subset \Omega^0_-(T)$.

Definition 3.4. We say that a given map V is *strictly dichotomicc with respect to (3.1)* if it is dichotomic with respect to that equation and, moreover, it satisfies the supplementary condition that $(\overline{\Omega'_+})^* \subset \Omega_-(T)$ and $\Omega_0(T) \cap \Omega'_0 = \{0\}$.

Thus, the above definitions extend to ordinary differential equations the concept of dichotomic and strictly dichotomic maps, formerly given for difference equations. As before, we see that Liapunov functions are automatically dichotomic with respect to the given equation. Note that there may exist points $y \in \Omega'_0$ which do not belong to $\overline{\Omega'_+}$.

Also, this time we put, for a given $y \in \mathbb{R}^N$,

$$c_j = \max\{V(x(t, y)) : (j-1)T \leq t \leq jT\}, j = 0, 1, 2, \ldots \tag{3.3}$$

and

$$t_j = \min\{t \in [(j-1)T, jT] : c_j = V(x(t_j, y))\} \ . \tag{3.4}$$

Then, we have:

Lemma 3.5. *If V is dichotomic and $x(t, y)$ is defined for $0 \leq t \leq jT$, then $c_{j-1} \geq c_j$.*

Proof. If $t_j = (j-1)T$, then by the definition of c_{j-1}, we have $c_j = V(x(t_j, y)) \leq c_{j-1}$. Now if $t_j > (j-1)T$, then set $f(t) = V(x(t, y))$, and note that $f(t) < f(t_j)$ for $(j-1)T \leq t < t_j$. By the mean value theorem there is $\tilde{t} \in (t, t_j)$ with $(f(t_j) - f(t))/(t_j - t) = f'(\tilde{t})$, and this gives $0 < f'(\tilde{t}) = \text{Grad}\, V(x(\tilde{t}, y)) \cdot g(x(\tilde{t}))$. Now $x(\tilde{t}, y) \to x(t_j, y)$ as $t \to t_j$, which immediately means that $x(t_j, y) \in \overline{\Omega'_+}$. Thus $V(x(t_j, y)) \leq V(x(t_j - T, y))$, so that $c_j = V(x(t_j, y)) \leq V(x(t_j - T, y)) \leq c_{j-1}$. □

By referring to the proof of Lemma 3.5, we now obtain.

Lemma 3.6. *If V is strictly dichotomic, $x(t,y)$ is defined for $0 \leq t \leq jT$, and $t_j > (j-1)T$, then $c_{j-1} > c_j$.*

The next lemma parallels Lemma 2.6.

Lemma 3.7. *If V is strictly dichotomic, $x(t,y)$ is defined for $0 \leq t \leq jT$, and $c_{j-1} \neq 0$, then $c_{j-2} > c_j$.*

Proof. If $t_j > (j-1)T$, then $c_{j-2} \geq c_{j-1} > c_j$. Now if $t_j = (j-1)T$, then either $c_{j-1} > c_j$, or $c_{j-1} = c_j$. In the first case we again have $c_{j-2} \geq c_{j-1} > c_j$. In the case that $c_{j-1} = c_j$, one notes that $\Omega_0(T) \cap \Omega_0' = \{0\}$ and $c_{j-1} \neq 0$, and sees that if one had $t_{j-1} = (j-2)T$, then for $f(t) = V(x(t,y))$, one could not have $f'(t_j) \neq 0$. In fact, for $t_{j-1} = (j-2)T$, one could not have $f'(t_j) < 0$, since that would immediately contradict $t_{j-1} = (j-2)T$, not could one have $f'(t_j) > 0$, since that would contradict $t_j = (j-1)T$. We conclude then that $t_{j-1} > (j-2)T$, so that $c_{j-2} > c_{j-1} = c_j$. □

Theorem 3.8. *If V is a positive definite map that is dichotomic with respect to equation (3.1), then the equilibrium $x(t,0)$ is stable.*

Proof. Take $R > 0$ with $\Omega \supset \{|y| \leq R\}$, and take any $r > 0$ with $\sup\{|x(t,y)| : |y| \leq r, 0 \leq t \leq T\} \leq R$. Set $\delta = \inf\{V(y) : r \leq |y| \leq R\}$, and note that $\delta > 0$. If $|y| \leq R$ and $V(y) < \delta$, then $|y| < r$. Noting continuity of V and of the map $(t,y) \rightarrow x(t,y)$, we can take $\mu \in (0,r)$ such that for $0 \leq t < T$ and $|y| < \mu$, one has $V(x(t,y)) \leq \delta/2, |x(t,y)| \leq R$. For such t,y one has $|x(t,y)| < r$. Now take $j \geq 1$, and note that if $|x(t,y)| < r$ for $(j-1)T \leq t \leq jT$, then $|x(t,y)| \leq R$ for $jT \leq t \leq (j+1)T$, and since $c_1(y) < \delta$, we have $c_{j+1}(y) < \delta$. Thus $V(x(t,y)) < \delta$ for $jT \leq t \leq (j+1)T$, and we find that $|x(t,y)| < r$ for $jT \leq t \leq (j+1)T$. From this inductive argument on the index $j \geq 1$ we see that $|x(t,y)| < r$ for all $t \geq 0$. □

The proof of the next theorem is nearly identical to the proof of the second part of the Metatheorem. For this reason we merely establish the notation and indicate the direction of proof.

Theorem 3.9. *If V is a positive definite map that is strictly dichotomic with respect to equation (3.1), then the equilibrium $x(t,0)$ is asymptotically stable.*

Proof. Let μ be as in Theorem 3.8 and take y with $|y| < \mu$. Let $c_j = c_j(y)$, note that $\{c_j\}$ is nonincreasing, and let $c_j \downarrow \alpha$ as $j \rightarrow \infty$. Again suppose $\alpha > 0$. Since $\{x(t_j,y)\}$ lies within a compact set, we have an increasing sequence $\{j_i\}$ of positive integers such that for $m_i = t_{ji}$, one has $x(m_i,y) \rightarrow y_0$ as $i \rightarrow \infty$. Note that $c_{j_i} = V(x(m_i,,y))$, take t with $0 \leq t \leq 4T$, and proceed just as in the proof of the Metatheorem, with $f^{m_i}(y)$ replaced by $x(m_i,y)$. To be specific we note that the difference equation $\tilde{x}_n = f^n(y_0)$ is replaced here by the differential equation $\dot{x}(t) = g(x(t))$, $x(0) = y_0$, and $\tilde{c}_j$ is again given by $\tilde{c}_j = c_j(y_0)$. □

It is clear from the above theorems that we can obtain sharp stability results for ordinary differential equations in the spirit of Liapunov's direct method [2] even when V' is not negative semi-definite . Theorems 3.8 and 3.9 can, therefore, be viewed as adaptations of Theorems 1.4 and 1.5, respectively, to the case of ordinary differential equations. The apparent drawback here is that, unlike Liapunov's theorems, we have to somehow integrate equation (3.1) over an interval $[-T,0]$ in order to check the dichotomic character of a given map V. In Example IV of the next section we indicate a numerical technique which in many cases allows one to overcome this difficulty.

4 Examples

I. Take $N = 2, g(x,y) = (y,-x)$ and $V(x,y) = x^2 + \frac{1}{2}y^2$. Then, V is positive definite in $\mathbb{R}^2$ and its variation with respect to the o.d.e. $(x'(t),y'(t)) = g(x(t),y(t)), x(0) = x_0, y(0) = y_0$ is given by $V'(x_0,y_0) = x_0y_0$, which is not negative semi-definite in any neighborhood of the origin. Hence, it is not a Liapunov function for this equation. The solution of this equation is $x(t) = x_0 \cos t + y_0 \sin t, y(t) = -x_0 \sin t + y_0 \cos t$. If we pick $T = 2\pi$, we obtain: $V(x(-2\pi),y(-2\pi)) - V(x_0,y_0) = 0$. Hence, we have for $\Omega = B_1$, for instance,

$$\overline{\Omega}'_+ = \{(x,y) \in \Omega : xy \geq 0\}$$

$$\Omega^0_-(2\pi) = B_1$$

so that $\overline{\Omega}'_+ \subset \Omega^0_-(2\pi)$, and Theorem 3.8 applies in order to prove the stability of the given equilibrium.

II. Everything as in example I, except that $g(x,y) = (y,-2x-2y)$. Then, $V'(x_0,y_0) = -2y_0^2$. Thus V is a Liapunov function to the given o.d.e. but, since V' is just negative semi-definite, it cannot prove (directly) the asymptotic stability of the null equilibrium. Since $x(t) = e^{-t}[x_0 \cos t + (x_0 + y_0) \sin t]$ and $y(t) = x'(t)$ is the solution through (x_0,y_0), we obtain

$$\overline{\Omega}'_+ = \emptyset$$

$$\Omega_-(2\pi) = B_1^* \ .$$

Hence, the null equilibrium is asymptotically stable because V is a strictly dichotomic map with respect to the given o.d.e. in B_1.

III. Everything as in example II, but $V(x,y) = \frac{1}{2}(3x^2+y^2)$. Then, $V'(x_0,y_0) = xy - 2y^2$. Thus, V is not a Liapunov function for the o.d.e. because V' is not negative semi-definite in Ω. A straightforward computation shows that

$$\Omega_-(2\pi) = \{(x,y) \in B_1 : \frac{1}{2}[3(1-e^{4\pi})x^2 + (1-e^{4\pi})y^2] < 0\} = \Omega^* \ .$$

Hence, we have that V is strictly dichotomic in Ω with respect to the given o.d.e., and its null equilibrium is asymptotically stable.

IV. If we discretize the o.d.e. of example III by, say, Euler's method, we obtain

(4.1) $$x_n = x_{n-1} + hy_{n-1} \qquad y_n = -2hx_{n-1} + (1-2h)y_{n-1}.$$

The matrix of (4.1) is

$$A(h) = \begin{bmatrix} 1 & h \\ -2h & 1-2h \end{bmatrix}$$

Given $T > 0$ and $m > 1$, let $h = T/m$. Then, one can diagonalize the matrix $A^m(h_m)$ and use L'Hopital's rule, and eventually find that $A^m(h_m) \to A_0(T)$ as $m \to \infty$, where

$$A_0(T) = e^{-T} \begin{bmatrix} \cos T + \sin T & \sin T \\ -2\sin T & \cos T - \sin T \end{bmatrix}$$

and this convergence is uniform in T for T in compact sets.

Using this feature, we shall again show that the map $V(x,y) = \frac{1}{2}(3x^2 + y^2)$ is strictly dichotomic for the differential equation of Example III.

In fact, note that for any $T > 0$, we have $h_m k_m \to T$ as $m \to \infty$. Now write $y_0 = \begin{pmatrix} u_0 \\ v_0 \end{pmatrix}$, and for $j \geq 1$ write $y_j = A^j y_0$. Then $y_{m+1} = \begin{pmatrix} u_{m+1} \\ v_{m+1} \end{pmatrix} = A(h_m)A^m(h_m)\begin{pmatrix} u_0 \\ v_0 \end{pmatrix}$, so that $\begin{pmatrix} u_{m+1} \\ v_{m+1} \end{pmatrix} \to A_0(T)\begin{pmatrix} u_0 \\ v_0 \end{pmatrix}$ as $m \to \infty$, i.e., $y_{m+1} \to e^{-T} y_0$ as $m \to \infty$.

In investigating whether V is dichotomic, one can first note for each $(u_0, v_0) \in R^2$ that $\frac{1}{h_m}\Delta_0^1 V(u_0, v_0) \to (u_0 v_0 - 2v_0^2)$ as $m \to \infty$, so that $V'(u_0, v_0) = u_0 v_0 - 2v_0^2$, as expected. Now this quadratic form $g(u_0, v_0) = u_0 v_0 - 2v_0^2$ is not sign definite. However, a simple calculation shows that $\Delta_0^{m+1} V(u_0, v_0) \to -\frac{1}{2}(1 - e^{-2T})(3u_0^2 + v_0^2)$ uniformly in compact neighborhoods of the origin as $m \to \infty$. If we write $3u_0^2 + v_0^2 = (u_0^2 + v_0^2) + 2u_0^2$, we see that for each nonzero $y_0 \in R^2$, there is $M > 0$ such that $\Delta_0^{m+1} V(y_0) \leq -\frac{1}{4}(1 - e^{-2T})|y_0|^2$ for all $m \geq M$. Since $\Delta_0^{m+1} V(y_0) \to V(x(T, y_0)) - V(y_0)$ as $m \to \infty$, we see that for all $y_0 \neq 0$, one has $V(x(T, y_0)) - V(y_0)) < 0$. We can now conclude that V is strictly dichotomic for the differential equation of Example III so that the null solution of this differential equation is asymptotically stable.

5 Acknowledgments

The author wishes to thank the referee for the various important suggestions and corrections which greatly improved the content and clarity of the paper.

This research was partially supported by FAPESP - Fundação de Amparo a Pesquisa do Estado de São Paulo - Brasil.

References

1. Carvalho, L.A.V., Ferreira, R.R. (1988): On a new extension of Liapunov's direct method to discrete equations. Quart. Appl. Math. **66**, No. 4, 778-788
2. Cech, E. (1969): Point Sets. Academia Publishing House of the Czechoslovak Academy of Sciences, Prague
3. Hale, J. K. (1969): Ordinary Differential Equations. Pure and Appl. Math. Series, **21**, Wiley Interscience, N. York, London, Sidney, Toronto

Systems Of Set-Valued Equations In Banach Spaces

G. Conti[1], *P. Nistri*[2], *P. Zecca*[2]

[1] Istituto di Matematica Applicata, Università di Firenze, Italy
[2] Dipartimento di Sistemi e Informatica, Università di Firenze, Italy

Introduction

In this paper we give sufficient conditions for the solvability of set-valued systems of the form

$$\begin{cases} 0 \in F(x,y) \\ 0 \in G(x,y) \end{cases} \tag{1}$$

where F and G are multivalued maps, defined in the following way

$$F(x,y) = y - \hat{F}(x,y)\ ,\ G(x,y) = x - \hat{G}(x,y)$$

where $x \in X$, $y \in Y$, with X, Y Banach spaces, and $\hat{F}$ and $\hat{G}$ are upper semicontinuous, compact multivalued maps, defined on the closure of a suitable open subset U of the $X \times Y$, taking values in X and Y respectively. For simplicity we will call F and G *multivalued compact vector fields*, even if in the literature this definition may be used also with different meaning in a different context. Roughly speaking, we solve the first equation in term of y as a function of "the parameter" x considering the application $x \multimap S(x) = \{y \in Y \ : \ 0 \in F(x,y)\}$ and hence we introduce the solution set $S(x)$ in the second one. The fixed points of the composite function $\hat{G}(x, S(x))$ are the solutions of system (1).

The paper is organized as follows. In Section 1 we give some definitions about set-valued maps and we recall some known results we will need in the sequel. In Section 2 we give a first result for solving system (1) by assuming F with convex, compact values and G single valued. Successively we solve system (1) under the hypotheses that G is a multivalued admissible map and the set of solutions $S(x)$ is acyclic. We want to note here that in general, in the convex case, the solution set $S(x)$ is not necessarily acyclic.

In Section 3 we give applications of our results. More precisely, we consider the problem of finding conditions for the solvability of two point boundary value problems for a multivalued differential system. We want to point out that while such problems can be formulated within the framework of the general theory of differential inclusions, they may also be viewed as models of problems in control theory.

1 Notations and definitions.

Definition 1.1. Let X and Y be topological spaces. A set-valued map M from X into Y is said to be *upper semicontinuous at* $x \in X$ if for every neighborhood V of $M(x)$ there exists a neighborhood U of x such that $M(x) \subset V$ for every $x \in U$. If, for every $x \in X$, M is upper semicontinuous at x and $M(x)$ is compact, then M is said to be upper semicontinuous on X. If M sends bounded sets into relatively compact sets, then it is said to be *compact*. M is said to be *proper* if, for each compact set K of Y, $M^{-1}(K)$ is compact. We will denote a multivalued map M from X to Y with the symbol $M : X \multimap Y$. By an r-neighborhood of a subset Ω of a metric space X we mean the set $B(\Omega, r) = \{y \in X : \exists\, x \in \Omega \text{ such that } d(x,y) < r\}$.

Definition 1.2. Let X and Y be Banach spaces and $M : X \multimap Y$ be a multivalued map. We say that a continuous map $\mu : X \to Y$ is a *ϵ-graph approximation of* M (shortly ϵ-approximation) if graph $\mu \subset B$ (graph M, ϵ). We will say that a map $\mu : X \to Y$ is a *ϵ-pointwise approximation for* $M : X \multimap Y$ if $\mu(x) \in B(M(x), \epsilon)$ for all $x \in X$.

The following result is due to Cellina [2].

Theorem 1.3. *Let X and Y be metric locally convex spaces and let $M : X \multimap Y$ be an upper semicontinuous map with compact and convex values. Then for any $\epsilon > 0$ there exists an ϵ-approximation of M.*

Using this result, in [3] Cellina and Lasota gave a definition of degree (we will denote it by Deg) for multivalued compact vector fields with convex values.

Definition 1.4. Let X and Y be topological Hausdorff spaces. An upper semicontinuous map, with a finite number of points as images, $M : X \multimap Y$ will be called a *weighted map* (shortly w-map) if to each x and $y \in M(x)$ a multiplicity or weight $m(y, M(x)) \in Z$ is assigned in such a way that the following property holds

a) if U is an open set in Y with $\partial U \cap M(x) = \emptyset$, then

$$\sum_{y \in M(x) \cap U} m(y, M(x)) = \sum_{y' \in M(x') \cap U} m(y', M(x'))$$

whenever x' is close enough to x, (see [5] and [11]).

Definition 1.5. The number

$$i(M(x), U) = \sum_{y \in M(x) \cap U} m(y, M(x))$$

will be called the *index or multiplicity* of $M(x)$ in U.

If U is a connected set, the number $i(M(x), U)$ does not depend on $x \in X$. In this case the number $i(M) = i(M(x), U)$ will be called the *index of the weighted map* M, (see e.g. [11]).

Definition 1.6. **([13])** Let X be a metric space. An upper semicontinuous set valued map $M : X \multimap X$ is *admissible* if there are maps $G_i : Y_i \to Y_{i+1}$, $i = 0, 1, ..., n$ (Y_i metric spaces, $Y_0 = Y_{n+1} = X$) satisfying
i) $F = G_n \circ ... \circ G_0$;
ii) G_i is upper semicontinuous with acyclic, compact values for each $i = 0, 1, ..., n$.
Each sequence $G_0, ..., G_n$ is called an admissible sequence for M.

We recall that the composition of upper semicontinuous maps is upper semicontinuous.

Definition 1.7. Let X be a Banach space and let $\overline{B}(0,r)$ be a closed ball in X of radius r. We will say that the upper semicontinuous map $M : \overline{B}(0,r) \multimap X$ verifies the *Borsuk-Ulam (B.U.) property on* ∂B if for all $x \in \partial B(0,r)$, $M(x)$ and $M(-x)$ are strictly separated by a hyperplane, i.e. for all $x \in \partial B(0,r)$ there exists a continuous functional $x^* \in X^*$, the dual space of X, such that $x^*(y) > 0$ for all $y \in M(x)$ and $x^*(y) < 0$ for all $y \in M(-x)$.

Definition 1.8. Let X, Y be metric spaces. Given $\mathcal{S} \subset X \times Y$ and $A \subset X$, denote by:

$$\mathcal{S}(x) = \{y \in Y : (x,y) \in \mathcal{S}\};$$
$$\mathcal{S}(A) = \{y \in Y : (x,y) \in \mathcal{S},\ x \in A\};$$
$$\mathcal{S}_x = \mathcal{S} \cap (\{x\} \times Y) \text{ and } \mathcal{S}_A = \mathcal{S} \cap (A \times Y) \text{ for } A \subset X.$$

Definition 1.9. Let X and Y be metric spaces. Let $U \subset X \times Y$ be open and locally bounded over X, i.e. for any $(x,y) \in U$ there exists a neighborhood $N \subset X$ of x such that U_N is a bounded set in $X \times Y$. We shall say that $F : \overline{U} \multimap Y$ is a *parametrized compact vector field* if $F(x,y) = y - \hat{F}(x,y)$ with $\hat{F}$ upper semicontinuous and $\hat{F}(D)$ relatively compact in Y for any bounded set D of $\overline{U}$. We shall denote by

$$S^F = \{(x,y) \in \overline{U} : y \in \hat{F}(x,y)\};$$
$$\mathcal{D}^F = \{x \in X : \mathcal{S}^F_x \cap \partial U = \emptyset\}.$$

Definition 1.10. Let $M : \overline{B}(0,r) \subset X \multimap X$ be an upper semicontinuous set valued map. The map M *satisfies the boundary condition "P"* if $x \in \partial B(0,r)$ and $\lambda x \in M(x)$ implies $\lambda \leq 1$.

Definition 1.11. A multivalued map $M : X \times Y \multimap Z$ is said to be *uniformly quasibounded with respect to x* if there exist α, $\beta \in R^+$ such that:

$$\| M(x,y) \| = \sup_{z \in M(x,y)} \| z \| \leq \alpha \; \| y \| \; + \beta \text{ for any } x \in X.$$

In the sequel by $\epsilon M(\epsilon p)$ we will denote the set $B(M(B(p,\epsilon)), \epsilon)$.

2 Results

We want to investigate now the existence of solutions for the system

$$\begin{cases} y \in \hat{F}(x,y) \\ x = \hat{g}(x,y) \end{cases} \text{ or } \begin{cases} 0 \in y - \hat{F}(x,y) = F(x,y) \\ 0 = x - \hat{g}(x,y) = g(x,y) \end{cases} \tag{2}$$

Theorem 2.1. *Let X, Y be Banach spaces, let $U \subset X \times Y$ be an open, locally bounded set over X. Let $\hat{F} : \overline{U} \multimap Y$ be an upper semicontinuous compact map with convex values and $\hat{g} : \overline{U} \to X$ be a continuous and compact map. Suppose that there exists r > 0 such that $\overline{B}(0,r) \subset \mathcal{D}^F$ and Deg $(F(0,\cdot), U(0), 0) \neq 0$. Let $T : \overline{B}(0,r) \multimap X$ be defined by $T(x) = x - \hat{T}(x)$, where $\hat{T}(x) = \hat{g}(x, S(x))$ and $S(x) = \{y \in Y : y \in \hat{F}(x,y)\}$. Let us suppose that for all $x \in \partial B$, such that $0 \notin T(x)$, $T(x)$ and $T(-x)$ are strictly separated by an hyperplane. Then there exists $x \in \overline{B}(0,r)$ such that $0 \in T(x)$. Hence system (2) has a solution.*

Before proving theorem 2.1 we need some preliminary results. First observe that the map $S : \overline{B}(0,r) \subset X \multimap Y$ defined by $x \multimap S(x)$ is upper semicontinuous on $\overline{B}(0,r)$. In fact the local boundedness of U implies that $S(x_0)$, is compact for every $x_0 \in X$. Moreover, if V is an open neighborhood of $S(x_0)$, then there exists an open neighborhood N of x_0, $N \subset \mathcal{D}^F$, such that $S(x) \subset V, \forall x \in N$. To see that let $y \in S(x_0)$ and let us consider neighborhoods of the form $N_{x_0} \times V_y$, with N_{x_0} a neighborhood of x_0 in $\mathcal{D}^F$ and V_y a neighborhood of y in Y, such that

$$V_y \subset U(x_0) \cap V \text{ and } N_{x_0} \times V_y \subset U.$$

Let S be the set $\{(x,y) \in X \times Y \; : \; y \in \hat{F}(x,y)\}$. By the compactness of $S_{x_0} = S \cap (\{x_0\} \times Y)$ there exists a finite number, say s, of neighborhoods of the previous form covering S_{x_0}. Let

$$N_0 = \bigcap_{i=1}^{s} N_i \text{ and } V' = \bigcup_{i=1}^{s} V_i$$

Clearly for each neighborhood N of x_0, with $N \subset N_0$, we have that $N \times V' \subset U$. Let us prove that there exists a neighborhood N of x_0 such that $S(x) \subset V'$ for all $x \in N$. Suppose not, then there exists a bounded sequence $\{(x_n, y_n)\}$ with $x_n \to x_0$, $y_n \in \hat{F}(x_n, y_n)$ and $y_n \notin V'$. As $\hat{F}$ is compact and upper semicontinuous, we may assume (by passing to a subsequence, if necessary) that

$y_n \to y_0$. Then $y_0 \in \hat{F}(x_0, y_0)$, that is $y_0 \in S(x_0)$, contradicting $S(x_0) \subset V'$. □

Note that $S(x) \neq \emptyset$ because of the hypothesis on the degree of F.

Lemma 2.2. *Let $X = R^n$, $Y = R^m$ and let $\overline{B}(0,r) \subset \mathcal{D}^F$. Then for each neighborhood W of S_B^F there exists $\epsilon > 0$ such that if $\hat{f} : \overline{U}_B \to R^m$ is an ϵ-approximation of $\hat{F}_{|\overline{U}_B}$, then $S_B^f \subset W$.*

Proof. S_B^F is a closed and compact set. In fact, let $\{(x_n, y_n)\} \subset S_B^F$ and $(x_n, y_n) \to (x_0, y_0)$. As $\{(x_n, y_n)\} \subset S_B^F$, we have that $y_n \in \hat{F}(x_n, y_n)$ for any $n \in N$. As $\hat{F}$ is upper semicontinuous, then $y_0 \in \hat{F}(x_0, y_0)$, that is $(x_0, y_0) \in S_B^F$. Then S_B^F is closed and obviously compact.

Let W be a neighborhood of S_B^F, V a ϵ_1-neighborhood of S_B^F, $V \subset W$, with $\partial V \cap \partial W = \emptyset$. Let $\epsilon_2 = d(\partial V, \partial W)$ and $A = \overline{U}_B \backslash V$. Since A is a compact set,we have that

$$\inf_{(\tilde{x},\tilde{y})\in A} \{\| s - \tilde{y} \|, s \in \hat{F}(\tilde{x},\tilde{y})\} = \epsilon_3 > 0.$$

Let $\epsilon = \min \{\epsilon_1, \epsilon_2, \epsilon_3\}$ and let $\hat{f} : \overline{U} \to Y$ be an ϵ-approximation of $\hat{F}_{|\overline{U}_B}$. Let $(x, y) \in \overline{U}_B \backslash W$ and let $y = \hat{f}(x, y)$, that is (x, y) in S_B^f. Then there exists $(\overline{x}, \overline{y}) \in \overline{U}_B$ such that:
$\| (x, y) - (\overline{x}, \overline{y}) \| + \| \hat{f}(x, y) - z \| < \epsilon$ for some $z \in \hat{F}(\overline{x}, \overline{y})$.
As $\| (x, y) - (\overline{x}, \overline{y}) \| < \epsilon$ it follows that $(\overline{x}, \overline{y}) \notin S_B^F$. Since $y = \hat{f}(x, y)$ we get that $\| y - z \| < \epsilon$. Hence $(\overline{x}, \overline{y}) \notin A$. Thus $(\overline{x}, \overline{y}) \in V \backslash S_B^F$.This is absurd, since $(x, y) \in \overline{U} \backslash W$ and $\| (x, y) - (\overline{x}, \overline{y}) \| < \epsilon$ and so $(\overline{x}, \overline{y}) \notin V$. □

Lemma 2.3. *Let $X = R^n$, $Y = R^m$ and let $\overline{B}(0,r) \subset \mathcal{D}^F$. There exists an $\epsilon_0 > 0$ such that for all $\epsilon < \epsilon_0$ there exists $\hat{f} : \overline{U} \to R^m$, such that $\hat{f}_{|\overline{U}_B}$ is an ϵ-approximation of $\hat{F}_{|\overline{U}_B}$ and*
a) S^f is a finite subset of $U(x), \forall x \in \overline{B}(0, r)$;
b) $Deg(F(0, \cdot), U(0), 0) = Deg(f(0, \cdot), U(0), 0)$.

Proof. The definition of degree given by Cellina and Lasota, [3], ensures the existence of a positive number, say ϵ_0, with the property that for any $\epsilon < \epsilon_0$ every ϵ-approximation of $F_{|\overline{U}_B}$ has the same degree as $F_{|\overline{U}_B}$. Fix $\epsilon < \epsilon_0$ and let $\tilde{f} : \overline{U}_B \to R^m$ be an $\epsilon/2$ approximation of $\hat{F}_{|\overline{U}_B}$. Using the same arguments as in Lemma 2.4 of [11] we can prove that there exists $\tilde{f}_1 : \overline{U}_B \to R^m$, $\epsilon/2$-pointwise approximation of $\tilde{f}$, that satisfies property a). Then $\tilde{f}_1$ is an ϵ-approximation of $\hat{F}_{|\overline{U}_B}$ that satisfies property a). The map $\hat{f}$ is then any continuous extension to $\overline{U}$ of $\tilde{f}_1$, therefore b) follows immediately from the choice of ϵ. □

Lemma 2.4. *Let X and Y be Banach spaces. Let U be a locally bounded, open subset in $X \times Y$. Let us suppose that for all $x \in \mathcal{D}^F$ the equation $y \in \hat{F}(x, y)$*

has only isolated solutions. Then the application $x \multimap S(x)$ *is a w-map and* $i(S) = Deg(F(x,\cdot),\ U(x),\ 0)$.

Proof. We have already seen that the map $x \multimap S(x)$ is an upper semicontinuous map. We want to prove that to any $y \in S(x)$ it is possible to associate an integer $m(y,\ S(x))$ with the property of Definition 1.4. If y is an isolated solution for $\hat{F}$ then there exists a neighborhood Ω of y such that $\Omega \cap S(x) = \{y\}$.

Let us define $m(y, S(x)) = Deg\ (\ F(x,\cdot),\ \Omega,\ 0)$. Using the excision property, $m(y,\ S(x))$ does not depend on the choice of Ω provided that $\Omega \cap S(x) = \{y\}$. Let W be an open set in Y such that $S(x) \cap \partial W = \emptyset$. As S is upper semicontinuous there exists a ball $\overline{B}(x,r)$ such that $\forall x' \in \overline{B}(x,r) \cap Y$ we get $S(x') \cap \partial W = \emptyset$.

Let us consider now the following homotopy, $H : [0,1] \times W \multimap Y$, defined by

$$H(t,y) = F(tx + (1-t)x', y).$$

As $tx + (1-t)x' \in \overline{B}(x,r) \cap Y$ for all $t \in [0,1]$, H is an admissible homotopy between $F(x,\cdot)_{|W}$ and $F(x',\cdot)_{|W}$. From this fact, using the additivity property of the degree, we get:

$$\sum_{y \in S(x) \cap W} m(y, S(x)) = Deg(\ F(x,\cdot),\ W, 0) =$$

$$Deg(\ F(x',\cdot),\ W,\ 0) = \sum_{y \in S(x') \cap W} m(y, S(x')).$$

□

The following lemmata, whose proofs can be found in [11], hold.

Lemma 2.5. *Let* $B = B(0,r)$ *be an open ball in* R^n *and let* $M : \overline{B} \multimap R^n$ *be a w-map with* $i(M) \neq 0$. *If* M *verifies the B.U. condition on* ∂B, *then there exists* $x \in \overline{B}(0,r)$ *such that* $0 \in M(x)$.

Lemma 2.6. *Let* $M\ :\ \overline{B} \multimap X$ *be a compact vector field satisfying the B.U. property. Then there exists* $\epsilon > 0$ *such that every* ϵ*-approximation of* M *satisfies the B.U. property.*

We can now give the proof of Theorem 2.1

Proof. The result is obviously true if there exists $x \in \partial B$, $B = B(0,r)$, such that $0 \in T(x)$. In any other case, assume that system (2) does not have solutions (x,y) with $y \in S(\overline{B})$, i.e. $0 \notin T(x)$ for all $x \in \overline{B}$. Since T is an upper semicontinuous and "compact" vector field, its image is closed. The assumption $0 \notin T(x)$ implies the existence of $\epsilon_1 > 0$ such that $B(0,\epsilon_1) \cap T(\overline{B}(0,r)) = \emptyset$. On the other hand from Lemma 2.6 there exists ϵ_2 such that every ϵ_2-approximation T' of T, $T' : \overline{B} \multimap X$ verifies the B.U. property.

Let $\delta = \min\ \{\epsilon_1, \epsilon_2\}$ and let $V \subset U$ be defined by: $V = \{(x,y) \in U\ :\ (x, g(x,y)) \in \delta\ \ \mathrm{Gr}\ T\}$, where Gr T stands for the graph of T and δA is the

δ-neighborhood of the set A. Clearly V is an open set being the inverse image of the open set δ Gr T, through the continuous map g.

We divide the proof into three parts.

First part. $X = R^n$, $Y = R^m$.

Let ϵ^* be that one given in Lemma 2.2, i.e. every ϵ^* approximation $\hat{f}$ of $\hat{F}$ has the property that $S_B^f \subset V$. Let ϵ_0 given by Lemma 2.3 and let $\epsilon' = \min\{\epsilon^*, \epsilon_0, \delta\}$. By Lemmata 2.3 and 2.4 there exists a continuous map $\hat{f} : \overline{V} \to R^n$ which is an ϵ'-approximation for $\hat{F}$ on $\overline{V} \backslash B$ and such that the set valued map $S' : \overline{B} \multimap R^n$ defined by $x \multimap S'(x) = S^f(x)$, $f = I - \hat{f}$, is a w-map such that $S'_B \subset V$. The index of the map is given by

$$i(S') = Deg(F(0,\cdot), V(0), 0) = Deg(F(0,\cdot), U(0), 0) \neq 0.$$

The set valued map $T'(x) = g(x, S'(x))$ is a w-map with index $i(T') = i(S') \neq 0$, (see [5]). As $S'_B \subset V$ we have that Gr $T' \subset \epsilon'$ Gr T. Since T' verifies the B.U. property then, by Lemma 2.5, there exists $x \in \overline{B}$ such that $0 \in T'(x)$. Then $0 \in \epsilon' T(\epsilon' x)$. This contradiction establishes the result.

Second part. $X = R^n$, *Y Banach space.*

Let $\tilde{f}$ be a $\frac{\epsilon'}{2}$-approximation of $\hat{F}$ on $\overline{U}_B$, and let $\hat{f}$ be a $\frac{\epsilon'}{2}$-pointwise approximation of $\tilde{f}$ whose range is contained in a finite dimensional subset K_1 of Y. By Lemma 2.2 and the properties of the degree, we get

$$\begin{aligned} 0 \neq Deg(F(0,\cdot), V(0), 0) &= deg(f(0,\cdot), V(0), 0) \\ &= deg(f(0,\cdot)_{|V_1}, V(0) \cap K_1, 0) \end{aligned}$$

with $V_1 = V \cap (X \times K_1)$ and $S_B^f \subset V_1$. Then $S_B^f \neq \emptyset$. Let $g_1 = g_{|V_1}$ and $\hat{f}_1 = \hat{f}_{|V_1}$. These two maps satisfy the hypotheses of Theorem 2.1. Then, for the first part of the proof, the set valued map $T''(x) = g_1(x, S^{f_1}(x))$ has a zero in $\overline{B}$. As $S_B^{f_1}(x) \subset V$ we have that Gr $T'' \subset \epsilon'$ Gr T, contradicting the fact that $0 \notin \epsilon_1 T(\epsilon_1 x)$.

Third part. X,Y *Banach spaces.*

Let $\hat{g}_2 : \overline{U}_B \to X$ be an ϵ-pointwise approximation of $\hat{g}$ on $\overline{U}_B$ with finite dimensional range. Let $X_1 \subset X$ be the subspace containing the range of $\hat{g}_2$ and let $g_2 = 1 - \hat{g}_{2|(X_1 \times Y) \cap \overline{U}_B}$ and $\hat{F}_2 = \hat{F}_{|(X_1 \times Y) \cap \overline{U}_B}$. Let $T' : \overline{B}' = \overline{B} \cap X_1 \multimap X_1$ defined by $T'(x) = g_2(x, S(x))$. T' is an ϵ-approximation of $T_{|\overline{B}}$ and by Lemma 2.6, T' satisfies the B.U. property on $\partial B'$. By the second part of the proof T' has a zero on $\overline{B}' \subset \overline{B}$, contradicting the fact that $0 \notin \epsilon\, T(\epsilon x)$. □

With a similar proof to the one of Theorem 1.4 of [11] we can prove the following

Theorem 2.7 *Let X and Y be as in Theorem 2.1. Let U be a locally bounded open set in* $X \times Y$ *and let* $\hat{F} : \overline{U} \multimap Y$ *be an upper semicontinuous convex valued mapping. Consider a compact, convex set* $Q \subset X$ *such that for every* $x \in Q$ *we*

have $y \notin \hat{F}(x,y)$ *on* $\partial U(x)$. *Assume that for some (and hence for all)* $x \in Q$ *we have* $Deg\ (F(x,\cdot),\ U(x),0) \neq 0$. *If* $\hat{g} : S_Q \to X$ *is any continuous map such that* $\hat{g}(x, S(x)) \subset Q$ *for any* $x \in Q$, *then there exists a solution* $(x,y) \in U$ *of (2) with* $x \in Q$ *and* $y \in S(x)$.

In the next theorem we give an existence result for system (2) under the assumption that F and G are both multivalued maps and G is admissible. However we have to assume that the set $S(x)$ is acyclic for every $x \in \mathcal{D}^F$. Notice that this assumption holds in many cases (see e.g. [7] and [8]).

Theorem 2.8 *Let X, Y be Banach spaces, let* $U \subset X \times Y$ *be an open, locally bounded set. Let* $\hat{F} : \overline{U} \multimap Y$ *be an upper semicontinuous, uniformly quasibounded with respect to* x, *compact map with closed values. Let us suppose that there exists* $r > 0$ *such that* $\overline{B}(0,r) \subset \mathcal{D}^F$ *and for any* $x \in \mathcal{D}^F$ *the set* $S(x) = \{y \in Y : y \in \hat{F}(x,y)\}$ *is non empty and acyclic. Let* $\hat{G} : \overline{U} \multimap X$ *be a compact, admissible map, and* $\hat{T} : \overline{B}(0,r) \multimap X$ *be the map defined by:*

$$x \multimap \hat{T}(x) = \hat{G}(x, S(x)).$$

Suppose that the map $\hat{T}(x)$ *satisfies property "P" of Definition 1.10. Then the system*

$$\begin{cases} y \in \hat{F}(x,y) \\ x \in \hat{G}(x,y) \end{cases} \tag{1}$$

has a solution.

Proof. We have already proved that the map S is upper semicontinuous. Therefore, $\hat{T}$ is an admissible map with compact values. Furthermore, being $\hat{G}$ compact and $\hat{F}$ uniformly quasibounded, we get that $\hat{T}$ is a compact map. We want to show that $\hat{T}$ has fixed point in $\overline{B}(0,r)$.

Using Lemma 2 in [10] we know that there exists a compact convex set $K \subset \overline{B}(0,r)$ such that $\overline{co}(\pi \circ \hat{T}(K)) = K$, where π is the radial projection of X on $\overline{B}(0,r)$. Since $\pi \circ \hat{T}$ is admissible , there exists $x \in K$ such that $x \in \pi \circ \hat{T}(x)$. Condition "P" implies that $x \in \hat{T}(x)$. □

3 Applications

In this section we present two applications of our results to problems involving convex-valued differential inclusions. While such problems can be formulated within the framework of the general theory of differential inclusions, a topic of independent interest, they may also be viewed as models of problems of a very different nature, for example: control theory. To be definite, consider a nonlinear control process described by a system of ordinary differential equations of the form:

$$\dot{x} = f(t, x, u) \tag{C}$$

where $f : [0,1] \times R^n \times R^m \to R^n$ satisfies Caratheodory's conditions and the control u is in Ω, with Ω a non-empty compact subset of R^m. If for $(t,x) \in [0,1] \times R^n$ we set

$$f(t,x,\Omega) = \Phi(t,x)$$

and assume that the multivalued map Φ is convex, then the trajectories of system (C) corresponding to controls from the set of functions $\mathcal{U} = \{u \in L^\infty((0,1),\ R^m) : u(t) \in \Omega \text{ for a.a. } t \in [0,1]\}$ are precisely those corresponding to the multivalued differential equation

$$\dot{x} \in \Phi(t,x).$$

One of the advantages of dealing with this equation rather than the original equation (C) is in the fact that in various situations it is easier to use differential inclusions in order to determine conditions sufficient to guarantee specified behaviour of the trajectories. On differential inclusions and their relationships with other fields see, for example, [1]. As mentionated above, the following two examples may also be viewed as control problems. Specifically, example 1 may be viewed as a problem of periodic controllability. Such problems have been treated in [9] and in [12] using degree theory (see also references therein). Example 2 may be considered as a reachability problem between two given sets. On this subject, see for example [4].

Example 1. Consider the initial value problem

$$\begin{cases} \dot{y} \in \Phi(t,y) \\ y(0) = y_0 \end{cases} \tag{E1}$$

where $\Phi : [0,1] \times R^n \to R^n$ is a Caratheodory function with compact, convex values, i.e. Φ satisfies the following conditions
(f1) Φ is a t-measurable, y-upper semicontinuous function;
(f2) for each $\rho > 0$ there exists $\alpha_\rho,\ \beta_\rho \in L^1((0,1),R)$ such that $|\Phi(t,p)| \leq \alpha_\rho(t) + \beta_\rho(t)\,|p|$, for a.a. $t \in [0,1]$ and every $p \in R^n$ with $|p| \leq \rho$, where

$$|\Phi(t,p)| = \sup_{z \in \Phi(t,p)} |z|.$$

Under our assumptions (E1) is equivalent to the integral form $y \in \hat{F}(y_0,y)$, where $\hat{F} : R^n \times (C)^n \to (C)^n$ is defined by:

$$(y_0,y) \multimap y_0 + \int_o^t \Phi(s,y(s))\,ds$$

where the integral is intended in the Aumann sense and $(C)^n$ stands for the Banach space $C([0,1],\ R^n)$.

Formulation of the problem: We want to give conditions on the map $\Phi(t,y)$ in order to prove, using Theorem 2.8, the existence of an initial state y_o corresponding to a 1-periodic solution of (E1).

First of all observe that, since Φ satisfies (f1) - (f2), for any $y_0 \in R^n$ the solution set $S(y_0)$ is a non empty, compact, acyclic subset of $(C)^n$. Moreover the solution map $S : R^n \to (C)^n$ is upper semicontinuous and sends bounded sets of R^n into bounded sets of $(C)^n$, which in turn, by (f2) are bounded in $AC([0,1], R^n) = (AC)^n$, and so they are compact in $(C)^n$.
Assume the following condition
(C1)-there exists $r > 0$ such that

$$\int_o^1 s(t)dt \leq 0,$$

where $s(t) = \sup_{|y| \geq r} \sigma(t,y)$ for almost all $t \in [0,1]$ and $\sigma(t,y) = \sup_{z \in \Phi(t,y)} < y, z >$.

Here $< \cdot, \cdot >$ denotes the standard Euclidean inner product in R^n.

In what follows we will rewrite our problem in a suitable form to apply Theorem 2.8. Since the set

$$\bigcup_{y_0 \in \overline{B}(0,r)} S(y_0)$$

is bounded in $(C)^n$, say by $\rho > 0$, we define a bounded, open set $U \subset R^n \times (C)^n$ by $U = B(0,r) \times B(0,\rho)$.
Let $\hat{G} : \overline{U} \multimap R^n$ be the map defined by $\hat{G}(y_0, y) = y(1)$. It is easy to see that $\hat{G}$ is a compact, continuous map.

Consider the map $\hat{T} : \overline{B}(0,r) \multimap R^n$, defined by

$$y_0 \multimap \hat{T}(y_0) = \hat{G}(y_0, S(y_0)).$$

Clearly $\hat{T}$ is an admissible map and a fixed point of $\hat{T}$ is the initial condition of a 1-periodic solution to (E1). We will prove the existence of a fixed point of $\hat{T}$ by showing that, via the condition (C1), $\hat{T}$ satisfies property "P". For this, let us prove the following

Proposition 3.1. *Assume condition (C1). For any $y \in S(y_0)$ with $|y_0| = r$ we have that $| y(1) | \leq r$.*

Proof :. Let $\tau_o = sup\{t \in [0,1] \ : | y(t) | \leq r\}$. If $\tau_0 = 1$, we are done.
On the other hand, if $\tau_0 < 1$, then $| y(t) | > r$ for any $t \in (\tau_0, 1)$. By integrating on $(\tau_0, 1)$ the inequality

$$\frac{d}{dt} \frac{| y(t) |^2}{2} = < y(t), \dot{y}(t) > \leq s(t) \text{ for a.a. } t \in [0,1],$$

and using (C1), we obtain that

$$\frac{1}{2} [| y(1) |^2 - | y(\tau_0) |^2] \leq 0 \text{ and so } | y(1) | \leq r.$$

□

Now it is immediate to see that Proposition 3.1 implies that the map $\hat{T}$ satisfies property "P", in fact from $\lambda y_0 \in \hat{T}(y_0)$ it follows $\lambda \leq 1$.

Notice that the problem treated in Example 1, under different assumptions, can be solved by using Theorem 2.1.

Example 2. We consider the following system:

$$\begin{cases} \dot{y} \in \Phi(t, y) \\ y(0) = y_0 \\ y(1) \in K \subset R^n \end{cases} \tag{E2}$$

where Φ is the map defined as in Example 1.

Formulation of the problem: Given $K \subset R^n$, K compact and acyclic, we want to give conditions ensuring the existence of y_0 in a suitable ball $B(0, r) \subset R^n$, such that (E2) is solvable.

Assume the following condition.

(C2)- there exists $r > 0$ such that

$$r^2 = \int_0^1 i(t, y_0)\, dt \geq \sup_{y_1 \in K} < y_0, y_1 > \quad \text{for any } | y_0 |= r,$$

$$\text{where } i(t, y_0) = \inf_{|y| \leq \rho} \inf_{z \in \Phi(t,y)} < y_0, z >,$$

for a.a.$t \in [0, 1]$, $|y_0| = r$, and ρ is determined as in Example 1.

Let $U = B(0, r) \times B(0, \rho) \subset R^n \times (C)^n$ and let $\hat{G} : \overline{U} \multimap R^n$ be the map defined by

$$\hat{G}(y_0, y) = \{y_0 + y_1 - y(1),\ y_1 \in K\}.$$

$\hat{G}$ is a compact, continuous map with compact, acyclic values.

Consider the map $\hat{T} : \overline{B}(0, r) \multimap R^n$ defined by $y_0 \multimap \hat{T}(y_0) = \hat{G}(y_0, S(y_0))$. Clearly $\hat{T}$ is an admissible map and a fixed point of $\hat{T}$ will be a solution to problem (E2).

We prove now the following proposition.

Proposition 3.2. *Assume condition (C2). Then* $< y_0,\ x > \geq 0$ for any $x \in T(y_0) = (I - \hat{T})(y_0)$.

Proof. Let $| y_0 |= r$. For a given $x \in T(y_0)$ we have,

$$< y_0,\ x >=| y_0 |^2 - < y_0,\ y_1 > + \\ < y_0, \int_0^1 \dot{y}(t)dt > \text{ for some } y \in S(y_0) \text{ and } y_1 \in K.$$

By using (C2) we obtain the assertion. □

This result ensures that the property "P" is satisfied by the map $\hat{T}$. In fact, if for some $| y_0 |= r$ we have that

$$\lambda y_0 = - \int_0^1 \dot{y}(t)\, dt + y_1, \quad \text{for some } y \in S(y_0) \text{ and } y_1 \in K,$$

or equivalently $(1-\lambda)y_0 \in (I - \hat{T})(y_0)$, then using Proposition 3.2 we obtain $(1 - \lambda)|y_0|^2 \geq 0$ and so $\lambda \leq 1$.

References

1. Aubin, J.P., Cellina, A. (1984): Differential Inclusions. Springer Verlag, Berlin-Heidelberg-New York
2. Cellina, A. (1969): A theorem on the approximation of compact multivalued mappings. Rend. Acc. Naz. dei Lincei, **47**, 429-433
3. Cellina, A., Lasota, A. (1969): A new approach to the definition of topological degree for multivalued mappings. Rend. Acc. Naz. Lincei, **47**, 434-440
4. Chukwu, E.N., Gronski, J.M. (1977): Approximate and complete controllability of nonlinear systems to a convex target set. J. Math. Anal. and Appl. **61**, 97-112
5. Darbo, G., (1958): Teoria dell' omologia in una categoria di mappe plurivalenti ponderate. Rend. Sem. Mat. Univ. Padova, **28**, 188-224
6. Dugundji, J. (1951): An extension of Tietze's Theorem. Pac. J. Math., **1**, 353-367
7. Gorniewicz, L. (1986): On the solution sets of differential inclusions. J.Math. Anal. and Appl., **113**, 235-244
8. Lasry, J.M., Robert, R. (1976): Analyse Non Lineaire Multivoque. Cahiers De Mathem. de la Decision, **7611**, Ceremade, Universite de Paris-Dauphine
9. Macki, J.W., Nistri, P., Zecca, P. (1988): Periodic solutions of a control problem via marginal maps. Annali di Mat. Pura e Appl. , **63**, 383-396
10. Martelli, M. (1975): A Rothe's type theorem for non-compact acyclic-valued maps. Boll. Un. Mat. Ital. , **11**, Suppl. Fasc. **3**, 70-76
11. Massabó, I., Nistri, P., Pejsachowicz, J., (1980): On the solvability of nonlinear equations in Banach spaces. In Fixed Point Theorems, E. Fadell and G. Fournier Eds., Lecture Notes in Math. **886**, 270-289, Springer Verlag, Berlin-Heidelberg-New York
12. Nistri,P., (1983): Periodic control problems for a class of nonlinear differential systems. Nonlinear Anal. T.M.A., . **7**, 79-80
13. Powers, M.J. (1972): Lefschetz fixed point theorems for a new class of multivalued maps. Pac. J. of Math., **42**, 211-220

Abstract Volterra Equations and Weak Topologies

C. Corduneanu

University of Texas at Arlington, Arlington, TX 76019

The equations involving abstract Volterra operators have been investigated since 1929 by many authors. We mention here the contributions made by L. Tonelli, S. Cinquini, D. Graffi, A.N. Tychonoff, L. Neustadt, which have been discussed in our recent book [6].

Roughly speaking, an abstract Volterra operator V (sometimes also called *causal* or *nonanticipative*) is an operator acting on a function space, the main feature consisting in the fact that $x(t) = y(t)$ for $t \leq \bar{t}$ implies $(Vx)(t) = (Vy)(t)$ for $t \leq \bar{t}$. Such operators have the distinction of introducing in the equations the past history of the phenomena governed by attached equations. They appear in many areas of investigation, and in [6] we have provided illustrations from Continuum Mechanics and the Dynamics of Nuclear Reactors. Applications of such operators/equations in Control Theory are contained in [4], [5], [8].

The existence problem, as well as other basic problems related to the abstract Volterra equation

$$x(t) = (Vx)(t), t \in J \subset R, \tag{1}$$

can be dealt with in various manners. The fixed point method is certainly one of the most inviting if one takes into account the form of equation (1). In our paper [3] (see also the book [6]), existence has been obtained mainly under the assumption of compactness of V on various function spaces $(C, L^p, 1 \leq p < \infty)$.

This paper has as main objective to provide similar results, but using the *weak topology* instead of the norm topology (consequently, *weak continuity* and *weak compactness* will apear as natural assumptions). This approach has the advantage that a Dunford-Pettis theorem, initially obtained for finite dimensional spaces, has been extended to infinite-dimensional spaces (see J.K. Brooks and N. Dinculeanu [1]). Therefore, results applicable to equations in Hilbert or Banach spaces can be obtained by this method. We shall consider here the case of Hilbert spaces, due to the fact that weak compactness and boundedness are equivalent concepts for such spaces.

As an illustration of the fact that weak topologies can be used with success in case of abstract Volterra equations, we shall prove a theorem similar to that given in [3] (see also [6]), concerning the existence of solutions. However, this time the restriction to the finite-dimensional case is not necessary. In the sequel, H will stand for a Hilbert space over the reals.

Theorem 1. *Consider the functional equation (1) and assume that V is an operator from the space $L^2([0,T],H)$ into itself $(0 < T < \infty)$, verifying the following conditions:*

1. *V is an operator of Volterra type, i.e., $x(t) = y(t)$ a.e. on $[0,\tau]$, implies $(Vx)(t) = (Vy)(t)$ a.e. on $[0,\tau]$, for any $\tau \leq T$;*
2. *V is weakly continuous on $L^2([0,T],H)$;*
3. *there exist two functions $A : [0,T] \to R_+$ continuous, and $B : [0,T] \to R_+$ integrable, such that for any $x \in L^2([0,T],H)$ satisfying*

$$\int_0^t |x(s)|^2 ds \leq A(t), \quad t \in [0,T], \tag{2}$$

one has

$$|(Vx)(t)|^2 \leq B(t), \quad \text{a.e. on} \quad [0,T], \tag{3}$$

while

$$\int_0^t B(s)ds \leq A(t), \quad t \in [0,T]. \tag{4}$$

Then there exists a solution $x \in L^2([0,T],H)$ of the equation (1), such that estimate (2) holds true.

Proof . Since a solution of (1) is a fixed point of V, we will apply the Tychonoff fixed point theorem in the space $L^2([0,T],H)$, endowed with the weak topology. This space is a locally convex space, and the usual Schauder fixed point theorem is not applicable.

The convex set on which the operator V is acting is defined by the inequality (2). Let us denote this set by S. We notice that S is convex, and it is closed in the norm topology. But the convex closure is the same in both norm and weak topologies. Therefore, the set S is closed in the weak topology of $L^2([0,T],H)$.

Taking into account (2), (3), and (4), there results

$$VS \subset S. \tag{5}$$

On the other hand, the set VS is obviously bounded in $L^2([0,T],H)$, and therefore it is weakly compact.

By hypothesis V is weakly continuous on S, which means that all conditions required by Tychonoff Theorem (see, for instance, [2]) are satisfied for the operator V, the set S, and the space $L^2([0,T],H)$ endowed with the weak topology.

Hence, the existence of (at least) a fixed point is guaranteed for the operator V (in S). This ends the proof of Theorem 1. □

Remark. Apparently, the fact that V is of Volterra type does not enter in the proof. Yet, conditions (2), (3) do not make sense if the operator V is not of Volterra type.

Remark. If $A(t) > 0$, which is a natural assumption, then condition (4) will be always verified if we restrict the interval $[0,T]$ to a sufficiently small interval

$[0,\delta] \subset [0,T]$. In other words, if we drop condition (4) from the statement of Theorem 1, then the remaining conditions assure the local existence of a solution. Furthermore, this local solution can be extended to a larger interval $[0,\delta_1] \supset [0,\delta]$, by using the same kind of argument. In a very standard way, the process of extension can be continued until a maximal interval of existence is obtained (see [6]).

Corollary *It is relatively easy to apply Theorem 1 to the special case of linear integral equations in H:*

$$x(t) = h(t) + \int_0^t k(t,s)x(s)ds, \quad t \in [0,T]. \tag{6}$$

In (6), $h \in L^2([0,T],H)$ while $k(t,s)$ stands for a family of bounded linear operators on H, $0 \leq x \leq t \leq T$, such that

$$\int_0^T dt \int_0^t \|k(t,s)\|^2 ds < +\infty. \tag{7}$$

Of course, the measurability (Bochner) of the map $(t,s) \to k(t,s) \in L(H,H)$, is assumed.

It can be easily seen that the operator

$$(Vx)(t) = h(t) + \int_0^t k(t,s)x(s)ds \tag{8}$$

is weakly continuous on $L^2([0,T],H)$, under assumption (7).

In order to get the existence for (6) in $L^2([0,T],H)$, it suffices to construct the functions $A(t)$ and $B(t)$ which appear in condition 3) of Theorem 1. This construction has been carried out in [3] (see also [6]), when we dealt with the finite-dimensional case. There are no differences at all when passing from R^n to a Hilbert space H. However, the discussion in [6] is conducted in a slightly more general situation, when $L^2([0,T],H)$ is substituted by $L^2_{\text{loc}}([0,T],H)$, with $T = +\infty$ a possible choice.

Another example of a classical integral equation for which existence can be obtained by applying Theorem 1 is the Volterra-Hammerstein equation

$$x(t) = h(t) + \int_0^t k(t,s)g(s,x(s))ds, \tag{VH}$$

in which h and k are as in the above Corollary, while g is a weakly continuous map on $L^2([0,T],H)$, satisfying an estimate of the form $|g(t,x)| \leq k_0(t,s)|x|$, with $k_0(t,s)$ like $k(t,s)$.

If instead of equation (1) we consider the functional differential equation

$$\dot{x}(t) = (Vx)(t), \quad t \in [0,T], \tag{9}$$

$$x(0) = x^0 \in H, \tag{10}$$

then we can reduce (by integrating both sides of (9) and taking (10) into account) this problem to the case dealt with in Theorem 1 above. Indeed, the right hand side of

$$x(t) = x^0 + \int_0^t (Vx)(s)ds \tag{11}$$

still represents an operator of Volterra type on $L^2([0,T],H)$, if V has this property.

Instead of proceeding to a direct application of Theorem 1 to the equation (11), we shall pursue a different approach, also based on the Tychonoff fixed point theorem.

We will assume that the oprator V in (9) can be represented in the form $V = L + N$, in which L is linear, and N is, in general, a nonlinear operator. Hence, the equation (9) will become

$$\dot{x}(t) = (Lx)(t) + (Nx)(t), \tag{12}$$

where L is a linear Volterra operator acting on the space $L^2([0,T],H)$. It is natural to assume the continuity of L on the space $L^2([0,T],H)$. Of course, the case when L is unbounded is also of interest, mainly in regard to applications to integrodifferential equations with partial derivatives. This case will be dealt with in subsequent papers.

Let us point out that equation (12), which appears as a perturbation of the linear equation

$$\dot{x}(t) = (Lx)(t) + f(t), \tag{13}$$

has been recently investigated in our paper [7], in which only the finite-dimensional case and, therefore, strong topologies have been considered.

Lemma *Consider equation (13), with the initial condition (10), and assume that*

$$L : L^2([0,T],H) \to L^2([0,T],H), \tag{14}$$

is a continuous linear operator of Volterra type, while

$$f \in L^2([0,T],H). \tag{15}$$

Then there exists a unique absolutely continuous solution of the problem (13), (10), defined on the interval $[0,T]$. *The following estimate is valid:*

$$|x(t)|_{L^2} \leq K(|x^0| + |f|_{L^2}), \tag{16}$$

where $K > 0$ *is a constant depending only of the operator* L.

Proof. The problem (13), (10) is equivalent to the functional-integral equation

$$x(t) = x^0 + \int_0^t f(s)ds + \int_0^t (Lx)(s)ds, \tag{17}$$

which can be treated by the method of successive approximations. The details can be found in [6], where the case $H = R^n$ is dealt with. However, no significant changes are required when passing from R^n to H.

After having the existence and uniqueness of a solution to (13), (10), in order to obtain the inequality (16) we can proceed as follows. We notice first that (17) and $(a+b+c)^2 \leq 3(a^2+b^2+c^2)$ imply

$$|x(t)|^2 \leq 3|x^0|^2 + 3T\int_0^T |f(t)|^2 dt + 3T\int_0^t |(Lx)(s)|^2 ds \tag{18}$$

for $0 \leq t \leq T$. Using the continuity of the operator L on $L^2([0,T],H)$, we can write for some constant $M > 0$

$$\int_0^t |(Lx)(s)|^2 ds \leq M\int_0^t |x(s)|^2 ds,$$

which taken into (18), and applying the Gronwall's integral inequality leads to (16). Obviously, we first get an estimate for $|x(t)|_C$, but we have

$$|x|_{L^2} \leq \sqrt{T}|x|_C,$$

and this ends the proof of the Lemma. □

We shall use now the Lemma and the Tychonoff (actually, Schauder's fixed point theorem for the weak topology, which is a particular case of the general Tychonoff result for locally convex spaces), in order to obtain existence for the equation (12), under initial conditions (10).

Theorem 2. *Assume the following conditions are satisfied for equation (12):*

1) L is as described in the Lemma;

2) N is a weakly continuous operator on $L^2([0,T],H)$, taking bounded sets into bounded sets, and such that

$$\phi(r) = \sup\{|Nx|_{L^2};\ \ |x|_{L^2} \leq r\} \tag{19}$$

verifies

$$\limsup_{r\to\infty} \frac{\phi(r)}{r} < K^{-1}. \tag{20}$$

Then there exists a solution $x(t)$ of the problem (12), (10), defined on $[0,T]$ and absolutely continuous. This solution verifies the estimate $|x|_{L^2} \leq r$, where $r > 0$ is the smallest solution of the equation $K(|x^0| + \phi(r)) = r$.

Proof. Let $r > 0$ be as described above. Such a number does exist because of the assumption (20). We shall apply the fixed point theorem to the operator V, $u \to x = Vu$, where

$$\dot{x}(t) = (Lx)(t) + (Nu)(t) \tag{21}$$

a.e. on $[0,r]$, under initial condition (10), with $u \in B_r = \{x : x \in L^2([0,T],H),$ $|x|_{L^2} \leq r\}$.

One needs to prove that the operator V is weakly continuous on $L^2([0,T],H)$, and takes bounded sets of $L^2([0,T],H)$ into bounded sets. Of course, we keep in mind here that bounded sets into a Hilbert space are weakly compact (relatively). Since V is the product of the operator N by the operator $f \to x$ from the Lemma, and N is by hypothesis weakly compact, it suffices to show that $f \to x$ is continuous. But inequality (16) shows even more than that, namely, that $(x^0, f) \to x$ is continuous from $H \times L^2$ into L^2. A fortiori, $f \to x$, is continuous from L^2 into itself. The linearity of $f \to x$ proves that this operator is also weakly continuous. Hence, the operator V is weakly continuous. It is almost obvious that V takes bounded sets of $L^2([0,T],H)$ into bounded sets.

The only property to be checked, before we apply the fixed point theorem is the inclusion

$$VB_r \subset B_r, \tag{22}$$

for the particular r precised at the beginning of the proof. This is obvious if we take into account condition (20).

To summarize, the operator V defined above is continuous and compact in the weak topology of $L^2([0,T],H)$, taking a closed convex set (i.e., B_r) into itself. This suffice to conclude, on behalf of the Tychonoff theorem, the existence of a fixed point in B_r.

This ends the proof of Theorem 2. □

Applications of Theorem 2 to equations of the form

$$\dot{x}(t) = \sum_{j=0}^{\infty} A_j x(t - t_j) + \int_0^t B(s)x(t-s)ds + f(t;x),$$

with $t_j \geq 0$ and $\sum_{j=0}^{\infty} ||A_j|| < \infty$, $||B|| \in L^1(R_+)$, can be easily obtained following the pattern in [6, Ch. VI].

References

1. Brooks, J.K., Dinculeanu, N. (1977): Weak compactness in spaces of Bochner integrable functions and applications. Adv. Math., **24**, 172-188
2. Corduneanu, C. (1973): Integral Equations and Stability of Feedback Systems. Academic Press, N.Y.
3. Corduneanu, C. (1986): An existence theorem for functional equations of Volterra type. Libertas Mathematica, VI, 117-124
4. Corduneanu, C. (1987): Control problems for abstract Volterra functional-differential equations. Modern Optimal Control (E. Roxin, Ed.), M. Dekker, N.Y., 41-48
5. Corduneanu, C.: Some control problems for abstract Volterra functional-differential equations. Proc. Int. Symp. MTNS-89, vol. III, Birkhäuser, 331-338
6. Corduneanu, C. (1990): Integral Equations and Applications. Cambridge Univ. Press
7. Corduneanu, C. (1990): Perturbation of abstract Volterra equations. J. Integral Equations and Applications, (in print)
8. Neustadt, L. (1976): Optimization (A theory of necessary conditions). Princeton Univ. Press

Semigroups and Renewal Equations on Dual Banach Spaces with Applications to Population Dynamics

Odo Diekmann[1], *Mats Gyllenberg*[2], *and Horst R. Thieme*[3]

[1] Centre for Mathematics and Computer Science, P.O. Box 4079, 1009 AB Amsterdam, The Netherlands and Institute of Theoretical Biology, University of Leiden, Kaiserstraat 63, NL-2311 GP Leiden, The Netherlands
[2] Department of Applied Mathematics, University of Luleå, S-951 87 Luleå, Sweden
[3] Department of Mathematics, Arizona State University, Tempe, Arizona 85287, USA

1 Introduction

Mathematical descriptions of the dynamics of structured populations take various forms. In many cases, most notably in linear age-dependent population dynamics, integral equations form a natural modelling tool. For an age structured population in a constant environment it is possible to derive a renewal equation (Lotka's equation) for the birth rate of the population from first principles.

An other approach to structured population dynamics is to start by writing down a population balance equation. This equation takes the form of a first order hyperbolic partial differential equation describing the continuous change of individual state as well as death. The birth process is described by a boundary condition supplementing the pde.

In the case of linear age-dependent population dynamics the two approaches described above are equivalent. Integrating the population balance equation (McKendrick's equation) along characteristics one obtains the age distribution of the population in terms of the birth rate. Substituting this into the boundary condition one obtains Lotka's renewal equation. On the other hand, once Lotka's equation has been solved, one can easily write down an explicit expression for the age distribution.

The purpose of this paper is to show that the equivalence of the renewal equation and pde approaches is not confined to age structured populations but has a much wider generality. A general structured population problem can usually be formulated as an abstract Cauchy problem in $M(\Omega)$, the space of all Borel measures on the individual state space Ω (see [7]). In the linear case this Cauchy problem generates a w^*-semigroup on $M(\Omega)$ (see [2, 5]). We show that a certain family of operators associated with the corresponding integrated semigroup satisfies a renewal equation on $M(\Omega)$ and conversely that the solution to this renewal equation uniquely determines the semigroup. All our results will be formulated in a general setting without any reference to population dynamics, but they will be illustrated by applications to age dependent population dynamics.

In section 2 we recall some facts from perturbation theory of dual semigroups as developed in [2-5] and we recall how this theory is related to Cauchy problems on dual Banach spaces. In section 3 we show how the perturbation problem gives rise to a renewal equation, the solutions of which can be used to define the solution semigroup of the problem. Finally in section 4 we take an abstract renewal equation as the starting point and we give conditions for when the solutions of this equation determine a semigroup on the dual Banach space. To achieve this goal which eventually yields the equivalence between the perturbation problem and renewal equation we introduce a new concept, that of a "multiplied integral of a semigroup". The relationship between this new notion and some more established ones like "integrated semigroup" is investigated.

2 Perturbation theory for dual semigroups

Let X be a Banach space. Recall that a w^*-semigroup on X^* is a family $T^\times = \{T^\times(t)\}_{t\geq 0}$ of bounded linear operators on X^*, which in addition to the semigroup properties $T^\times(0) = I$, $T^\times(t+s) = T^\times(t)T^\times(s)$, satisfies the continuity condition that $t \to \langle x, T^\times(t)x^*\rangle$ is continuous for any $x \in X, x^* \in X^*$. The w^*-generator $A^\times$ of $T^\times$ is defined by

$$A^\times x^* = weak^* - \lim_{h\downarrow 0} \frac{1}{h}[T^\times(h)x^* - x^*]$$

the domain $D(A^\times)$ being defined as the set of all $x^* \in X^*$ for which the above limit exists. In general there is not a unique correspondence between semigroup and generator (see [6] where properties (i) and (ii) of Theorem 2.1 below are motivated).

Let $T_0 = \{T_0(t)\}_{t\geq 0}$ be a strongly continuous semigroup of linear operators on X with infinitesimal generator A_0. Then $T_0^* = \{T_0^*(t)\}_{t\geq 0}$ is a w^*-semigroup on X^* with w^*-generator A_0^*. Define $X^\odot := \overline{D(A_0^*)}$. Then $T_0^\odot$, the restriction of T_0^* to $X^\odot$, is a strongly continuous semigroup whose generator $A_0^\odot$ is the part of A_0^* in $X^\odot$, that is $A_0^\odot x^\odot = A_0^* x^\odot, D(A_0^\odot) = \{x^\odot \in D(A_0^*): A_0^* x^\odot \in X^\odot\}$.

Let C be a bounded linear operator from $X^\odot$ into X^*. The basic perturbation result is given by the following theorem.

Theorem 2.1 *Let $A^\times := A_0^* + C, D(A^\times) = D(A_0^*)$. Then $A^\times$ generates a w^*-semigroup $T^\times = \{T^\times(t)\}_{t\geq 0}$ on X^* with the properties*

(i) $x^* \in D(A^\times)$ and $A^\times x^* = y^*$
if and only if
$\frac{d}{dt}\langle x, T^\times(t)x^*\rangle = \langle x, T^\times(t)y^*\rangle \; \forall x \in X, \forall t \geq 0$

(ii) For all $x^* \in X^*$ and all $t \geq 0$

$$\int_0^t T^\times(\tau)x^* d\tau \in D(A^\times) \text{ and } A^\times \int_0^t T^\times(\tau)x^* d\tau = T^\times(t)x^* - x^*.$$

Note that the integral in (ii) is defined as a *weak**-integral, i.e.

$$\langle x, \int_0^t T^\times(\tau)x^* \, d\tau\rangle := \int_0^t \langle x, T^\times(\tau)x^*\rangle d\tau.$$

Theorem 2.1 can be proved in several different ways. In [6] we used a general Hille-Yosida type characterization of w^*-generators of w^*-semigroups. A perhaps more appealing approach is to first construct a strongly continuous semigroup $T^\odot = \{T^\odot(t)\}_{t\geq 0}$ on the smaller space $X^\odot$ by the variation of constants formula

$$T^\odot(t)x^\odot = T_0^\odot(t)x^\odot + \int_0^t T_0^*(t-\tau)CT^\odot(\tau)x^\odot d\tau, \; x^\odot \in X^\odot, \tag{1}$$

and then extend $T^\odot$ to all of X^*. That the variation of constants formula (1) indeed defines a strongly continuous semigroup $T^\odot$ on $X^\odot$ was proved in [2]. The extension of $T^\odot$ to X^* can be performed in two different ways, either through the intertwining formula

$$T^\times(t) := (\lambda I - A^\times)T^\odot(t)(\lambda I - A^\times)^{-1}, \tag{2}$$

or by duality. In the latter case one obtains, after taking adjoints of $T^\odot(t)$, a w^*-semigroup $T^{\odot *}$ on $X^{\odot *}$ and after restricting to the space $X^{\odot\odot} := \overline{D(A^{\odot *})}$ of strong continuity, a strongly continuous semigroup $T^{\odot\odot}$ on $X^{\odot\odot}$. There is a duality pairing $[\,,\,]$ between $X^{\odot\odot}$ and X^* and hence $T^\times(t)$ can be defined on X^* by

$$[x^{\odot\odot}, T^\times(t)x^*] = [T^{\odot\odot}(t)x^{\odot\odot}, x^*] \tag{3}$$

for all $x^{\odot\odot} \in X^{\odot\odot}$ (in particular all $x^{\odot\odot} \in X$), $x^* \in X^*$. The details and equivalence of these approaches are explained in [5].

Observe that a variation of constants formula like (1) is not possible for $T^\times$ since the perturbation C is defined on $X^\odot$ only. (A more involved variant which also holds for $T^\times$ on X^* is derived in [8]). However, we will be able to write down a renewal equation for an associated family of operators on X^* from which the semigroup $T^\times$ can be recovered. This leads to yet another method of defining the solution semigroup on all of X^*.

We close this section by illustrating the abstract setting with an example from population dynamics.

The classical age-dependent population problem is usually formulated in the population state space $L^1[0,\infty)$ by the McKendrick equation supplemented by boundary and initial conditions

$$\begin{aligned} &\frac{\partial n}{\partial t}+\frac{\partial n}{\partial a}=-\mu(a)n, \\ &n(t,0)=\int_0^\infty \beta(a)n(t,a)da, \\ &n(0,a)=\phi(a). \end{aligned} \tag{4}$$

Here the age-specific fertility function β is assumed to belong to $L^\infty[0,\infty)$. The problem (4) can be put in the abstract framework by considering the birth process as a perturbation of the process of aging and dying. On the predual space $X = C_0[0,\infty)$ the unperturbed semigroup, i.e, the solution semigroup for the problem with $\beta \equiv 0$, is given by

$$[T_0(t)f](a) = e^{-\int_a^{a+t}\mu(\alpha)d\alpha} f(a+t). \tag{5}$$

The population state space is $X^* = M[0,\infty)$. T_0^* is strongly continuous on $X^\odot = \overline{D(A_0^*)}$, the subspace of absolutely continuous (with respect to Lebesque measure) measures on $[0,\infty)$. $X^\odot$ can of course be identified with $L^1[0,\infty)$ by the Radon-Nikodym theorem, thus yielding the classical state space. The birth process is described by the bounded perturbation $C: X^\odot \to X^*$, where $C: L^1[0,\infty) \to M[0,\infty)$ is defined by

$$C\phi = \langle\beta,\phi\rangle\delta = \int_0^\infty \beta(a)\phi(a)da\,\delta, \tag{6}$$

where δ is the Dirac measure concentrated at the origin. Thus the problem can be abstractly reformulated as the Cauchy problem

$$\begin{aligned} &\frac{dn}{dt} = (A_0^* + C)n, \\ &n(0) = n_0 \end{aligned} \tag{7}$$

on $X^* = M[0,\infty)$. If $n_0 \in L^1[0,\infty)$ then the variation of constants formula (1) yields the usual mild solution in $L^1[0,\infty)$.

3 The renewal equation

In the last section we showed how the classical McKendrick formulation of age dependent population dynamics could be viewed as an abstract perturbation problem on a dual Banach space. In order to motivate the subsequent derivation of a renewal equation associated with the perturbation problem, we start by considering the same example again.

Let the semigroup T_0 on $X = C_0[0,\infty)$ be given by (5) and let the perturbation C be given by (6). Applying the linear functional induced by $\beta \in L^\infty[0,\infty)$ to both sides of the variation of constants formula (1) one obtains the equation

$$b(t) = b_0(t) + \langle\beta, \int_0^t T_0^*(t-\tau)\,\delta\, b(\tau)d\tau\rangle, \tag{8}$$

where

$$b(t) := \langle \beta, T^{\odot}(t)x^{\odot} \rangle, \tag{9}$$

and

$$b_0(t) := \langle \beta, T_0^{\odot}(t)x^{\odot} \rangle. \tag{10}$$

The function $b(t)$ may be interpreted as the instantaneous birth rate of the population and $b_0(t)$ as the instantaneous birth rate of offspring of parents present in the initial population. We want to transform equation (8) into a renewal equation. To this end we note that the function K defined by

$$K(t) := \langle \beta, \int_0^t T_0^*(\tau)\delta d\tau \rangle \tag{11}$$

is locally Lipschitz continuous and therefore is differentiable almost everywhere, the derivative k belonging to $L_{loc}^{\infty}[0,\infty)$:

$$K(t) = \int_0^t k(\tau)d\tau. \tag{12}$$

It is now easily seen (for details, see [3]) that b satisfies the renewal equation

$$b(t) = b_0(t) + \int_0^t k(t-\tau)b(\tau)d\tau. \tag{13}$$

Equation (13) is nothing but Lotka's integral equation. Once b is solved from (13), the solution is obtained on $X^{\odot} = L^1[0,\infty)$ by the explicit formula

$$T^{\odot}(t)x^{\odot} = T_0^{\odot}(t)x^{\odot} + \int_0^t T_0^*(t-\tau)\,\delta\, b(\tau)d\tau. \tag{14}$$

As a matter of fact we obtain the solution on all of $X^* = M[0,\infty)$, not in terms of the instantaneous birth rate, but in terms of the cumulative number of births. Exactly as in the case of K above, we see that

$$B_0(t) := \langle \beta, \int_0^t T_0^*(\tau)x^*\, d\tau \rangle \tag{15}$$

defines a locally Lipschitz continuous function B_0. Therefore the renewal equation

$$B(t) = B_0(t) + \int_0^t K(t-\tau)dB(\tau) \tag{16}$$

has a unique solution B, which is again locally Lipschitz continuous. It follows that B has a derivative $b \in L_{loc}^{\infty}[0,\infty)$ almost everywhere:

$$B(t) = \int_0^t b(\tau)d\tau. \tag{17}$$

If $x^* = x^\odot \in X^\odot$, then b is of course the continuous function defined by (9). It follows that $B(t)$ is the cumulative number of births in the time interval $[0, t]$. We have the following explicit representation of the perturbed semigroup $T^\times$ on X^*:

$$T^\times(t)x^* = T_0^*(t)x^* + \int_0^t T_0^*(t-\tau)\delta dB(\tau). \tag{18}$$

The crucial "trick" in the derivation above was integration with respect to time. We will now extend this idea to the general theory. It will therefore come as no surprise that integrated semigroups will play a key role.

If one integrates the variation of constants formula (1) from 0 to t one obtains

$$\begin{aligned}\int_0^t T^\odot(\tau)x^\odot d\tau &= \int_0^t T_0^\odot(\tau)x^\odot d\tau \\ &+ \int_0^t T_0^*(t-u)C\int_0^u T^\odot(\tau)x^\odot d\tau du.\end{aligned} \tag{19}$$

Applying the operator $C: X^\odot \to X^*$ to both sides of equation (19) one obtains

$$\begin{aligned}C\int_0^t T^\odot(\tau)x^\odot d\tau &= C\int_0^t T_0^\odot(\tau)x^\odot d\tau \\ &+ C\int_0^t T_0^*(t-u)C\int_0^u T^\odot(\tau)x^\odot d\tau du.\end{aligned} \tag{20}$$

Observe that since $\int_0^t T_0^*(\tau)x^* d\tau \in X^\odot$ and $\int_0^t T^\times(\tau)x^* d\tau \in X^\odot$ for all $x^* \in X^*$, all terms in equations (19) and (20) still make sense if $x^\odot \in X^\odot$ is replaced by $x^* \in X^*, T_0^\odot$ by T_0^* and $T^\odot$ by $T^\times$. Introducing the integrated semigroups

$$S_0^*(t) = \int_0^t T_0^*(\tau)d\tau, \tag{21}$$

$$S^\times(t) = \int_0^t T^\times(\tau)d\tau, \tag{22}$$

mapping X^* into $X^\odot$ and the operators

$$V_0(t) = CS_0^*(t), \tag{23}$$

$$V(t) = CS^\times(t), \tag{24}$$

mapping $X^\odot$ into X^* and integrating by parts we arrive at the representation

$$S^\times(t) = S_0^*(t) + \int_0^t S_0^*(t-\tau)dV(\tau) \tag{25}$$

of $S^\times(t)$, where $V(t)$ is the solution of the abstract renewal equation

$$V(t) = V_0(t) + \int_0^t V_0(t-\tau)dV(\tau). \tag{26}$$

Once $S^\times(t)$ has been obtained from (26) one gets the semigroup $T^\times(t)$ by differentiation with respect to the *weak**-topology:

$$T^\times(t)x^* = \frac{d^*}{dt}S^\times(t)x^* = T_0^*(t)x^* + \int_0^t T_0^*(t-\tau)d[V(\tau)x^*]. \tag{27}$$

The Stieltjes integrals in equations (25) and (26) are in the operator norm, whereas the Stieltjes integral in (27) must be interpreted in the *weak**-sense.

In [8] we show how the formal manipulations above can be made rigorous. There we develop a convolution calculus on a Fréchet algebra of Lipschitz continuous operators. Resolvent theory (see [9]) then implies that the renewal equation (26) has a unique solution in this algebra and that it is given by the generation expansion

$$V(t) = \sum_{n=0}^{\infty} V_0^{n\star}(t), \tag{28}$$

where the terms in the series are powers with respect to the Stieltjes convolution

$$(U \star V)(t) = \int_0^t U(t-\tau)dV(\tau). \tag{29}$$

It follows that the integrated semigroup $S^\times(t)$ and the semigroup $T^\times(t)$ also have representations in terms of generation expansions:

$$S^\times(t) = \sum_{n=0}^{\infty} S_n^\times(t), \tag{30}$$

where

$$S_{n+1}^\times(t) = \int_0^t S_n^\times(t-\tau)dV_0(\tau), \tag{31}$$

and

$$T^\times(t) = \sum_{n=0}^{\infty} T_n^\times(t), \tag{32}$$

where

$$T_{n+1}^\times(t)x^* = \int_0^t T_n^\times(t-\tau)d[V_0(\tau)x^*], x^* \in X^*. \tag{33}$$

Finally we note that (the integrated version of) Lotka's integral equation (16) is a special case of the abstract renewal equation (26). In age dependent population dynamics the perturbation C is a rank one operator and hence so are $V_0(t)$ and $V(t)$. Equation (16) i simply the scalar component of (26), with $V_0(t) = B_0(t)\delta$ and $V(t) = B(t)\delta$.

4 Multiplied integrals of semigroups

The main result of the last section was that if the perturbation $C: X^{\odot} \to X^*$ is given, then the perturbed semigroup $T^{\times}(t)$ on X^* is obtained in terms of solutions of an associated renewal equation on $\mathcal{B}(X^*)$, the space of bounded linear operators on X^*. In the McKendrick model of age dependent population dynamics it was clear how to define C, but this is not longer true in more general models of structured populations. If for instance the individual state space Ω is multidimensional (a subset of R^n with $n \geq 2$), then there is no reasonable representation of $X^{\odot}$ as a space of functions or measures, and hence it is not clear where exactly the *instantaneous* birth rate operator C should be defined. In some concrete models it is not even clear from biological arguments how to define C, see e.g. [12]. However, it is usually possible to define directly the *cumulative* birth function of a given population in $M(\Omega)$. In the case of multidimensional individual state space this is not a scalar, but a measure-valued function. Moreover, cumulative births, not rates, are what one can actually measure.

We are thus led to the following converse problem : Given a w^*-semigroup T_0^* on X^* and a family $\{V_0(t)\}_{t\geq 0}$ of locally Lipschitz continuous operators in $\mathcal{B}(X^*)$, under what conditions does the formula

$$T^{\times}(t)x^* = T_0^*(t)x^* + \int_0^t T_0^*(t-\tau)d[V(\tau)x^*], \tag{34}$$

where V is the solution of the renewal equation

$$V(t) = V_0(t) + \int_0^t V_0(t-\tau)dV(\tau), \tag{35}$$

define a w^*-semigroup $T^{\times}$ on X^*? Does there exist a bounded linear operator $C: X^{\odot} \to X^*$ such that $V(t) = CS^{\times}(t)$, where $S^{\times}(t) = \int_0^t T^{\times}(\tau)d\tau$? The rest of this section is concerned with these and related questions. We start with some definitions.

Definition 4.1 (see [6]) A w^*-semigroup $T^{\times}$ on X^* is called an *integral w^*-semigroup* if

$$T^{\times}(t)\int_0^s T^{\times}(\tau)x^* d\tau = \int_0^s T^{\times}(t+\tau)x^* d\tau \tag{36}$$

for all $x^* \in X^*$ and all $s, t \geq 0$.

Definition 4.2 (see [1, 10, 11, 13]) A family $S = \{S(t)\}_{t\geq 0}$ of bounded linear operators on a Banach space is called an *integrated semigroup* if

$$S(0) = 0 \tag{37}$$

$$t \to S(t) \text{ is strongly continuous} \tag{38}$$

$$S(s)S(t) = \int_0^s [S(t+\tau) - S(\tau)]d\tau, \; s,t \geq 0. \tag{39}$$

It is clear from the above definitions that $T^\times$ is an integral w^*-semigroup if and only if $S^\times$ defined by $S^\times(t) := \int_0^t T^\times(\tau)d\tau$ is an integrated semigroup.

Definition 4.3 Let $T^\times$ be an integral w^*-semigroup on X^* and let $S^\times$ be the corresponding integrated semigroup. A family $V = \{V(t)\}_{t\geq 0}$ of bounded linear operators on X^* is called a *multiplied integral* of $T^\times$ if it satisfies the following conditions

$$V(0) = 0 \tag{40}$$

$$t \to V(t) \text{ is locally Lipschitz continuous with respect to the operator norm} \tag{41}$$

$$V(s)S^\times(t) = \int_0^t [V(\tau+s) - V(\tau)]d\tau, t, s \geq 0. \tag{42}$$

Remark 4.4 Formal differentiation of (42) with respect to t yields

$$V(t+s) = V(t) + V(s)T^\times(t). \tag{43}$$

In the context of population dynamics condition (43) has a clear biological interpretation. $V(t)$ then stands for cumulative births. Let x^* be the initial population state. In the time interval [0,t] there are $V(t)x^*$ births while the population itself evolves to $T^\times(t)x^*$. In the time interval $[t, t+s]$ there are therefore $V(s)T^\times(t)x^*$ births. This should equal $V(t+s)x^*$, that is, equation (43) should hold.

Note that (42) is equivalent to

$$V(s)S^\times(t) = \int_0^s [V(t+\tau) - V(\tau)]\, d\tau,$$

which shows that the integrated semigroup corresponding to an integral w^*-semigroup is also a multiplied integral of that semigroup.

The terminology "multiplied integral of $T^\times$ " introduced in Definition 4.3 is justified by the following proposition.

Proposition 4.5 *Let $T^\times$ be an integral w^*-semigroup with w^*-generator $A^\times$ on the Banach space X^*. Let $X^\odot = \overline{D(A^\times)}$ and let C be a bounded linear operator from $X^\odot$ into X^*. Then*

$$V(t) = CS^\times(t), t \geq 0 \tag{44}$$

defines a multiplied integral of $T^\times$. *Conversely, if* V *is a multiplied integral of* $T^\times$, *then there exists a unique bounded linear operator* $C: X^\odot \to X^*$ *such that* (44) *holds. One has*

$$Cx^\odot = \lim_{t\downarrow 0} \frac{1}{t} V(t)x^\odot \tag{45}$$

for all $x^\odot \in X^\odot$.

Proof. Let V be defined by (44). Then it is obvious that V satisfies (40) and (41). Also,

$$\begin{aligned}\int_0^t [V(\tau+s) - V(\tau)]d\tau &= \int_0^t C[S^\times(\tau+s) - S^\times(\tau)]d\tau \\ &= C\int_0^t [S^\times(\tau+s) - S^\times(\tau)]d\tau \\ &= CS^\times(s)S^\times(t) \\ &= V(s)S^\times(t),\end{aligned}$$

that is, (42) holds.

Conversely, let V satisfy (40) - (42). Let $x^* \in X^*, h > 0$. Then

$$\begin{aligned}\frac{1}{h}V(h)S^\times(t)x^* &= \frac{1}{h}\int_0^t [V(\tau+h)x^* - V(\tau)x^*]d\tau \\ &\to V(t)x^* \text{ as h} \downarrow 0.\end{aligned} \tag{46}$$

In fact, $\frac{1}{h}V(h)S^\times(t) \to V(t)$ in the operator norm as $h \downarrow 0$. Since $x^\odot = \lim_{h\downarrow 0} \int_0^h T^\times(\tau)\frac{1}{h}x^\odot d\tau, x^\odot \in X^\odot$, the set of elements of the form $S^\times(t)x^*$ is dense in $X^\odot$. Since moreover $\frac{1}{h} \parallel V(h) \parallel \leq L < \infty$, $0 < h \leq 1$, it follows that the limit in (45) exists for all $x^\odot \in X^\odot$ and defines a bounded linear operator $C: X^\odot \to X^*$. It follows from (46) that (44) holds. Uniqueness is clear.

Remark 4.6 The limit in (45) agrees with the notion of instantaneous birth rate as the derivative of cumulative number of births.

We now turn our attention to the main question of this section. Let $T_0 = \{T_0(t)\}_{t\geq 0}$ be a strongly continuous semigroup on X and let V_0 be a multiplied integral of T_0^*. Let $S_0^*(t) := \int_0^t T_0^*(\tau)d\tau, t \geq 0$. Let $V(t)$ be the unique solution of the renewal equation

$$V(t) = V_0(t) + \int_0^t V_0(t-\tau)dV(\tau), \tag{47}$$

or equivalently,

$$V(t) = V_0(t) + \int_0^t V(t-\tau)dV_0(\tau), \tag{48}$$

and let $S^\times(t)$ be defined by

$$S^\times(t) = S_0^*(t) + \int_0^t S_0^*(t-\tau)dV(\tau). \tag{49}$$

Then $S^\times(t)$ is the unique solution of the equation

$$S^\times(t) = S_0^*(t) + \int_0^t S^\times(t-\tau)dV_0(\tau) \tag{50}$$

(for details, see [8]). It follows from (50) that the mapping $t \to S^\times(t)x^*$ is differentiable in the *weak**-sense. So we can define

$$T^\times(t)x^* := \frac{d^*}{dt}S^\times(t)x^* = T_0^*(t)x^* + \int_0^t T_0^*(t-\tau)d[V(\tau)x^*] \tag{51}$$

for $x^* \in X^*$ and $t \geq 0$. It is obvious that

$$S^\times(t)x^* = \int_0^t T^\times(\tau)x^* d\tau. \tag{52}$$

Proposition 4.6 *Let $C: X^\odot \to X^*$ be the unique bounded linear operator satisfying $V_0(t) = CS_0^*(t), t \geq 0$. Then $V(t) = CS^\times(t), t \geq 0$.*

Proof. By (47) and (49) we have

$$\begin{aligned} CS^\times(t) &= CS_0^*(t) + C\int_0^t S_0^*(t-\tau)dV(\tau) \\ &= V_0(t) + \int_0^t V_0(t-\tau)dV(\tau) \\ &= V(t). \end{aligned}$$

Proposition 4.6 shows that we are exactly in the situation described in section 3 and rigorously treated in [8]. However, the point of this section consists in deriving the main result — that V is a multiplied integral of the integral w^*-semigroup $T^\times$ — without any reference to the operator C.

Theorem 4.7 *$T^\times$ is an integral w^* -semigroup and V is a multiplied integral of $T^\times$.*

Proof. Since $T^\times$ and $S^\times$ are related by (52) we have to show that $S^\times$ is an integrated semigroup. It follows then immediately that $T^\times$ is an integral w^*-semigroup. To this end, note that it follows from (50) that

$$S^\times(t+\tau)-S^\times(\tau)=S_0^*(t+\tau)-S_0^*(\tau)+\int_0^{t+\tau}S^\times(t+\tau-\sigma)dV_0(\sigma)-\int_0^{\tau}S^\times(\tau-\sigma)dV_0(\sigma)$$

and hence that

$$\begin{aligned}\int_0^s[S^\times(t+\tau)-S^\times(\tau)]d\tau &= \int_0^s[S_0^*(t+\tau)-S_0^*(\tau)]d\tau\\ &+\int_0^s\int_\tau^{\tau+t}S^\times(t+\tau-\sigma)dV_0(\sigma)\,d\tau\\ &+\int_0^s\int_0^\tau[S^\times(t+\tau-\sigma)-S^\times(\tau-\sigma)]dV_0(\sigma)d\tau\end{aligned} \tag{53}$$

Since S_0^* is an integrated semigroup the first term on the right hand side of (53) is of course equal to $S_0^*(t)S_0^*(s)$. Using the fact that V_0 is a multiplied integral of T_0^*, one finds that the second term to the right of (53) equals

$$\begin{aligned}\int_0^s\int_0^t S^\times(t-\sigma)d_\sigma[V_0(\sigma+\tau)]d\tau &= \int_0^s\int_0^t S^\times(t-\sigma)d_\sigma[V_0(\sigma+\tau)-V_0(\tau)]d\tau\\ &= \int_0^t S^\times(t-\sigma)d_\sigma[\int_0^s\{V_0(\sigma+\tau)-V_0(\tau)\}d\tau]\\ &= \int_0^t S^\times(t-\sigma)d_\sigma[V_0(\sigma)S_0^*(S)],\end{aligned}$$

while the third term can be written as

$$\int_0^s\int_0^{s-\sigma}[S^\times(\tau+t)-S^\times(\tau)]d\tau\,dV_0(\sigma).$$

(The subscript σ in d_σ indicates the integration variable). Defining

$$U_t(s):=\int_0^s[S^\times(\tau+t)-S^\times(\tau)]d\tau$$

it follows from (53) that

$$U_t(s)=S_0^*(t)S_0^*(s)+\int_0^t S^\times(t-\sigma)d_\sigma[V_0(\sigma)S_0^*(s)]+\int_0^s U_t(s-\sigma)dV_0(\sigma), \tag{54}$$

or, by (50),

$$U_t(s) = S^\times(t)S_0^*(s) + \int_0^s U_t(s-\sigma)dV_0(\sigma). \tag{55}$$

But (47) and (48) show that V is the resolvent kernel of V_0 and hence U_t, as the solution of (55), has the representation

$$\begin{aligned} U_t(s) &= S^\times(t)S_0^*(s) + \int_0^s S^\times(t)S_0^*(s-\sigma)dV(\sigma) \\ &= S^\times(t)[S_0^*(s) + \int_0^s S_0^*(s-\sigma)dV(\sigma)] \\ &= S^\times(t)S^\times(s), \end{aligned} \tag{56}$$

which shows that $S^\times$ is an integrated semigroup.

It remains to be shown that V is a multiplied integral of $T^\times$. That (40) and (41) hold is obvious. The proof that V and $S^\times$ satisfy (42) is completely analogous with the proof that $S^\times$ is an integrated semigroup. Again using the fact that V_0, S_0^* satisfy (42) one derives an equation similar to (55) for $U_s(t) := \int_0^t [V(\tau+s) - V(\tau)]d\tau$, after which the resolvent representation yields the desired result.

Acknowledgement

Horst R. Thieme is supported by a Heisenberg scholarship from Deutsche Forschungsgemeinschaft.

References

1. W. Arendt (1987): Vector-valued Laplace transforms and Cauchy problems. Israel J. Math. **59**, 327-352
2. Ph. Clément, O. Diekmann, M. Gyllenberg, H.J.A.M. Heijmans and H.R. Thieme (1987): Perturbation theory for dual semigroups. I. The sun-reflexive case. Math. Ann. **277**, 709-725
3. Ph. Clément, O. Diekmann, M. Gyllenberg, H.J.A.M. Heijmans and H.R. Thieme (1988): Perturbation theory for dual semigroups. II. Time-dependent perturbations in the sun-reflexive case. Proc. Roy. Soc. Edinburgh **109A**, 145-172
4. Ph. Clément, O. Diekmann, M. Gyllenberg, H.J.A.M. Heijmans and H.R. Thieme (1989): Perturbation theory for dual semigroups. III. Nonlinear Lipschitz continuous perturbations in the sun-reflexive case. In Volterra integrodifferential equations in Banach spaces and applications, G. Da Prato and M. Iannelli (Eds.), Pitman Research Notes in Mathematics Series 190, Longman, Harlow, 67-89
5. Ph. Clément, O. Diekmann, M. Gyllenberg, H.J.A.M. Heijmans and H.R. Thieme (1989): Perturbation theory for dual semigroups. IV. The intertwining formula and the canonical pairing. In Semigroup theory and applications, Ph. Clément, S. Invernizzi, E. Mitidieri and I.I. Vrabie (Eds.), Lecture Notes in Pure and Applied Mathematics 116, p.95-116, Marcel Dekker, New York

6. Ph. Clément, O. Diekmann, M. Gyllenberg, H.J.A.M. Heijmans and H.R. Thieme (1989): A Hille-Yosida type theorem for a class of *weakly** continuous semigroups. semigroup Forum **38**, 157-178
7. O. Diekmann (1987): On the mathematical synthesis of physiological and behavioural mechanisms and population dynamics. In: Mathematical topics in population biology, morphogenesis and neurosciences, E. Teramoto and M. Yamaguli (Eds.), Springer Lecture Notes in Biomath. **71**, 48-52
8. O. Diekmann, M. Gyllenberg, and H.R. Thieme: Perturbation theory of dual semigroups. V. Variation of constants formulas. submitted
9. G. Gripenberg, S-O. Londen, O. Staffans (1990): Volterra Integral and Functional Equations, Cambridge University Press, Cambridge
10. H. Kellermann and M. Hieber (1989): Integrated semigroups, J. Funct. Anal. **84**, 160-180
11. F. Neubrander (1988): Integrated semigroups and their applications to the abstract Cauchy problem. Pac. J. Math. **135**, 111-155
12. H.R. Thieme (1988): Well-posedness of physiologically structured population models for Daphnia magna. How biological concepts can benefit by abstract mathematical analysis. J. Math. Biol. **26**, 299-317
13. H.R. Thieme (1990): 'Integrated semigroups' and integrated solutions to abstract Cauchy problems. J. Math. Anal. Appl. **152**, 416-447

Estimates for Spatio-Temporally Dependent Reaction Diffusion Systems

W.E. Fitzgibbon[1], *J.J. Morgan*[2], *R.S. Sanders*[3], *and S.J. Waggoner*[4]

[1] University of Houston, Houston, TX
[2] Texas A & M, College Station, TX
[3] University of Houston, Houston, TX
[4] Furman University, Greenville, SC

1 Introduction

In this paper we are concerned with a priori bounds for globally defined solutions to a class of weakly coupled semilinear parabolic equations which may include spatio-temporal inhomogeneity. Specifically we consider an m-component system of the form:

$$\partial u_i/\partial t = \mathcal{L}_i(x,t)u_i + f_i(x,t,u) = \nabla \cdot (a_i(x,t) \nabla u_i) + f_i(x,t,u) \qquad (1.1a)$$

$$x \in \Omega, \quad t > 0, \quad i = 1 \text{ to } m,$$

$$u_i(x,t) = 0 \quad x \in \partial\Omega, \quad t > 0, \quad i = 1 \text{ to } m \qquad (1.1b)$$

$$u_i(x,0) = u_{0_i}(x) \quad x \in \Omega, \quad i = 1 \text{ to } m, \qquad (1.1c)$$

where Ω is a bounded domain in $\mathbf{R}^n$ whose boundary is an $(n-1)$ dimensional $C^{2+\alpha}$ manifold for some $\alpha \in (0,1)$ such that Ω lies locally on one side of $\partial\Omega$. The closure of Ω and the Lebesgue measure of Ω shall be given by $\overline{\Omega}$ and $|\Omega|$ respectively. We let $\mathbf{R}^m_+$ denote the positive orthant of $\mathbf{R}^m$, i.e., $\mathbf{R}^m_+ = \{u \in \mathbf{R}^m | u_i \geq 0\}$.

To put things in perspective we consider a biological model investigated in [19] and [20]. The system was developed to portray the interaction of oxygen O_2, carbon dioxide CO_2 and hemoglobin H_b as blood travels through a pulmonary capillary. The relevant reaction scheme is:

$$H_b + O_2 \longleftrightarrow H_bO_2 \qquad H_b + CO_2 \longleftrightarrow H_bCO_2$$

If we add diffusion to the associated kinetics we obtain the following semilinear parabolic system for all $x \in \Omega, \ t \geq 0$:

$$\partial u_1/\partial t - \nabla \cdot (a_1(x,t) \nabla u_1) = k_2u_2 - k_1u_1u_5 \qquad (1.2a)$$

$$\partial u_2/\partial t - \nabla \cdot (a_2(x,t) \nabla u_2) = -k_2 u_2 + k_1 u_1 u_5 \tag{1.2b}$$

$$\partial u_2/\partial t - \nabla \cdot (a_3(x,t) \nabla u_3) = k_4 u_4 - k_3 u_3 u_5 \tag{1.2c}$$

$$\partial u_2/\partial t - \nabla \cdot (a_4(x,t) \nabla u_4) = -k_4 u_4 + k_3 u_3 u_5 \tag{1.2d}$$

$$\partial u_2/\partial t - \nabla \cdot (a_5(x,t) \nabla u_5) = k_2 u_2 + k_4 u_4 - k_1 u_1 u_5 - k_3 u_3 u_5 \tag{1.2e}$$

where Ω is bounded domain with sufficiently smooth boundary. We impose homogeneous Dirichlet boundary conditions $u_i(x,t) = 0$, $i = 1$ to 5 on $\partial\Omega$. If the initial date is nonnegative straightforward maximum principle arguments imply that the solutions remain nonnegative. If we set

$$H(u) = \sum_{i=1}^{5} h_i(u_i) = u_1 + 2u_2 + u_3 + 2u_4 + u_5,$$

we may multiply the i-th component by $h_i'(u_i)$, integrate in space time and sum the result to produce the formal a priori L_1 estimate

$$\int_\tau^T \int_\Omega H(u)dxdt \leq \left\{ \int_\Omega H(u,(x,0)dx \right\} (T - \tau) \tag{1.3}$$

for $0 < \tau < T$.

The question becomes that of bootstrapping the L_1 estimate 1.3 (or more generally a L_p estimate for (1.1a-c)) to L_∞ estimate. We shall exploit the existence of a separable convex function to guarantee the existence of uniform bounds for globally defined solutions. Here, we are extending recent work of Morgan [14]. We do not address the question of existence of solutions for all time. Criteria which guarantee this existence for (1.1) (including 1.3) are provided in [6].

We point out that for reasons of simplicity we have chosen not to include transport terms in our differential operators. Such terms will be treated within a more general context by the authors in forthcoming work [7].

2 Preliminary estimates

We shall assume that the reader is familiar with standard Sobolev spaces. We use $Q(\tau,\ T)$ for the space time cylinder $\Omega \times (\tau,\ T)$. If $p \in [1,\ \infty]$ we introduce $W_p^{2,1}(Q(\tau,\ T))$ as the Banach space of functions $\varphi \in L_p(Q(\tau,\ T))$ having generalized derivatives of the form $\partial\varphi/\partial t$, $\partial\varphi/\partial x_i$ and $\partial^2\varphi/\partial x_j \partial x_k$ as elements of $L_p(Q(\tau,\ T))$. The norm for $W_p^{2,1}(Q(\tau,\ T))$ is given by

$$\begin{aligned} \| \varphi \|_{p,\ Q(\tau,\ T)}^{(2)} = \| \varphi \|_{p,\ Q(\tau,\ T)} + \| \varphi_t \|_{p,Q(\tau,\ T)} + \| D_x\varphi \|_{p,Q(\tau,\ T)} \\ + \| D_x^2\varphi \|_{p,Q(\tau,\ T)}. \end{aligned} \tag{2.1}$$

We place the following conditions on our initial data, coefficients and nonlinearities.

C_1 $\qquad u_o = (u_{o_i})_{i=1}^m \in C^2(\Omega,\ R_+^m) \cap C_o(\overline{\Omega},\ R_+^m)$

C_2 $f = (f_i(\,))_{i=1}^m : \overline{\Omega} \times [0,\infty) \times R_+^m \to R^m$ is locally Lipschitz in each variable.

C_3 $a = (a_i(\,))_{i=1}^m \in C^{2,1}(\overline{\Omega} \times [0,\infty), R_+^m)$ and there exists $\epsilon > 0$ so that for

$$i = 1 \text{ to } m \text{ and all } x,\ t \in \overline{\Omega} \times [0,\infty),\ a_i(x,t) > \epsilon.$$

C_4 There exists $M > 0$ so that sup $\{a_i, (x,t),\ |\partial_t a_i(x,t)|,\ |a_i(x,t)_{x_j}|\} < M$.

C_5 For each $i = 1$ to $m, x \in \overline{\Omega},\ t \geq 0,\ v \in R_+^m,\ f_i(x,t,v) \geq 0$ whenever $v_i = 0$.

We remark that it is possible to consider more general cases of elliptic operators; however, we restrict ourselves to the present case to avoid additional notational and technical difficulties. The extension of our theory to boundary conditions of Robin or Neumann type is more difficult and it will be the subject of future work.

We shall need the concept of strong L_p solutions for parabolic equations. By a strong L_p solution to a parabolic equation of the form,

$$\partial v/\partial t = \mathcal{L}(x,t)v + f(t,x), \quad x \in \Omega \tag{2.3a}$$

$$v = 0, \quad x \in \partial\Omega \tag{2.3b}$$

$$v(x,0) = v_0(x), \quad x \in \Omega \tag{2.3c}$$

on $\overline{\Omega} \times [0, T)$ we shall mean a function $v \in W_p^{2,1}(Q(0,T))$ which satisfies the boundary conditions, the initial condition and satisfies the differential equation a.e. We shall now produce several estimates for parabolic equations and results from semigroup theory which will serve to develop our analysis of the relevant adjoint equations.

We shall consider equations of the form

$$\partial\varphi/\partial t - \nabla \cdot (d(x,\ t) \nabla \varphi) + \varphi = \theta \qquad x \in \Omega,\ \tau \leq t \leq T \tag{2.4a}$$

$$\varphi(x,t) = 0 \qquad x \in \partial\Omega,\ \tau \leq t < T \tag{2.4b}$$

$$\varphi(x,\tau) = 0 \qquad x \in \Omega \tag{2.4c}$$

where $\theta \in L_p(Q(\tau,\ T)), \theta \geq 0$, and $\|\ \theta\ \|_{p,Q(\tau,T)} = 1$. The diffusivities satisfy conditions C_3 and C_5.

Lemma 2.5 *If* $p \in (1,\infty)$, *and* $\theta \in L_p(Q(r,\ T))$ *there exists a unique strong* L_p *solution to* (2.4 $a - c$) *such that:*

(*i*) $\qquad \varphi \geq 0.$

(*ii*) $\qquad$ There exists $C_p > 0$ such that $\|\ \varphi\ \|^{(2)}_{p,Q(\tau,\ T)} \leq C_p$.

(*iii*) If $1 < p < \dfrac{n+2}{2}$ and $p \le q \le \dfrac{p(n+2)}{n+2-2p}$ then there exists $\overline{C}_p$ such that

$$\| \varphi \|_{q,Q(\tau,\, T)} \le \overline{C}_p.$$

Proof. Part (i) is a maximum principle consequence. Part (ii) follows from Ladyzenskaja, Solonnikov and Uralceva [Theorem 9.1, 13] and Part (iii) follows from Ladyzenskaja, Solonnikov and Uralceva [Lemma 3.10,13]

We shall also consider globally defined parabolic equations of the form

$$\partial\varphi/\partial t - d\Delta\varphi + \varphi = \theta \quad x \in \mathbf{R}^n,\ \tau \le t \le T \tag{2.6a}$$

$$\varphi(x,\tau) = 0 \quad x \in \mathbf{R}^n \tag{2.6b}$$

where $\theta \in L_p(\mathbf{R}^n \times (\tau,\, T))$ and $d > 1$. For each $p \in (1,\infty)$ we define A_p on $L_p(\mathbf{R}^n)$ by

$$A_p w \;=\; d\Delta w - w \tag{2.7a}$$

with

$$D(A_p) = \{w | w \in W_p^2(\mathbf{R}^n)\}. \tag{2.7b}$$

It is well known, cf. Pazy [16], that A_p is the infinitesimal generator of an analytic semigroup on $L_p(\mathbf{R}^n)$, $\{T_p(t) | t \ge 0\}$. Moreover, $0 \notin \sigma(A_p)$ and there exists $\lambda_o > 0$ so that

$$\| T_p(t) \|_{p,\mathbf{R}^n} \le e^{-\lambda_o t} \tag{2.8}$$

Fractional powers of $-A_p$ exist. For $0 < \gamma < 1$ we define, cf [16],

$$B_p^\gamma = (-A_p)^\gamma. \tag{2.9}$$

Each $-B_p^\gamma$ is the infinitesimal generator of an analytic semigroup on $L_p(\mathbf{R}^n)$. For $\gamma_1 > \gamma_2 \ge 0$, $D(B_p^{\gamma_1}) \subseteq D(B_p^{\gamma_2})$. Because $D(B_p^0) = L_p(\mathbf{R}^n)$ we can define the graph norm on $X_{p,\gamma} = D(B_p^\gamma)$ by setting $|w|_{p,\gamma} = \| B_p^\gamma w \|_{p,\, \mathbf{R}^n}$. We shall need the following result.

Lemma 2.10 *Suppose* $\{T_p(t) | t \ge 0\}$ *and* B_p^γ *are as above. The following are true:*

(*i*) $T_p(t) \;:\; L_p(\mathbf{R}^n) \rightarrow D(B_p^\gamma)$ for all $t > 0$.

(*ii*) $\| B_p^\gamma\, T_p(t) w \|_{p,\mathbf{R}^n} \le C_{\gamma,\,p} t^{-\gamma} \| w \|_{p,\, \mathbf{R}^n}$ for some $C_{\gamma,p} > 0$ and all $w \in L_p(\mathbf{R}^n)$ and $t > 0$.

(*iii*) For all $t \ge 0$, if $w \in D(B_p^\gamma)$, $T_p(t) B_p^\gamma\, w = B_p^\gamma T_p(t) w$.

(*iv*) If $\gamma > n/2p$ then $D(B_p^\gamma) \subseteq L_\infty(\mathbf{R}^n)$ and $\| w \|_{\infty,\, \mathbf{R}^n} \le M_{\gamma,p} \| B_p^\gamma w \|_{p,\mathbf{R}^n}$

for some constant $M_{\gamma\ ,p} > 0$ and all $w \in D(B_p^\gamma)$

(*v*)
If $\mu > n(q-p)/2pq$ then $D(B_p^\mu) \subseteq L_q(\mathbf{R}^n)$ and $\| w \|_{q,\mathbf{R}^n} \le N_{\mu,p} \| B_p^\mu w \|_{p,\mathbf{R}^n}$

for some constant $N_{\mu,p}$ and all $w \in D(B_p^\mu)$

Proof. Parts (i-iii) are standard estimates from the theory of analytic semigroups, cf. Pazy [16]. Part (iv) may be also found in Pazy and the proof of (v) is contained in Henry [p.40,9].

We now return to the spatio-temporal dependent equation (2.4 a-c).

Lemma 2.11 *Assume the hypotheses of Lemma 2.5 are satisfied. If $\varphi(\ )$ is the stong L_p solution to (2.4a-c) then the following are true.*

(*i*) If $p \in (1, \infty)$ then $\| \varphi(., T) \|_{p,\ \Omega} \le p^{\frac{1}{p}}$.

(*ii*) If $p \in (1, \infty)$ then there exists $a\ d > 0$ and $N > 0$ so that

$$\| \varphi \|_{p,\ Q(\tau,\ T)} \le \frac{N}{N + d(p-1)} \left[\left(1 - exp(-(T-\tau) \frac{N + d(p-1)}{N} \right] \right).$$

(iii)) If $p > \frac{n+2}{2}$ then there exists $K_{p(T-\tau)} > 0$ such that

$$\| \varphi \|_{\infty,k\ Q(\tau,\ T)} < K_{p(T-\tau)}$$

.

(iv) If $1 < p < \frac{n+2}{2}$ and $1 < q < \frac{np}{n-2(p-1)}$ then there exists $K_{p,q,(T-\tau)} > 0$ such that

$$\| \varphi(,\ T) \|_{q,\Omega} \le K_{p,q,(T-\tau)}.$$

Proof. Let $\overline{d}(x,t)$ be an extension of $d(x,t)$ to R^n which preseves the positivity, smoothness and boundedness properties of $d(x,t)$ and its first partials. Similarly let $\tilde{\theta}$ be an extension of θ which is identically zero on $R^n - \overline{\Omega}$. We consider the Cauchy initial value problem

$$\partial\tilde{\varphi}/\partial t - \nabla \cdot (\tilde{d}(x.t)\nabla\tilde{\varphi}) + \tilde{\varphi} = \tilde{\theta} \qquad x \in \mathbf{R}^n,\ t \in (\tau, T) \qquad (2.12a)$$

$$\tilde{\varphi}(x,\ \tau) = 0 \qquad x \in \mathbf{R}^n \qquad (2.12b)$$

Standard maximum principle arguments show that if $(x,\ t) \in Q(\tau,\ T)$ then $0 \le \varphi(x,t) \le \tilde{\varphi}(x,t)$. Powerful estimates of Aronson [2], [3] and Ladyzenskaja, Uralceva and Solonnikov [13] permit us to estimate the fundamental solution $\gamma(x, \eta, t, \tau)$ associated with $\partial\eta/dt - \nabla \cdot (\tilde{d} \nabla \eta) = 0$ by means of an appropriately chosen heat kernel. In particular there exists $c > 0$ and a heat kernel $\rho(x-\xi, t-\tau)$ associated with some $a\Delta u = \partial u/\partial t (a > 0)$ so that if $x \ne \xi,\ t \ne \tau$

$$0 \le \gamma(x, \xi,\ t,\ \tau) \le c\ \rho(x - \xi,\ t - \tau) \qquad (2.13)$$

Consequently,

$$0 \le e^{(t-\tau)}\gamma(x,\ \xi,\ t,\ \tau) \le c\, e^{-(t-\tau)}\rho(x-\xi,\ t-\tau) \tag{2.14}$$

and

$$\int_0^t \int_\Omega e^{-(t-s)}\gamma(x,\xi,t,s)\tilde{\theta}(\xi,s)d\xi ds \le c \int_o^t \int_\Omega e^{-(t-s)}\rho(x-\xi,t-s)\tilde{\theta}(\xi,s)d\xi ds. \tag{2.15}$$

However the left hand side of (2.15) is a strong solution to

$$\begin{aligned} \partial\tilde{\varphi}/\partial t - \nabla\,(\tilde{d}\,\nabla\,\tilde{\varphi}) + \tilde{\varphi} = \tilde{\theta} \qquad & x\ \in\ \mathbf{R}^n,\ t\ \in (\tau,T) \\ \varphi(x,\tau) = 0 \qquad & x\ \in\ \mathbf{R}^n \end{aligned} \tag{2.16}$$

while the right hand side satisfies

$$\partial w/\partial t - a\Delta w + w = \tilde{\theta} \qquad x\ \in\ \mathbf{R}^n,\ t\ \in (\tau,T) \tag{2.17a}$$

$$w(x,\tau) = 0 \qquad x\ \in\ \mathbf{R}^n \tag{2.17b}$$

in the sense of a strong solution. Therefore

$$0 \le \varphi(x,t) \le \tilde{\varphi}(x,t) \le cw(x,t) \tag{2.18}$$

for $x \in \Omega$ and $\tau \le t \le T$.

The estimates (i), (ii) are a straightforward adaptions of a lemma in Morgan [14]. In a forthcoming paper on globally defined reaction diffusion systems [7] the authors establish the inequality of Part (iii) for $w(x,t)$; hence, by virtue of (2.18) the same estimates hold for $\varphi(x,t)$. To obtain Part (iv) we write (2.17a-b) as abstract ordinary differential equation in $L_p(\mathbf{R}^n)$.

$$dw/dt + A_p w = \tilde{\theta} \tag{2.19a}$$

$$w(\tau) = 0 \tag{2.19b}$$

Solutions to (2.19 a-b) have variation of parameters representation

$$w(t) = \int_\tau^t T_p(t-s)\ \tilde{\theta}(s)da. \tag{2.20}$$

We mention that the constant c depends on $\tilde{d}$. However the facts that $\tilde{d}(x,t)$ is bounded above and below and satisfies a uniform Lipschitz condition in x_i and t imply that a uniform c can be chosen for all x and t, cf [13, p360]. Our hypotheses imply that there exists $0 < \mu < 1$ such that $\mu > n(q-p)/2pq$ and $\mu p/(p-1) < 1$. Thus from Lemma 2.10 we have

$$\begin{aligned} \|\ \varphi(.,T)\ \|_{q,\Omega} &\le N_{\mu,p}\ \|\ B_p^\mu w(.,T)\ \|_{p,\mathbf{R}^n} \\ &\le N_{\mu,p}C_{\mu,p} \int_0^{T-\tau} (T-\tau-s)^{-\mu}\ \|\ \theta(\ ,s)\ \|_{p,\mathbf{R}^n}\ ds \end{aligned} \tag{2.21}$$

and from Holder's inequality

$$\int_0^{T-\tau} (T-\tau-s)^{-\mu} \parallel \theta(\cdot,s) \parallel_{p,\mathbf{R}^n} ds \leq \left[\frac{(T-\tau)^{1-\frac{\mu p}{p-1}}}{1-\frac{\mu p}{p-1}} \right]^{\frac{p-1}{p}} \tag{2.22}$$

The argument establishing Part (ii) uses Lemma 2.10 in much the same way. We note that if $2p > n+2$ then the choice $q=\infty$ is admissible.

The final next lemma which appears in Morgan [14] provides some necessary numerical estimates.

Lemma 2.23 *If* $1 \leq r < k$ *and there exists* $0 < \mu < 2$ *such that* $r + \frac{2-\mu}{n+2} < 1 + \frac{2k}{n+2}$ *then there exist* $\delta > 1$ *and* $1 < p < \frac{n+2}{2}$ *such that*

(i)
$$k = \frac{np}{(n+2)(p-1)}\delta,$$

(ii)
$$\frac{k}{k-r} \leq \frac{p(n+2)}{n+2-2p},$$

(iii)
$$\frac{p}{p-1} \geq \frac{n+2}{n+\mu}k.$$

Our subsequent results are predicated on the existence foa priori cylinder bounds. These a priori bounds are bootstrapped to higher L_p spaces by use of duality arguments. We now formulate the system of adjoint equations used in the duality argument.

We let φ_i, $i=1$ to m, be the unique strong solution to

$$\partial\varphi_i/\partial t - \nabla \cdot (a_i(x,t) \nabla \varphi_i) + \varphi_i = \theta \qquad x \in \Omega,\ t \in (\tau,\ T) \tag{2.24a}$$
$$\varphi_i = 0 \qquad x \in \partial\Omega,\ t \in (\tau,\ T) \tag{2.24b}$$
$$\varphi_i(x,\tau) = 0 \qquad x \in \Omega \tag{2.24c}$$

where $\theta \in L_p(Q(\tau,\ T))$, $\theta \geq 0$, and $\parallel \theta \parallel_{p,Q(\tau,T)} = 1$. We set $\tilde{\theta}(x,\ t) = \theta(x,\ T+\tau-t)$ and $\psi_i(x,t) = \psi_i(x, T+\tau-t)$ on $\overline{\Omega} \times [\tau, T]$ for $i=1$ to m and observe ψ_i, satisfies

$$\partial\psi_i/\partial t = -\nabla \cdot (a,(x,t) \nabla \psi_i) + \psi_i - \tilde{\theta} \qquad x \in \Omega,\ t \in (\tau, T) \tag{2.25a}$$
$$\psi_i(x,\ ) = 0 \qquad x \in \partial\Omega,\ t \in (\tau,\ T) \tag{2.25b}$$
$$\psi_i(x,\ T) = 0 \qquad x \in \Omega. \tag{2.25c}$$

We hereby obtain a system formally adjoint to (2.24 a-c). The estimates of this section hold for ψ_i as well.

We conclude this section by detailing the hypotheses needed for the separable convex functional alluded to in the introduction. We assume that there exist $H(\ ) \in C^2(R_+^m,\ R_+^m)$ and $h_i(\ ) \in C^2(R_+,\ R_+)$ for $i = 1$ to m such that:

H_1 $$H(v) = \sum_{i=1}^{m} h_i(v_i) \text{ for } v \in R_+^m$$

H_2 $$H(z) = 0 \text{ if and only if } z = (0, ..., 0)^T.$$

H_3 $$\partial^2 H(v) \text{ is nonnegative definite for } v \in R_+^m$$

We now introduce the intermediate sums condition.
H_4 There exists $A = (a_{ij}) \in R^{m \times m}$ satisfying $a_{ij} > 0,\ a_i j \geq 0$ for all $1 \leq i \leq m$ such that for each $1 \leq j \leq m$ there exists $r,\ K_1,\ K_2 > 0$ independent of j so that

$$\sum_{i=1}^{j} a_{ji} h_i'(v_i) f_i(x, t, v) \leq K_1 (H(v))^r + K_2.$$

Finally we need,

H_5 There exist $q_1,\ K_5,\ K_4$ such that for $1 \leq i \leq m$ and $v \in R_+^m$

$$h_i'(v_i) f_i(x, t, v) \leq K^3 (H(v))^{q1} + K^4$$

Some remarks are in order. We point out that H^5 places a polynomial growth bound on the vector field. Moreover it will insure that condition H_4 holds for some exponent. However the exponent produced in this fashion is generally too large to be useful. Smaller exponents for H_4 are produced by careful examination of the additive cancellation of terms of the form $h_i'(v_i) f_i(x, t, v)$. Morgan [15] was the first to recognize the importance of the role played by the intermediate sums condition. It is subsequently used in [14], [6], [7], [17].

3 Main result-boundness and decay

Our development continues techniques constructed by Hollis, Martin and Pierre [10]. The initial extension to m-components systems was done by Morgan [14]. We repeat for emphasis that we are assuming the existence of globally defined solutions to (1.1a-c). The results of their section are predicated upon the following lemma.

Lemma 3.1 *Suppose that C_1-C_6, H_1-H_3 and H_5 hold, $p > (n + 2)/2$ and there exists $T > 0$, a sequence $\{t_i\}$ and $\tilde{g} \in C([0, \infty))$ such that* (*i*) $t_1 > 0$ and $T/4 < t_{i+1} - t_i < T/2$ for all $i \geq 1$,

(*ii*) $||H(u)||_{p,Q(\tau,\tau+T)}$, $||(H(u))^{q_1}||_{p,Q(\tau,\tau+T)} \leq \tilde{g}(\tau)$ for all $\tau \geq 0$,

(*iii*) $||h_j(u_j(\cdot,t_i))||_{p,\Omega} \leq \tilde{g}(t_i)$ for all $i \geq 1, 1 \leq j < m$.

If there exists $K > 0$ such that $\tilde{g}(t) \leq K$ for all $t \geq 0$, then there exists $N > 0$ such that $||u_i(\cdot,t)||_{\infty,\Omega} \leq N$ for all $t \geq 0$, $1 \leq i \leq m$. Furthermore, if $K_4 = 0$ and $\lim_{t\to\infty} \tilde{g}(t) = 0$ then $\lim_{t\to\infty} ||u_i(\cdot,t)||_{\infty,\Omega} = 0$.

Proof. Suppose $t > t_3$. There exists $i \geq 1$ so that $t_i < t_{i+1} < t \leq t_{i+2}$; furthermore, $t - t_i < T$. Also there exist $\delta > 0$ such that $p = (n+2+\delta)/2$. Hence if $\mu = n/(n+2)$ then $n/2p < \mu < 1$.

If we multiply (1.1a) by $h_i'(u_i)$ we may use the convexity of $h_i(u_i)$, the ellipticity and H_5 to observe that

$$\partial_t(h_i(u_i)) \leq \mathcal{L}_i(x,t)h_i(u) + K_3(H(u))^{q_1} + K_4 \text{ on } Q(0,\infty), \tag{3.2a}$$

$$h_i(u_i(x,0)) = 0 \text{ on } \partial\Omega \times [0,\infty), \tag{3.2b}$$

$$h_i(u_i(x,0)) = h_i(u_{0_i}(x,0)) \text{ on } \Omega, \tag{3.2c}$$

for $i = 1$ to m. Hence,

$$\partial_t(h_i(u_i)) \leq \mathcal{L}_i(x,t)h_i(u) - h_i(u_i) + H(u) + K_3(H(u))^{q_1} + K_4 \tag{3.3}$$

on $Q(0,\infty)$ with the same boundary and initial conditions. We let $\tilde{\mathcal{L}}_i(x,t)$ extend $\mathcal{L}_i(x,t)$ to all of $\mathbf{R}^n$ and preserve all boundedness and smoothness properties. Additionally we define

$$G_i(x,t) = \begin{cases} H(u) + K_3(H(u))^{q_1} + K_4 & \text{if } (x,t) \in \overline{\Omega} \times [0,\infty) \\ 0 & \text{if } (x,t) \in (\mathbf{R} - \overline{\Omega}) \times [0,\infty) \end{cases} \tag{3.4}$$

and

$$v_{o_i}(x) = \begin{cases} h_i(u_{o_i}(x)) & x \in \overline{\Omega} \\ 0 & x \in \overline{\Omega} \end{cases} \tag{3.5}$$

Standard maximum principle arguments insure that if $v_i(\ ,\ )$ is a solution to

$$\partial v_i/\partial t = \tilde{\mathcal{L}}_i(x,t)v_i - v_i + G_i \qquad x \in \mathbf{R}^n, t > 0 \tag{3.6a}$$

$$v_i(x,0) = v_{o_i}(x) \qquad x \in \mathbf{R}^n \tag{3.6b}$$

then for all $(x,t) \in \Omega \times [0,\infty)$, $v_i(x,t) \geq h_i(u_i(x,t)) \geq 0$. Let $\gamma_i(x,\xi,t,\tau)$, cf Aronson [2], be the fundamental solution associated with $\partial y_i/\partial t = \tilde{\mathcal{L}}_i(x,t)y_i$. Then the solution to (3.6a-b) has the Duhamel representation

$$v_i(x,t) = e^{-t}\int_{\Omega}\gamma_i(x,\xi,t,0)v_{o_i}(\xi)d\xi + \int_0^t\int_{\Omega}e^{-(t-s)}\gamma_i(x,\xi,t,s)G_i(\xi,s)d\xi ds. \tag{3.7}$$

The aforementioned work of [2], [3], [13] established the existence of $c > 0$ and a heat kernel of $\rho_i(x,t)$ associated with an appropiated chosen heat equation $\partial z/\partial t = d_i \Delta z$ such that if $x \neq \xi, t \neq \tau$, then $0 \leq \gamma_i(x,\xi,t,\tau) \leq c\rho_i(x-\xi,t-\tau)$. Thus we may observe that

$$\begin{aligned} 0 \leq v_i(x,t) \leq ce^{-t}&\int_{\Omega}\rho_i(x-\xi,0)v_{o_i}(\xi)d\xi \\ &+ \int_0^t\int_{\Omega}ce^-t\rho_i(x-\xi,t-s)G_i(\xi,s)d\xi ds. \end{aligned} \tag{3.8}$$

However if $w_i(x,t)$ is defined to be the right hand side 3.8 we see that

$$\partial w_i/\partial t = d_i\Delta w_i - w_i + G_i \quad (x,t) \in \mathbf{R}^n \times [0,\infty) \tag{3.9}$$

$$w_i(x,0) = v_{o_i}(x) \quad x \in \mathbf{R}^n. \tag{3.10}$$

Moreover $cw_i(x,t) \geq v_i(x,t) \geq h_i(u_i(x,t))$ for $(x,t) \in \Omega \times [0,\infty)$ and $i = 1$ to m. As we have seen in Lemma 2.11 a uniform c can be chosen.

Referring back to (2.17 a-b) and (2.18) we cast (3.9 a-b) as a system of ordinary differential equations in $L_p(\mathbf{R}^n)$,

$$\dot{w}_i(t) + A_{ip}w(t) = G_i(t) \tag{3.10a}$$

$$w(0) = w_{o_i} \tag{3.10b}$$

which has componentwise variation of parameters solution

$$w_i(t) = T_{i,p}(t)w_{o_i} + \int_0^t T_{i,p}(t-s)G_i(s)ds. \tag{3.11}$$

By virtue of Lemma 2.11, part (iii) we have

$$\begin{aligned} ||w_j(t)||_{\infty,\mathbf{R}^n} &\leq M_{\mu,p}C_{\mu,p}\left[(t-t_i)^{-\mu}||h_j(u_j(\cdot,t_i))||_{p,\Omega}\right. \\ &\quad \left. + ||(t-s)^{-\mu}||_{\frac{p}{p-1},(t_i,t)}((K_3+1)\tilde{g}(t) + K_4(t-t_i)^{\frac{1}{p}})\right] \\ &\leq M_{\mu,p}C_{\mu,p}\left(\frac{T}{4}\right)^{-\mu}\quad[\tilde{g}_i(t_i) \\ &\quad + ||(T-s)^{-\mu}||_{\frac{p}{p-1},(0,\frac{T}{4})}(K_3+1)\tilde{g}(t_i) + K_4T^{\frac{1}{p}}] \end{aligned} \tag{3.12}$$

Because $p\mu/(p-1) < 1$ there exists $N_\mu > 0$ such that

$$||(T-s)^{-\mu}||_{\frac{p}{p-1},(0,\frac{T}{4})} \leq N_\mu$$

Becasue we have produced a uniform bound for $||w_i(t)||_{\infty,\Omega}$ we have a uniform bound for $||h_i(u_i(\cdot,t))||_{\infty,\Omega}$ and the coerciveness of h_i () implies the existence of $N > 0$ so that

$$||u_i(\cdot,t)||_{\infty,\Omega} \le N \tag{3.14}$$

To obtain the final assertion we suppose $\lim_{t\to\infty} \tilde{g}(t) = 0$ and $K_4 = 0$. From (3.12) and (3.13) we obtain:

$$||h_j(u_j(\cdot,t))||_{\infty,\Omega} \le ||w_j(t)||_{\infty,\mathbf{R}^n} \le M_{\mu,p}C_{\mu,p}\left[\left(\frac{T}{4}\right)^{-\mu} + N_\mu K_5 + 1\right] \tag{3.15}$$

Hence $\lim_{t\to\infty} ||h_j(u_j(\cdot,t))||_{\infty,\Omega} = 0$ and H_2 implies $\lim_{t\to\infty} ||u_j(\cdot,t)||_{\infty,\Omega} = 0$.

Our boundedness result is obtained by bootstrapping known as a priori cylinder bounds by an iterative process to sufficently high cylinder bounds via utilization of duality arguments in conjunction with the intermediate sums condition. We have the following theorem.

Theorem 3.16 *Let $C_1 - C_5$ hold and $H_1 - H_3$ and H_5 hold. Let $u = (u_i)_{i=1}^m$ be the globally defined classical solution to (1.1 a-c). If H_4 holds with exponent r satisfying $1 \le r \le a$ or $1 \le r = a < (n+2)/2$ then there exists an $N > 0$ such that $||u_i(\cdot,t)||_{\infty,Q(0,\infty)} \le N$ for all $1 \le i \le m$.*

Proof. We get $M_1 = g(3)$ and observe that $||H(u)||_{a,Q(T-\tau)} \le g(T-\tau)$ for all $0 \le \tau \le T$ implies that there exists a sequence $\{t_{1,i}\}$ such that $t_{1,1} > 0$, $0 < t_{1,i+1} - t_{1,i} < 3$ and $||H(u(\cdot,t_{1,i}))||_{a,\Omega} \le M_1$ for all $i \ge t$. Thus if $T > t_{1,i}$ there exists $i \ge 1$ such that $t_{1,i} < T \le t_{1,i+1}$. Set $\tau = t_{1,i}$ following Morgan [14] we subdived our argument into two cases.

Case 1. Suppose $1 \le r < a$. Because $r < 1 + (2a)/n + 2$ there exists $0 < \mu < 2$ such that $r + (2-\mu)/(n+2) < 1 + (2a)/n + 2$. Lemma 2.21 implies there exist $\delta > 1$ and $1 < p < (n+2)/2$ such that $a = np\delta/(n+2)(p-1)$, $a/(a-r) \le p(n+2)/(n+2-2p)$ and $p/(p-1) \ge a(n+2)/(n+\mu)$. We set $p_1 = p/(p-1)$. We now find ourselves able to invoke the intermediate sums hypothesis. Suppose that $1 \le j \le m$ and there exists $M_2 > 0$ such that for all $l \ge 1$ and $1 \le i < j$, $||h_i(u_i)||_{p_1,Q(t_{1,l},t_{1,l+1})} \le M_2$. Assume there exists $\delta_{p_1} \in (0,1)$ and constants K_{5p_1} and K_{6p_1} so that if $j \le k \le m$ then

$$||h_i(u_i)||_{p_1,Q(\tau,T)} \le K_{5p_1} + K_{6p_1}||H(u)||_{Q(\tau,T)}^{\delta_{p_1}} \tag{3.17}$$

is satisfied for $j \le i < k$ but not for $i = k$. Here we let $\tau = t_{1,l} < T \le t_{1,l+1}$. Our immediate objective will be to use H_4 to show via contradiction that an inequality of the form (3.17) must hold for all $j \le i \le m$.

Let $\tilde{\theta} \in L_p(Q(\tau,T))$ be such that $\tilde{\theta} \ge 0$ and $||\tilde{\theta}||_{p,Q(\tau,T)} = 1$ and suppose that ψ_k is the solution to the adjoint equation $(2.6a - c)$ with $i = k$. We integrate $h_i(u_i)$ for $1 \le i \le k$ against $\tilde{\theta}$ on the space time cylinder $Q(\tau,T)$ and obtain via integration by parts and the convexity of $h_i()$

$$\int_\tau^T \int_\Omega h_i(u_i)\tilde{\theta}dxdt \leq \int_\tau^T \int_\Omega h_i(u_i)(\mathcal{L}_i(\psi_k) - \mathcal{L}_i(\psi_k))dxdt$$
$$+ \int_\Omega h_i(u_i(x,\tau))\psi_k(x,\tau)dx \tag{3.18a}$$
$$+ \int_\tau^T \int_\Omega \psi_k \left[h_i(u_i) + h'(u_i)f_i(x,t,u)\right] dxdt.$$

We use H_4, to sum the integrals from $i = 1$ to k and obtain

$$\int_\tau^T \int_\Omega \sum_{i=1}^{k} a_{ki}h_i(u_i)\tilde{\theta}dxdt \leq \sum_{i=1}^{k-1} a_{ki} \int_\tau^T \int_\Omega h_i(u_i)(\mathcal{L}_i(\psi_k) - \mathcal{L}_k(\psi_k))dxdt$$
$$+ \sum_{i=1}^{k} a_{ki} \int_\Omega h_i(u_i(x,\tau))\psi_k(x,\tau)dx$$
$$+ \int_\tau^T \int_\Omega \psi_k \left[K_1(H(u))^r + K_2\right] dxdt.$$
$$+ \int_\tau^T \int_\Omega \psi_k \sum_{i=1}^{k} a_{ki}h_i(u_i)dxdt.$$
$$\tag{3.18b}$$

We proceed to estimate the right hand side of (3.18b) termwise. For the first term we have

$$\sum_{i=1}^{k-1} a_{ki} \int_\tau^T \int_\Omega h_i(u_i)(\mathcal{L}_i(\psi_k) - \mathcal{L}_k(\psi_k))dxdt$$
$$= \sum_{i=1}^{j-1} a_{ki} \int_\tau^T \int_\Omega h_i(u_i)(\mathcal{L}_i(\psi_k) - \mathcal{L}_k(\psi_k))dxdt \tag{3.19}$$
$$= \sum_{i=j}^{k-1} a_{ki} \int_\tau^T \int_\Omega h_i(u_i)(\mathcal{L}_i(\psi_k) - \mathcal{L}_k(\psi_k))dxdt.$$

We now observe that

$$\left| \int_\tau^T \int_\Omega h_i(u_i)(\mathcal{L}_i(\psi_k) - \mathcal{L}_i(\psi_k))dxdt \right|$$
$$\leq \|h_i(u_i)\|_{p_1,Q(\tau,T)} \|\mathcal{L}_i(\psi_k) - \mathcal{L}_i(\psi_k)\|_{p,Q(\tau,T)} \tag{3.20}$$

From Lemma 2.5 we are guaranteed the existence of a constant $\tilde{C}_{p,T-\tau}$ so that

$$\|\mathcal{L}_i(\psi_k) - \mathcal{L}_i(\psi_k)\|_{p,Q(\tau,T)} \leq \tilde{C}_{p,T-\tau} \tag{3.21}$$

Therefore, we may obtain

$$\sum_{i=1}^{k-1} a_{ki} \int_{\tau}^{T} \int_{\Omega} h_i(u_i)(\mathcal{L}_i(\psi_k) - \mathcal{L}_i(\psi_k))dxdt$$
$$\leq \sum_{i=1}^{k-1} a_{ki}\tilde{C}_{p,T-\tau} \left[M_2 + g(3) + g(3)\|H(u)\|_{p_1,Q(\tau,T)}^{\delta_{p_1}}\right] \tag{3.22}$$

The succeeding terms are estimated in exactly same manner as in [14]. We therefore simply list the estimates and refer the reader to [14] for the details. We have:

$$\int_{\tau}^{T} \int_{\Omega} \psi_k \left[K_1(H(u))^r + K_2\right] dxdt \leq C_{p,(T-\tau)} \left[K_1(g(3))^r + K_2(3 \mid \Omega \mid)^{\frac{r}{a}}\right] \tag{3.23}$$

$$\int_{\Omega} \psi(x,\tau) h_i(u_i(x,\tau))dx \leq K_{p,a/a-1,(T-\tau)} M_1 \tag{3.24}$$

$$\sum_{i=1}^{k-1} a_{ki} \int_{\tau}^{T} \int_{\Omega} \psi_k(h_i(u_i))dxdt \leq$$
$$\sum_{i=1}^{k-1} a_{ki}\tilde{C}_{p,T-\tau} \left[M_2 + g(3) + g(3)\|H(u)\|_{p_1,Q(\tau,T)}^{\delta p_1}\right] \tag{3.25}$$

and

$$\int_{\tau}^{T} \int_{\Omega} a_{kk}\psi_k h_k(u_k)dxdt \leq \left[1 - e^{-(T-\tau)}\right] \|h_k(u_k)\|_{p_1,Q(\tau,T)} \tag{3.26}$$

Combining these estimates we obtain

$$\begin{aligned}\int_{\tau}^{T} \int_{\Omega} a_{kk}h_k(u_k)\tilde{\theta}dxdt \leq \sum_{i=1}^{k-1} a_{ki}(2\tilde{C}_{p,T-\tau}) & [M_2+ \\ & +g(3)\|H(u)\|_{p_1,Q(\tau,T)}^{\delta_1}\Big] + K_{p,a/a-1,(T-\tau)}M_1 \\ & + C_{p,T-\tau}\left[K_1(g(3))^r + K_2(3 \mid \Omega \mid)^{\frac{r}{a}}\right] \\ & + a_{kk}(1 - e^{-3})\|h_k(u_k)\|_{p_1,Q(T,\tau)}\end{aligned} \tag{3.27}$$

Therefore using (3.27) and viewing $h_k(u_k)$ as a linear function on $L_p(Q(\tau,T))$ we are guaranteed the existence of K_7, $K_8 > 0$ such that

$$\|h_k(u_k)\|_{p_1,Q(\tau,T)} \leq K_7 + K_8\|H(u)\|_{p_1,Q(\tau,T)}^{\delta_1} \tag{3.28}$$

for all $\tau = t_{1,l} < T \leq t_{1,l+1}$ independent of $l \geq 1$. Thus K_{5p_1} and K_{6p_1} may be chosen so that (3.17) holds for k. This leads to a contradiction and hence (3.17) holds for all $j \leq i \leq m$. Because $\|h_i(u_i)\|_{p_1,Q(\tau,T)} \leq M_2$ for $1 \leq i < j$ we have for $i \geq 1$

$$\|H(u)\|_{p_1,Q(t_{1,i},t_{1,i+1})} \le \epsilon^1 m(M_2 + g(3)) + (m(g(3))^{\epsilon^1} = \tilde{M}_2 \qquad (3.29)$$

where $\epsilon^1 = 1/(1-\delta_p)$. Recall that $p_1 \ge \frac{n+2}{n+\mu}a, \frac{n+2}{n+\mu} > 1$ and $1 < t_{1,i+1} - t_{1,i} < 3$. Therefore there exists a sequence $\{t_{2,1}\}_{i=1}^{\infty}$ such that $t_{i,2i-1} < t_{2,i} < t_{1,2i}$ for all $i \ge 1$ and $\|H(u(\cdot,t_{2,i}))\|_{p_1,\Omega} \le \tilde{M}_2$ for all $i \ge 1$. We note that $1 < t_{2,i+1} - t_{2,i} < 9$ for all $i \ge 1$ and that there exists $\tilde{\tilde{M}}_2 > 0$ such that $\|H(u)\|_{p_1,Q(t_{2,i},t_{2,i+1})} \le \tilde{\tilde{M}}_2$ for all $i \ge 1$. We now iterate the foregoing argument. We replace a by p_1, chose the same vaule of μ, a corresponding value of $p > 1$ and set $p_2 = p/(p-1) \ge \left(\left(\frac{n+2}{n+\mu}\right)^2\right) a$. Following the preceeding argument we find there exists $M_3 > 0$ such that $\|H(u)\|_{p_2,Q(t_{2,i},t_{2,i+1})} \le \tilde{M}_3$ for an $i \ge 1$

If we proceed inductively, the for $k \ge 2$, there exists $\tilde{M}_{k+1} > 0, p_k > \left(\left(\frac{n+2}{n+\mu}\right)^k\right) a$ and a sequence $\{t_{k,i}\}_{i=1}^{\infty}$ such that

(i) $t_{k,1} > 0$ and $1 < t_{k,i+1} - t_{k,i} < 3^k$for all $i \ge 1$
(ii) $\|H(u(,t_{k,i}))\|_{p_{k-1},\Omega} \le \tilde{M}_k$ for all $i \ge 1$
(iii) $\|H(u)\|_{p_k,Q(t_{k,i},t_{k,i+1})} \le \tilde{M}_{k+1}$ for all $i \ge 1$

Note that $\lim_{k\to\infty} p_k = \infty$ because $(n+2)/(n+\mu) > 1$. Hence if we can take k sufficiently large, we may apply Lemma 3.1 to guarantee the existence of $N > 0$ so that for $1 \le i \le m$, $\|u_i\|_{\infty,Q(0,\infty)} < N$.

The second case assumes that $1 \le r = a < (n+2)/n$. Then there exists $0 < \epsilon < 2$ such that $a < \frac{n+2}{n+\epsilon}$ and we can set $p = \frac{n+2}{2-\epsilon} > \frac{n+2}{2}$. Once again we let $\theta \epsilon L_p(Q(\tau,T))$ be such that $\theta \ge 0$ and $\|\theta\|_{p,Q(\tau,T)} = 1$. The second case is reduced to the first case via arguments totally analogous to those of [14, Theorem 2.5].

We also obtain a decay result.

Theorem 3.30 *Let the hypotheses of Theorem 3.18 hold and $K_2 = K_6 = 0$. If there exists $T > 0$ such that for all $p \ge 1$*

$$\lim_{\tau\to\infty} [\|H(u)\|_{a,Q(\tau,\tau+T)}] = 0$$

then $\lim_{t\to\infty} \|u_i(\cdot,t)\|_{\infty,\Omega} = 0$ for all $1 \le i \le m$.

Proof. One argues that there exists $M_1 \in C([0,\infty))$ and a sequence $\{t_{1,i}\}_{i=1}^{\infty}$ such that $\lim_{t\to\infty} M_1(t) = 0$, $t_1 > 0$, $1 < t_{1,1+i} - t_{1,i} < 3$ and $\|H(u(,t_{1,i}))\|_{a,\Omega} \le M_1(t_{1,i})$. Proceeding as in the proof Therom 3.16 with $K_2 = K_4 = 0$ one shows the existence of $M_2 \in C([0,\infty))$ such that $\lim_{t\to\infty} M_2(t) = 0$ and $\|H(u)\|_{p_1,Q(t_{1,i},t_{1,i+1})} \le M_2(t_{1,i})$ for $i \ge 1$ and $p_1 = \frac{n+2}{n+\mu}$ if $r < a$ and $p_1 = \frac{na+n+2}{2n}$ if $1 \le r = a < (n+2)/2$. One observes that $p_1 > r \ge 1$ and can proceed as in Theorem 3.16 and establish for $k \ge 2$ the existence of $M_{k+1} \in C([0,\infty))$, $p_k \ge a\left(\frac{n+2}{n+\mu}\right)^{k-1}$ and a sequence $\{t_{k,i}\}_{i=1}^{\infty}$ such that

(i) $\lim_{t\to\infty} M_{k+1}(t) = 0$,
(ii) $t_{k,1} > 0$ and $1 < t_{k,i+1} - t_{k,i} < 3^k$ for all $i \ge 1$,
(iii) $\|H(u(\cdot,t_{k,i}))\|_{p_{k-1},\Omega} \le M_k(t_{k-1,i})$ for all $i \ge 1$,

(iv) $\quad \|H(u)\|_{p_k,Q(t_{k,i},t_{k,i+1})} \leq \tilde{M}_{k+1}(t_{k-1,i})$ for all $i \geq 1$
We note that if $(n+2)/(n+\mu) > 1$ we can apply the second assertion of Lemma 3.2 to conclude that for each $1 \leq i \leq m$, $\lim_{t\to\infty} \|u_i(\cdot,t)\|_{\infty,\Omega} = 0$.

We conclude this section with a simple result wich guarantees the existence of L_1 cylinder bounds which satisfy the hypotheses of Theorem 3.16.

Proposition 3.31 *Suppose that $C_1 - C_5$ hold and that $H_1 - H_3$ hold. If for all $v \in \mathbf{R}_+^m$ $\Delta H(v)f(x,t,v) = \sum_{i=1}^m h_i'(v_i)f_i(x,t,v) \leq 0$ then for all $0 \leq \tau \leq T$ we have $\|H(u)\|_{1,Q(\tau,T)} \leq [\|H(u_0)\|_{1,\Omega}](T-\tau)$.*

Proof. We multiply each component by $h_i'(u_i)$ and obtain

$$h_i'(u_i)\nabla \cdot (a_i(x,t)\nabla u_i) + h_i'(u_i)f_i(x,t,u). \tag{3.32}$$

The convexity of $h_i()$ yields

$$\partial h_i(u)/\partial t \leq \Delta \cdot (a_i(x,t)\Delta u_i) + h_i'(u_i)f_i(x,t,u) \tag{3.33}$$

Integrating (3.33) on $Q(\tau,T)$ and summing we obtain

$$\|H(u(,t))\|_{1,\Omega} \leq \|H(u(,\tau))\|_{1,\Omega} \tag{3.34}$$

Consequently, for all $\tau \geq 0$

$$\|H(u(\cdot,\tau))\|_{1,\Omega} \leq \|H(u_0)\|_{1,\Omega} \tag{3.35}$$

If we integrate (3.20) on (τ,T) and use (3.21), we have

$$\|H(u)\|_{1,Q(\tau,T)} \leq \|H(u_0)\|_{1,\Omega}(T-\tau)$$

We return to example (1.2 a-e) of the first section. We have demonstrated the existence of a priori $L_1(Q(\tau,T))$ bounds for the functional $H(u) = \sum_{i=1}^5 h_i(u_i) = u_i + 2u_2 + u_3 + 2u_4 + u_5$ applied along the trajectories of the solutions. These bounds depend only upon $L_1(\Omega)$ norm of the intial data and upon $(T-\tau)$. A cursory examination of the vector field reveals the intermediate sums condition holds for matrix

$$A = \begin{bmatrix} 1 & 0 & 0 & 0 & 0 \\ \frac{1}{2} & \frac{1}{2} & 0 & 0 & 0 \\ 1 & 1 & 1 & 1 & 0 \\ \frac{1}{2} & \frac{1}{2} & \frac{1}{2} & \frac{1}{2} & 0 \\ 1 & 1 & 1 & 1 & 1 \end{bmatrix} \tag{3.36}$$

and exponent $r = 1$. The question of global existence of solutions is settled in [6]. Because $1 < (n+2)/n$, Theorem 3.16 may be invoked to produce uniform L_∞ bounds. We refer the reader to [14], [15], [5], [6] for further examples of vector fields and convex functionals $H()$ which satisfy our hypotheses. We remark that additional arguments can be constructed to guarantee the precompactness of trajectories.

Acknowledgments

W.E. Fitzgibbon's work was supported in part by NSF Grant #DMS8803151, ONR and by Grant #N0014-89-J-1011 Texas Advanced Research Program Grant #1100. J.J Morgan's work was supported in part by NSF Grant #DMS8813071. R.S. Sanders' work was supported in part by NSF Grant #DMS 8703383

Bibliography

1. Amann H. (1986): Quasilinear evolution equations and parabolic systems. Trans. Amer. Math. Soc., Providence, 191-227
2. Aronson, D.G. (1967): Bounds for the fundamental solution of a parabolic equation. Bulletin American Mathematical Society, **73**, 890-896
3. Aronson, D. G. (1968): Non-negative solutions of linear parabolic equations. Annali Scuola Norm. Sup. Pisa, **22**, 607-694
4. Farr, W., Fitzgibbon, W., Morgan, J., Waggoner, S.: Asymptotic convergence for a class of autocatalytic chemical reactions. Partial Differential Equations and Applications, Marcel Dekker, to appear
5. Fitzgibbon, W., Morgan, J., Waggoner, S. (1990): Generalized Lyapunov structure for a class of semilinear parabolic systems. JMAA, **152**, 109-130
6. Fitzgibbon, W., Morgan, J., Waggoner, S.: Weakly coupled semilinear parabolic evolution systems. Annali Mat. Pura Appl., to appear
7. Fitzgibbon, W., Morgan, J, Sanders, R.: Global existence and boundedness for a class of inhomogeneous semilinear parabolic systems. University of Houston, Technical Report UH/MD-87
8. Gray, P. and Scott, S. (1985): Sustained oscillations in a CSTR. J. Phys. Chem., **89**, 22
9. Henry, D., (1981): Geometric Theory of Semilinear Parabolic Equations, Lecture Notes in Mathematics **840**, Springer Verlag, Berlin-Heidelberg-New York
10. Hollis, S., (1986): Globally bounded solutions of reaction-diffusion systems. Dissertation, North Carolina State University
11. Hollis, S., Martin, R., Pierre, M. (1987): Global existence and boundedness in reaction-diffusion systems. SIAM J. Math. Anal., **18**, 744-761
12. Kanel, Y.I. (1984): Cauchy's problem for semilinear parabolic equations with balance conditions. Trans. Diff. Urav., **20**, No. 10, 1753-1760
13. Ladyzenskaja, O., Solonnikov, V., Uralceva, N., (1968): Linear and Quasilinear Equations of Parabolic Type. Translations of Mathematical Monograph **23**, American Mathematical Society, Providence
14. Morgan, J. (1990): Boundedness and decay results for reaction diffusion systems. SIAM J. Math Anal., to appear
15. Morgan, J. (1989): Global existence for semilinear parabolic systems. SIAM J. Math Anal., **20**, No.5, 1128-1144
16. Pazy, A. (1983): Semigroups of Linear Operations and Applications to Partial Differential Equations. Applied Mathematical Science, **44**, Springer Verlag, Berlin-Heidelberg-New York
17. Waggoner, S. (1988): Global existence for solutions of semilinear and quasilinear parabolic systems of partial differential equations. Dissertation, University of Houston

18. Haraux A., Youkana, A., (1988): On a Result of K. Masuda concerning reaction diffusion equations. Tohoku J. Math. **40**, 159-183
19. Singh, M., Khetarpal, K., Sharan (1980): A theoretical model for studying the rate of oxygenation of blood in pulmonary capillaries. J. Math Biology **9**, 305-330
20. Feng, W. (1988): Coupled systems of reaction-diffusion equations and applications. Dissertation, North Carolina State University

The Mountain Circle Theorem

G. Fournier[1] *and M. Willem*[2]

[1] University of Sherbrooke, Sherbrooke, Canada
[2] University of Louvain, L.L.N., Belgium

Introduction

In this paper, we present notions of Lusternik-Schnirelman relative category and some applications to differential equations. There are many notions of the relative category see Fadell [5], Fournier-Willem [6], and Szulkin [13]. In this work, we shall attempt to show that, provided one uses the strong relative category as the basic count of critical points and the (weak) relative category to calculate it, the relative category almost behaves like a degree or a fixed point index theory. We shall attempt to do so by proving the mountain circle theorem, a modification of the well known mountain pass theorem [1], but which gives the existence of one more critical point than the latter.

1 Lusternik-Schnirelman category

Let A be a closed subset of a topological space X.

The Lusternik-Schnirelman category of A in X, $cat_X(A)$, is the least integer n such that A can be covered by n closed subsets of X each of which is contractible in X [A is contractible in X if there exists a continuous h:$A \times I \to X$, where I=[0,1], such that h(x,0)=x $\forall x \in A$ and $\exists y \in X$ such that h(x,1)=y $\forall x \in A$].

The following properties are easy consequences of the definition.
(1.1) if $X \supset B \supset A$ then $cat_X(A) \leq cat_X(B)$,
(1.2) $cat_X(A \cup B) \leq cat_X(A) + cat_X(B)$,
(1.3) if A is closed and if h$\in$C$([0,1] \times A, X)$ is such that h(0,x)=x for every x$\in$A, then $cat_X(A) \leq cat_X(h_1(A))$.

Definition 1.4 (Palais-Smale condition; see [11]) A map $\phi : M \to \Re$, where M is a C^1 Finsler manifold and ϕ is C^1, satisfies P-S if every sequence $\{s_n\}$ of elements of M such that $\{\phi(s_n)\}$ is bounded and $||\phi'(s_n)|| \to 0$ as $n \to \infty$, has a convergent subsequence.

The following result is due to Palais [11].

Theorem 1.6 *Let M be a complete C^2 Finsler manifold and $\phi \in C^1(M, \Re)$ be a map satisfying the Palais-Smale condition. Then if ϕ is bounded from below, ϕ has at least $cat(M)$ critical points.*

2 Relative category

For the definitions of relative category given here see [6], for other definitions see [5] [13].

Definition 2.1 Let X be a topological space and Y a closed subset of X. A closed subset A of X is of the n-th category relative to Y (we write $cat_{X,Y}(A) = n$) if and only if n is the least positive integer such that

$$A = \bigcup_{i=0}^{n} A_i$$

and

$$(1) \quad \forall i \geq 1 \;\; A_i \text{ is contractible in } X,$$

$$(2) \quad A_0 \text{ is strongly deformable into } Y \text{ in } X.$$

[A is strongly deformable into Y in X if there exists h:$A \times I \to X$, such that h(x,0)=x and $h(x,1) \in Y \quad \forall x \in A$ and $h(x,t) = x \quad \forall x \in A \cap Y$].

We say that A is of the n-th strong category relative to Y (we write $Cat_{X,Y}(A) = n$) if and only if n is the least positive integer such that

$$A = \bigcup_{i=0}^{n} A_i$$

and

$$(1) \quad \forall i \geq 1 \;\; A_i \text{ is contractible in } X \backslash Y,$$

$$(2) \quad A_0 \text{ is touch and stop deformable into } Y \text{ in } X.$$

[A is touch and stop deformable into Y in X if there exists h:$A \times I \to X$, such that h(x,0)=x and $h(x,1) \in Y \quad \forall x \in A$ and if $h(x,t) = y \in Y$ implies $h(x,s) = y \; \forall s \geq t$].

Remark 2.2.
(1) We have that $cat_{X,Y}(A) \leq Cat_X(A)$.
(2) If $Y = \emptyset$ then $A_0 = \emptyset$ and $cat_{X,\emptyset}(A) = Cat_X(A)$.

Definition 2.3 Let Z, Z' be subsets of X. Then $Z <_Y Z'$ if and only if there exists h:$Z \times I \to X$ such that
(1) $h_0 = i_Z : Z \to X$ is the inclusion
(2) $Z' \supset h_1(Z)$

(3) $h(x,t) = x \ \forall x \in Y, \ \forall t \in I$.

We have the following properties most of which are generalizations of properties of the category itself.

Proposition 2.4 *Let* A, B, Y *be closed subsets of* X.

(i) *if* $B \supset A$ *then* $cat_{X,Y}(A) \leq cat_{X,Y}(B)$

(ii) $A <_Y B$ *implies* $cat_{X,Y}(A) \leq cat_{X,Y}(B)$

(iii) $A <_Y B$ *and* $B <_Y A$ *imply* $cat_{X,Y}(A) = cat_{X,Y}(B)$

(iv) $cat_{X,Y}(A \cup B) \leq cat_{X,Y}(A) + cat_X(B)$

(v) $cat_{X,Y}(X) \geq cat_X(X) - cat_Y(Y)$

(vi) $cat_{X,Y}(A) = 0 \Leftrightarrow A <_Y Y$.

Furthermore, in each of the above properties we may replace cat by Cat, provided we replace $<_Y$ *by the obvious corresponding relation* $\ll_Y$ *and that in (iv) we replace* $cat_X(B)$ *by* $cat_{X \setminus Y}(B)$ *with the added condition that* $X \setminus Y \supset B$.

Proposition 2.5 *Let* $X' \supset X \supset A$ *and* $X' \supset Y' \supset Y$ *and* $X' \supset A' \supset A$ *and* $X \supset Y$. *Then,*

(a) $cat_{X',Y}(A) \leq cat_{X,Y}(A)$,

(b) $cat_{X,Y'}(A) \geq cat_{X,Y}(A)$, *provided that* $cat_{X,Y}(Y') = 0$,

(c) $cat_{X',Y'}(A') \geq cat_{X,Y}(A)$, *provided that there is a retraction* $r: X' \rightarrow X$ *(i.e.* $r(x) = x \ \forall x \in X$*) such that* $A' \supset A$, *and* $r^{-1}(Y) = Y'$,

(d) $cat_{X',Y'}(A') \leq cat_{X,Y}(A)$ *if* $A' \setminus A = Y' \setminus Y = X' \setminus X$ *and* X *is closed in* X'.

Furthermore, in each of the above properties, we may replace cat by Cat.

In the following, we may not replace cat by Cat.

Proposition 2.6 *Let* $X' \supset X \supset A$, $X \supset Y$ *and* $X' \supset Y' \supset Y$. *If there exists a retraction* $r: X' \rightarrow X$ *such that* $r(Y') <_Y Y$ *in* X, *then* $cat_{X',Y'}(A') \geq cat_{X,Y}(A)$ *provided that* $A \subset A'$.

In the following, we may not replace Cat by cat.

Proposition 2.7 (Excision) *Let* $X \supset A$ *and* $X \supset Y$, *then*

$$Cat_{X,Y}(A) = Cat_{X \setminus V, Y \setminus V}(A \setminus V) = Cat_{X \cap F, Y \cap F}(A \cap F),$$

for any V *contained in the interior of* Y *and any closed* F *such that* $F \cup Y = X$.

The next proposition is evident but is useful for applications; in it Cat can be replaced by cat.

Proposition 2.8 *Let* $\{X_j\}_{j \in J}$ *be a finite set of disjoint closed non-empty subsets whose union is* X, *then*

$$Cat_{X,Y}(A) = \sum_{j \in J} Cat_{X_j, Y \cap X_j}(A \cap X_j).$$

3 Connection with the cup length

For the results presented here see [7]; see also [5,3,13]. In the following Y is a closed subset of an ANR X and we use the singular cohomology over the real field.

Definition 3.1 cuplength(X,Y)=n iff n is the maximum number such that there are $\alpha_0 \in H^k(X,Y)$, with $k \geq 0$, and $\alpha_m \in H^k(X)$, with $k > 0$, for $m = 1, \ldots, n$, such that: $\alpha_0 \cup \alpha_1 \cup \ldots \cup \alpha_n \neq 0$. If such an α_0 does not exist put cuplength(X,Y)$=-\infty$. [cuplength X=cuplength (X,$\emptyset$).]

Theorem 3.2 *If X is an ANR and $Y \subset X$ is closed non empty, then*

$$cat_{X,Y}(X) \geq 1 + cuplength(X,Y).$$

Theorem 3.3 *If Y is closed in X and $H^*(X,Y)$ or $H^*(Z)$ is of infinite type, then $cuplength(X \times Z, Y \times Z) \geq cuplength(X,Y) + cuplength(Z)$.*

Corollary 3.4 *If T is a topological space then $cat_{T \times B_{n+1}, T \times S_n}(T \times B_{n+1}) \geq cuplength(T) + 1$.*

4 Application to critical point theory

Let M be a complete C^2 Finsler manifold i.e. a C^2 Banach manifold with a Finsler structure on its tangent bundle (Important examples are complete Riemannian manifolds and Banach spaces). Let $\phi \in C^1(M, \Re)$. Set

$$\phi^c = \{u \in M | \phi(u) \leq c\}$$
$$K_c = \{u \in M | \phi(u) = c,\ d\phi(u) = 0\}.$$

We shall use the following variation of the deformation lemma due to Clark [4] for Banach spaces and to Ni [10] for Finsler manifolds.

Lemma 4.1 *If $\phi \in C^1(M, \Re)$ satisfies the Palais-Smale condition and if U is an open neighbourhood of K_c, then, for every $e' > 0$ there exists $e \in]0, e'[$ and a map $f{:}M \to M$ isotopic to id_M such that for all $d \in [0, e]$, $\phi^{c-d} \supset f(\phi^{c+e} \backslash U)$.*

For the following see [6], see also [5] [13].

Theorem 4.2 *If $\phi \in C^1(M, \Re)$ satisfies the Palais-Smale condition and if $-\infty < a < b < +\infty$ and $K_a = K_b = \emptyset$, then*

$$\#\{x \in \phi^{-1}([a,b]) | d\phi = 0\} \geq Cat_{\phi^b, \phi^a}(\phi^b) \geq Cat_{M,\phi^a}(\phi^b).$$

5 The mountain pass theorem

These results can be proved using the relative category [8], they can also be proved using the Minimax principle (Palais [12]). Note also that the homological approach to the mountain pass theorem was pointed out in Tian [14] and in Chang [2].

Theorem 5.1 *Let $\phi \in C^1(M, R)$ satisfy the Palais-Smale condition on $\phi^{-1}[a, b]$, when M is a complete C^2 Finsler manifold, with $-\infty < a < b < +\infty$. If ϕ^a is disconnected and ϕ^b is a connected set, then ϕ has at least one critical point in $\phi^b \backslash \phi^a$.*

Theorem 5.2 (The Ambrosetti-Rabinowitz mountain pass theorem [1]) *Let $\phi \in C^1(B, R)$ satisfy the Palais-Smale condition on $\phi^{-1}[a, \infty)$, where B is a Banach space and $a \in R$. If there exist $x, y \in B$ and $r \in R$ such that $||x - y|| > r$ and, $\phi(z) > b$ for all z with $||x - z|| = r$, for some $b > a = \max\{\phi(x), \phi(y)\}$, then ϕ has a critical point in $B \backslash \phi^a$.*

6 The mountain circle theorem

The following is similar to the mountain pass theorem [1], but gives the existence of two critical points instead of one. The method of proof is to first apply the excision property (to augment cat but not Cat) and secondly to obtain a better lower bound for Cat by calculating the new cat, using the results of Sect. 3. Unfortuately, the method seems to use, in an essential way, the finite dimensionality of the circle or sphere of its hypotheses.

Theorem 6.1 (The mountain circle theorem) *Let $\phi \in C^1(B \times R^n, R)$ satisfy the Palais-Smale condition on $\phi^{-1}[a, \infty)$, where B is a Banach space and $a, b \in R$. If there exists $x \in R^n$ such that, for some $r > s > t \geq 0$, $\phi(u, y) > b > a \geq \phi(u, z)$ for all $u \in B$ and all $||x - y|| = s$ and $||x - z|| = r, t$, then ϕ has two critical points in $(B \times R^n) \backslash \phi^a$.*

Proof. Choose $b > c > a$. By (4.2) we need only to show that the strong category of $(B \times R^n)$ relative to ϕ^c, is greater than or equal to 2. By excision twice and by (2.8), if F is the complement of the interior of ϕ^c and if $Z = F \cap B \times A$, then, since $(B \times \partial A) \cap F = \emptyset$,

$$Cat_{B\times R^n,\phi^c}(B\times R^n) = Cat_{F,F\cap\phi^c}(F) \geq Cat_{Z,Z\cap\phi^c}(Z)$$
$$= Cat_{B\times A,\phi^c\cap(B\times A)}(B\times A) \geq cat_{B\times A,\phi^c\cap(B\times A)}(B\times A),$$

where $A = \{x \in R^n | r \geq ||x|| \geq t\}$.

By (2.5b), if ∂A is the boundary of A in R^n, we get that,

$$cat_{B\times A,\phi^c\cap(B\times A)}(B\times A) \geq cat_{B\times A,B\times\partial A}(B\times A).$$

By (3.2),

$$cat_{B\times A,B\times\partial A}(B\times A) \geq 1 + cuplength(B\times A, B\times\partial A).$$

Finally, by (3.3) we get that

$$1 + cuplength(B\times A, B\times\partial A) \geq 1 + cuplength(B) + cuplength(A,\partial A) \geq 1+0+1 = 2.$$

That is the conclusion. □

Theorem 6.2 (The mountain circle theorem) *Let $\phi \in C^1(B\times R^n, R)$ satisfy the Palais-Smale condition on $\phi^{-1}[a,\infty)$, where B is a Banach space and $a,b \in R$. If there exist $x \in R^n$ and $r,s,t : B \to R$ continuous and such that $r(u) > s(u) > t(u) \geq 0$ for all $u \in B$, and $\phi(u,y) > b > a \geq \phi(u,z)$ for all $||x-y|| = s(u)$ and $||x-z|| = r(u), t(u)$, then ϕ has two critical points in $(B\times R^n)\backslash\phi^a$.*

Proof. This theorem is a corollary of the proof of the preceding theorem. In fact it is sufficient to notice that the relative categories are invariant under homeomorphism preserving the subspaces (the two whose categories are calculated and the two relative to which the categories are calculated). And to notice that obviously such a homeomorphism exists. □

7 Example

Consider the following system of equations

$$I \begin{cases} x'' = -ax + by + Ax^{1/3} + F_x(x,y) \\ y'' = bx - cy + By^{1/3} + F_y(x,y) \\ x(0) = x(T),\ y(0) = y(T),\ x'(0) = x'(T),\ y'(0) = y'(T) \end{cases}$$

where $a, c, A, B > 0$ and F is continuously differentiable.

For simplicity let us assume that F_x and F_y are continuous and bounded by M and that F is bounded by N. Then the solutions of I are the critical points of ϕ, where

$$\phi(x,y) = \frac{1}{T}\int_0^T \left[\frac{1}{2}(x'^2 + y'^2 - ax^2 - cy^2 + 2bxy) + \frac{3Ax^{4/3} + 3By^{4/3}}{4} + F(x,y)\right] dt$$

Here ϕ is defined on the space $H = H_T^1 \times H_T^1$ where

$$H_T^1 = \{y : [0,T] \to R | y' \in L_2,\ y(0) = y(T)\}$$

is given by the scalar product

$$\ll x, y \gg = \frac{1}{T}\int_0^T (xy + x'y')dt.$$

Let us define $\bar{x} = \frac{1}{T}\int_0^T x\,dt$. Then $x = \bar{x} + \tilde{x}$ where $\bar{\tilde{x}} = 0$. Thus we can write

$$\begin{aligned}\phi(x,y) = \frac{1}{2T}\int_0^T \left[x'^2 + y'^2 - a\tilde{x}^2 - c\tilde{y}^2 + 2b\tilde{x}\tilde{y}\right] dt + \frac{1}{2}\left[2b\bar{x}\bar{y} - a\bar{x}^2 - c\bar{y}^2\right] \\ + \frac{3}{4T}\int_0^T \left[Ax^{4/3} + By^{4/3}\right] dt + \frac{1}{T}\int_0^T F(x,y)\,dt \quad (7.1) \\ = \alpha(\tilde{x},\tilde{y}) + \beta(\bar{x},\bar{y}) + \gamma(x,y) + \delta(x,y) \quad \text{in short.}\end{aligned}$$

Proposition 7.1. *If $(a+c)T^2 < 4\pi^2$ and $b^2 < ac$ then*
a) $\alpha(\tilde{x},\tilde{y}) \geq 0\ \forall (x,y) \in H$
b) ϕ satisfies tha P-S condition on H.

Proof.
a) By the Wirtinger inequality ($\omega\|\tilde{x}\|_2 \leq \|x'\|_2$ where $\omega = 2\pi T^{-1}$), $\alpha(\tilde{x},\tilde{y})$ dominates

$$\frac{1}{2T}\int_0^T \left[(1-s)(x'^2 + y'^2) + ((s\omega^2 - a)\tilde{x}^2 + 2b\tilde{x}\tilde{y} + (s\omega^2 - c)\tilde{y}^2)\right] dt \quad (7.2)$$

for any $s < 1$. Our conditions imply that $a + c < \omega^2$ and $b^2 < (\omega^2 - a)(\omega^2 - c)$ so we may choose $0 < s < 1$ close enough to 1 such that $a + c < s\omega^2$ and $b^2 < (\omega^2 s - a)(\omega^2 s - c)$. Thus (7.2) can be split into squares and so it dominates

$$\frac{1-s}{2T}\int_0^T (x'^2 + y'^2)\,dt \geq 0. \quad (7.3)$$

b) Assume that $S_n = (x_n, y_n) \in H$ where $f(S_n)$ is bounded and $\|\nabla\phi(S_n)\| \to 0$.

i) Let us show that $\bar{x}_n$ and $\bar{y}_n$, $\tilde{x}_n$, $\tilde{y}_n$, $\dot{x}_n$, $\dot{y}_n$ are bounded.
In fact if we denote $< x, u >= \frac{1}{T}\int_0^T xu\,dt$ and $\|x\|_2^2 =< x, x >$ we have that

$$\begin{aligned}\ll \nabla\phi(x,y),(u,v) \gg =< \dot{x},\dot{u} > + < \dot{y},\dot{v} > + < -ax + by + Ax^{1/3} \\ + F_x(x,y), u > + < -cy + bx + By^{1/3} + F_y(x,y), v > . \quad (7.4)\end{aligned}$$

We may assume that $|\phi(x_n,y_n)| \leq K$ and $|\nabla\phi(x_n,y_n)| \leq K\ \ \forall n \geq 0$, for some $K > 0$.

α) Let us put $(x,y) = (x_n,y_n)$ and $(u,v) = (\bar{x}_n,\bar{y}_n)$ in (7.4). We get

$$K(|\bar{x}_n| + |\bar{y}_n|) \geq | - a\bar{x}_n^2 + b\bar{y}_n\bar{x}_n + A\bar{x}_n \ < x^{1/3}, 1 > + \bar{x}_n \ < F_x, 1 > \\ - c\bar{y}_n^2 + b\bar{y}_n\bar{x}_n + B\bar{y}_n \ < y^{1/3}, 1 > + \bar{y}_n \ < F_y, 1 > |,$$

and since

$$| < x^{1/3}, y > | = |\frac{1}{T}\int_0^T yx^{1/3}dt| \\ \leq \frac{1}{T}\Big[\int_0^T |x^{1/3}|^2 dt\Big]^{1/2}\Big[\int_0^T y^2\Big]^{1/2} \leq \Big(\frac{1}{T}\int_0^T x^{2/3}dt\Big)^{1/2} ||y||_2 \\ \leq \Big(\frac{1}{T}\Big(\int_0^T (x^{2/3})^3\Big)^{1/3}\Big(\int_0^T 1^{3/2}\Big)^{2/3}\Big)^{1/2} ||y||_2 \leq ||x||_2^{1/3}||y||_2$$

we obtain that

$$a\bar{x}_n^2 - 2b|\bar{x}_n|\,|\bar{y}_n| + c\bar{y}_n^2 \leq (K+M)(|\bar{x}_n| + |\bar{y}_n|) + \\ A|\bar{x}_n|\,||x_n||_2^{1/3} + B|\bar{y}_n|\,||y_n||_2^{1/3}. \quad (7.5)$$

β) Let us put $(x,y) = (x_n, y_n)$ and $(u,v) = (\tilde{x}_n, \tilde{y}_n)$ in (7.4). We get

$$K\big(||\tilde{x}_n||_2 + ||\tilde{y}_n||_2 + ||\dot{x}_n||_2 + ||\dot{y}_n||_2\big) \\ \geq \Big| ||\dot{x}_n||_2^2 + ||\dot{y}_n||_2^2 - a||\tilde{x}_n||_2^2 + \frac{2b}{T}\int_0^T \tilde{x}_n\tilde{y}_n\,dt - c||\tilde{y}_n||_2^2 \\ + A < x_n^{1/3}, \tilde{x}_n > + B < y_n^{1/3}, \tilde{y}_n > + < F_x, \tilde{x}_n > + < F_y, \tilde{y}_n > \Big|.$$

Since

$$\Big|\frac{1}{T}\int_0^T \tilde{x}\,F_x\Big| \leq \frac{1}{T}\Big(\int_0^T \tilde{x}^2\Big)^{1/2}\Big(\int_0^T (F_x)^2\Big)^{1/2} = ||\tilde{x}||_2 M$$

we get

$$K\big(||\tilde{x}_n||_2 + ||\tilde{y}_n||_2 + ||\dot{x}_n||_2 + ||\dot{y}_n||_2\big) + A||x_n||_2^{1/3}||\tilde{x}_n||_2 \\ + B||y_n||_2^{1/3}||\tilde{y}_n||_2 + M\big(||\tilde{x}_n||_2 + ||\tilde{y}_n||_2\big) \geq ||\dot{x}_n||_2^2 + ||\dot{y}_n||_2^2 \\ - a||\tilde{x}_n||_2^2 - c||\tilde{y}_n||_2^2 + \frac{2b}{T}\int_0^T \tilde{x}_n\tilde{y}_n\,dt \quad (7.6)$$

γ) Let us put together (7.5) and (7.6). We obtain

$$(K+M)(|\bar{x}_n| + |\bar{y}_n|) + A||x_n||_2^{1/3}(|\bar{x}_n| + ||\tilde{x}_n||_2) \\ + B||y_n||_2^{1/3}(|\bar{y}_n| + ||\tilde{y}_n||_2) + (M+K)(||\tilde{x}_n||_2 + ||\tilde{y}_n||_2) \\ + K(||\dot{x}_n||_2 + ||\dot{y}_n||_2) \quad (7.7) \\ \geq a\bar{x}_n^2 - 2b|\bar{x}_n|\,|\bar{y}_n| + c\bar{y}_n^2 + ||\dot{x}_n||_2^2 + ||\dot{y}_n||_2^2 \\ -a||\tilde{x}_n||_2^2 - c||\tilde{y}_n||_2^2 + \frac{2b}{T}\int_0^T \tilde{x}\tilde{y}\,dt \\ \geq a(1-t)\bar{x}_n^2 + c(1-t)\bar{y}_n^2 + (1-s)(||\dot{x}_n||_2^2 + ||\dot{y}_n||_2^2) \\ \geq a(1-t)\bar{x}_n^2 + c(1-t)\bar{y}_n^2 + (1-s)\omega^2(||\tilde{x}_n||_2^2 + ||\tilde{y}_n||_2^2)$$

(where s is as in a), $0 < t < 1$, and $act^2 = b^2$. This implies $at\bar{x}^2 \pm 2b\bar{x}\bar{y} + ct\bar{y}^2 = (\sqrt{at}\bar{x} \pm \sqrt{ct}\bar{y})^2 \geq 0$.

That is there exists positive constants, K_1, K_2, K_3 such that

$$K_1(X_n)^{4/3}+K_3 \geq K_2(X_n)^2\ ,\ X_n = max\{||\tilde{x}_n||_2\, , ||\tilde{y}_n||_2\, , |\bar{x}|\, , |\bar{y}|\, , ||\dot{x}_n||_2\, , ||\dot{y}_n||_2\}.$$

Thus X_n is bounded, i.e. $||\tilde{x}_n||_2\, , ||\tilde{y}_n||_2\, , |\bar{x}|\, , |\bar{y}|\, , ||\dot{x}_n||_2\, , ||\dot{y}_n||_2$ are bounded.

ii)Let us prove that $(x_{nj}, y_{nj}) \to (x_0, y_0)$ for some $(x_0, y_0) \in H$ and some subsequence n_j. By passing to a subsequence, we may assume that $(x_n, y_n) \to (x_0, y_0)$ weakly in H for some $(x_0, y_0) \in H$. So $(x_n, y_n) \to (x_0, y_0)$ in $C[0,T]$ thus strongly in L_2. It remains to show that $||x_n' - x_0'||_2 \to 0$ and $||y_n' - y_0'||_2 \to 0$. In fact

$$\begin{aligned}&\ll \nabla\phi(x_n, y_n), (x_n - x_0, y_n - y_0) \gg - \ll \nabla\phi(x_0, y_0), (x_n - x_0, y_n - y_0) \gg \\ &\quad =< x_n' - x_0', x_n' - x_0' > + < y_n' - y_0', y_n' - y_0' > \\ &+ < Ax_n^{1/3} - Ax_0^{1/3} - a(x_n - x_0) + b(y_n - y_0) + F_x(x_n, y_n) - F_x(x_0, y_0), x_n - x_0 > \\ &+ < By_n^{1/3} - By_0^{1/3} - c(y_n - y_0) + b(x_n - x_0) + F_y(x_n, y_n) - F_y(x_0, y_0), y_n - y_0 >\end{aligned}$$

Now since $\nabla\phi(x_n, y_n) \to 0$ strongly and $(x_n - x_0, y_n - y_0) \to 0$ weakly in H_T' and so is bounded, the first term goes to 0; evidently the second one also goes to zero. The last two terms of the right side also go to zero since for those the convergences are strong. We are left with

$$||x_n' - x_0'||_2^2 + ||y_n' - y_0'||_2^2 \to 0 \text{ as } n \to \infty. \qquad \square$$

Proposition 7.2 *If the conditions of proposition 7.1 are satisfied, then*

a) $\phi(0,0) \leq N$

b) $\phi(x,y) \geq \frac{3}{4}(k-\epsilon)^{4/3}D - k^2(a+c) - N = S_\epsilon$ *if* $\bar{x}^2 + \bar{y}^2 = (k-t)^2$ *for all* $0 \leq t \leq \epsilon$ *where* $D = min\{A, B\}$

c) $\phi(x,y) \leq f(\tilde{x},\tilde{y}) + g(\bar{x},\bar{y})$, *where* $g(\bar{x},\bar{y}) = \beta(\bar{x},\bar{y}) + \frac{3}{4}(A|\bar{x}|^{4/3} + B|\bar{y}|^{4/3}) + N$, *and* f *is continuous.*

Proof.

a) $\phi(0,0) = \frac{1}{T}\int_0^T F(0,0)\,dt \leq N$

b) $\phi(x,y) = \alpha(\tilde{x},\tilde{y}) + \rho(\bar{x},\bar{y}) + \gamma(x,y) + \delta(x,y) \geq 0 + \beta(\bar{x},\bar{y}) + \alpha(x,y) - N$ but $\beta(\bar{x},\bar{y}) = \frac{1}{2}[2b\bar{x}\bar{y} - a\bar{x}^2 - c\bar{y}^2] \geq -a\bar{x}^2 - c\bar{y}^2 \geq -k^2[a+c]$ since $|\bar{x}| \leq k$ and $|\bar{y}| \leq k$, and

$$\begin{aligned}\gamma(x,y) &= \frac{3}{4T}\int_0^T [Ax^{4/3} + By^{4/3}]dt \\ &\geq \frac{3}{4}[A\bar{x}^{4/3} + B\bar{y}^{4/3}] \geq \frac{3D}{4}(\bar{x}^2 + \bar{y}^2)^{2/3} \geq \frac{3D}{4}(k-\epsilon)^{4/3}\end{aligned}$$

since for any positive numbers x and y we have that $x^{2/3} + y^{2/3} \geq (x+y)^{2/3}$ and since

$$\pm\bar{x} = \frac{\pm 1}{T}\int_0^T x\,dt \leq \frac{1}{T}\left(\int_0^T (\pm x)^{4/3}\right)^{3/4}\left(\int_0^T 1^4\right)^{1/4} = \left(\frac{1}{T}\int_0^T x^{4/3}\right)^{3/4}$$

which implies that $\frac{1}{T}\int_0^T x^{4/3} \geq |\bar{x}|^{4/3}$.

c) Let

$$f(\tilde{x},\tilde{y}) = \alpha(\tilde{x},\tilde{y}) + \frac{3}{4}[A||\tilde{x}||_2^{4/3} + B||\tilde{y}||_2^{4/3}]$$
$$g(\bar{x},\bar{y}) = \beta(\bar{x},\bar{y}) + \frac{3}{4}(A\bar{x}^{4/3} + B\bar{y}^{4/3}) + N$$

since $\gamma(x,y) \leq N$ and

$$\frac{1}{T}\int_0^T x^{3/4}dt \leq \frac{1}{T}\Big(\int_0^T (x^{3/4})^{3/2}dt\Big)^{2/3}\Big(\int_0^T 1^3\Big)^{1/3}$$
$$\leq \Big(\frac{1}{T}\int_0^T x^2 dt\Big)^{2/3} = (||\tilde{x}||_2^2 + \bar{x}^2)^{2/3} \leq ||\tilde{x}||_2^{4/3} + \bar{x}^{4/3}. \quad \square$$

Proposition 7.3 *If in addition to the conditions of (7.1) we have that* $\frac{3}{4}D - 2N > a + c$, *and if* $S = S_0$ *is given by (7.2b) with* $\epsilon = 0$, *then, for some* $k > 1$, ϕ *has a critical point in*

$$V = \{(x,y)|\phi(x,y) < \frac{S+N}{2} \text{ and } R^2 = \bar{x}^2 + \bar{y}^2 < k^2\}.$$

Proof.

By 7.2 and our condition, for any $k > 1$ close enough to 1 we have that $\frac{3}{4}k^{4/3}D > 2N + k^2(a+c)$ and so $N < S$ which gives $\phi(0,0) \leq N < \phi(x,y)$ for all $(x,y) \in \partial V$.

But if $(x,y) \in V$ then $|\bar{x}^2| + |\bar{y}^2| < k^2$ so

$$\phi(x,y) = \alpha(\tilde{x},\tilde{y}) + \beta(\bar{x},\bar{y}) + \gamma(x,y) + \delta(x.y) \geq 0 - \frac{k^2}{2}[2|b| + a + c] + 0 - N$$

So ϕ is bounded below on V which is a sub-manifold of H. If ϕ satisfies P-S in V, since V is not empty, $cat(V) > 0$ and, by (1.6), ϕ has a critical point in V. If ϕ does not satisfy P-S in V (and this for any k close enough to 1), it does satisfy the P-S in H so there must exist a sequence on which ϕ is bounded with vanishing gradient and converging to a point in $H\backslash V$. This point must then be a critical point of ϕ contained in ∂V, that is contained in the V corresponding to any k bigger than the one previously considered. In any case we may assume that V contains a critical point of ϕ. $\square$

Remark 7.4. Using the mountain pass theorem, one finds another critical point of critical value $\geq s$. But we would like to find one more critical point.

Lemma 7.5 *If in addition to the conditions of 7.3 we have that and* $\omega > (a + c + 2M) < D$ *then if we denote* $R^2 = \bar{x}^2 + \bar{y}^2$, *there exists* $k > 1$ *and* $\epsilon > 0$ *and* $\Gamma : H = H_T^1 \times H_T^1 \to \Re$ *a* C^1 *functional such that:*

a) Γ *satisfies P-S on* $R \geq k - 2\epsilon$

b) ϕ, Γ *have no critical points on* $k - 2\epsilon \leq R \leq k$

c) $\Gamma(x,y) = \phi(x,y)$ *if* $R \geq k$
d) if $R \leq k - 2\epsilon$ *then* $\Gamma(x,y) < 0$.

Proof. We may assume that there exist $k > 1$ satisfying $\frac{3}{4}k^{4/3}D > 2N + k^2(a+c)$ and such that $D > (a+c+2M)k^{2/3}$. Choose $\epsilon > 0$ such that $k - 2\epsilon > 1$ and

$$D(k-2\epsilon)^{4/3} > (a+c+2M)k^2 \text{ and } \phi(x,y) \geq S_{2\epsilon} > 0 \text{ if } k - 2\epsilon \leq R \leq k.$$

This is possible by (7.3) and (7.2b) since S_ϵ is a continuous function of ϵ and $S_0 > N$. Choose $L > N$, $(A+B)k^{4/3}$.

a) $\ll \nabla\phi(x,y), (x,y) \gg\, > 0$ if $(k-2\epsilon)^2 \leq \bar{x}^2 + \bar{y}^2 \leq k^2$ and ϕ has no critical points on $k - 2\epsilon \leq R \leq k$.

In fact,

$$\begin{aligned}
&\ll \nabla\phi(x,y), (x,y) \gg \\
&\quad = \frac{1}{T}\int_0^T \Big[x'^2 + y'^2 - ax^2 - cy^2 + 2bxy + Ax^{1/3}x + By^{1/3}y \\
&\qquad\qquad + F_x(x,y)x + F_y(x,y)y \Big]\, dt \\
&\quad \geq \omega^2\|\tilde{x}\|_2^2 + \omega^2\|\tilde{y}\|_2^2 - a\|\tilde{x}\|_2^2 - c\|\tilde{y}\|_2^2 \\
&\qquad - 2b\|\tilde{x}\|_2\|\tilde{y}\|_2 - 2M\|\tilde{x}\|_2^2 - 2M\|\tilde{y}\|_2^2 \\
&\qquad + A\bar{x}^{4/3} + B\bar{y}^{4/3} - a\bar{x}^2 - c\bar{y}^2 - 2b|\bar{x}|\,|\bar{y}| - 2M\bar{x}^2 - 2M\bar{y}^2
\end{aligned}$$

since

$$\begin{aligned}
\Big|\frac{1}{T}\int_0^T F_x(x,y)x\,dt\Big| &\leq \Big(\frac{1}{T}\int_0^T |F_x(x,y)|^2 dt\Big)^{1/2}\Big(\frac{1}{T}\int_0^T x^2 dt\Big)^{1/2} \\
&\leq M(\|\tilde{x}\|_2^2 + \bar{x}^2)^{1/2},
\end{aligned}$$

because $a^{1/2} + b^{1/2} \leq 2(a+b)$ provided that $a + b \geq 1$, $a, b \geq 0$ and $\bar{x}^2 + \bar{y}^2 \geq (k-2\epsilon)^2 > 1$.

So we get

$$\begin{aligned}
\ll \nabla\phi(x,y), (x,y) \gg\, &\geq (\omega^2 - a - c - 2M)\|\tilde{x}\|_2^2 + (\omega^2 - a - c - 2M)\|\tilde{y}\|_2^2 \\
&\quad + D(\bar{x}^{4/3} + \bar{y}^{4/3}) - (a+c+2M)(\bar{x}^2 + \bar{y}^2) \\
&\geq D(k-2\epsilon)^{4/3} - (a+c+2M)k^2 > 0,
\end{aligned}$$

by our choice of ϵ.

b) Define

$$\psi(x,y) = f(R)\Big[\gamma(x,y) - \gamma(\bar{x},\bar{y}) + L + \delta(x,y)\Big] + \alpha(\tilde{x},\tilde{y}) + \beta(\bar{x},\bar{y}) + \gamma(\bar{x},\bar{y}) - L$$

where $f \in C^\infty(\Re,\Re)$ is 0 on $(-\infty, k-\epsilon]$, is 1 on $[k,\infty)$ and $f'(t) > 0$ $\forall t \in (k-\epsilon, k)$. Then ψ satisfies P-S, has no critical points on $k - 2\epsilon \leq R \leq k$ and $\psi(x,y) = \phi(x,y)$ if $R > k$.

In fact, since

$$\begin{aligned}\ll \nabla\psi(x,y),(u,v) \gg &= < \dot{x},\dot{u} > + < \dot{y},\dot{v} > \\ &+ < -ax+by+f(R)(Ax^{1/3}+F_x(x,y)),u > \\ &+ < -cy+bx+f(R)(By^{1/3}+F_y(x,y)),v > \\ &+(1-f(R))[A < \bar{x}^{1/3},\bar{u} > +B < \bar{y}^{1/3},\bar{v} >] \\ &+\frac{1}{R}f'(R)[\gamma(x,y)-\gamma(\bar{x},\bar{y})+L+\delta(x,y)][< \bar{x},\bar{u} > + < \bar{y},\bar{v} >]\end{aligned}$$

For $R > 0$, if we give the proof that ϕ satisfies P-S (7.1b), the first two lines of the above equation would give the same bounds as before; the last would add only a term of the form $C_1 + C_2||\tilde{x}||_2^{4/3} + C_3\bar{x}^{4/3} + C_2||\tilde{y}||_2^{4/3} + C_3\bar{y}^{4/3}$, since the last line is zero if $R > k$. This would still permit us to obtain that $\bar{x}_n$, $\tilde{x}_n$, $\dot{x}_n$ are bounded (the same for y). In ii) the same proof, with the added terms, still gives that ψ satisfies P-S. As for the fact that $\ll \nabla\psi(x,y),(x,y) \gg > 0$ if $(k-2\epsilon)^2 \leq \bar{x}^2+\bar{y}^2 \leq k^2$ we need only repeat the proof of a) obtaining the same right member with the exception of the addition of the following positive term,

$$f'(R)R[\gamma(x,y)-\gamma(\bar{x},\bar{y})+L+\delta(x,y)].$$

Thus we get the conclusion of this section.

c) Define

$$\begin{aligned}\Gamma(x,y) = f(R)\Big[\gamma(x,y)-\gamma(\bar{x},\bar{y})+L+\delta(x,y)\Big] \\ + f(R+\epsilon)\alpha(\tilde{x},\tilde{y})+\beta(\bar{x},\bar{y})+\gamma(\bar{x},\bar{y})-L.\end{aligned}$$

Then $\Gamma(x,y) = \psi(x,y)$ if $R \geq k-\epsilon$ and Γ satisfies P-S in $R \geq k-2\epsilon$ and has no critical points on $k-2\epsilon \leq R \leq k$.

In fact if $k-2\epsilon \leq R \leq k-\epsilon$ we have that

$$\begin{aligned}\ll \nabla\Gamma(x,y),(\bar{x},\bar{y}) \gg &= f'(R+\epsilon)R\alpha(\tilde{x},\tilde{y})+2b\bar{y}\bar{x} \\ &\quad -a\bar{x}^2-c\bar{y}^2+A\bar{x}^{4/3}+B\bar{y}^{4/3} \\ &\geq DR^{4/3}-(a+c)R^2 \geq D(k-2\epsilon)^{4/3}-(a+c)k^2 > 0\end{aligned}$$

so

$$||\nabla\Gamma(x,y)|| \geq \frac{D(k-2\epsilon)^{4/3}-(a+c)k^2}{k} > 0$$

since $||(\bar{x},\bar{y})|| = R \leq k$.

Thus $\Gamma(x,y)$ satisfies P-S in $R \geq k-2\epsilon$ and has no critical points on $k > R \geq k-2\epsilon$.

d) If $R \leq k-2\epsilon$ then $\Gamma(x,y) < 0$.

In fact if $R \leq k-2\epsilon$ then

$$\begin{aligned}\Gamma(x,y) &= \beta(\bar{x},\bar{y})+\gamma(\bar{x},\bar{y})-L \\ &= \frac{3}{4}(A\bar{x}^{4/3}+B\bar{y}^{4/3})-L+\frac{1}{2}[2b\bar{x}\bar{y}-a\bar{x}^2-c\bar{y}^2] \\ &\leq \frac{3}{4}(A+B)k^{4/3}-L < 0,\end{aligned}$$

by our choice of L. □

Proposition 7.6 *Assume that the conditions of Lemma 7.5 are satisfied, then ϕ has two critical points in $R > k$.*

Proof. By (6.2) and our hypothesis, it is sufficient to prove that the hypotheses of (6.2) are satisfied for Γ with $a = 0$, $b = S$, $t = k - 2\epsilon$, $s = k$ and $r(u)$ to be determined later, since by (7.5b) and (7.5d) the critical points of Γ that we find must also be critical points of ϕ.

By (7.5a) and (7.5d), Γ satisfies P-S on $\Gamma^{-1}[a, \infty)$. Since t is a constant function, t is continuous. By (7.5d) it also satisfies its required condition. The same is true for s, by (7.5c) and (7.2b). As for r, by(7.5c) and (7.2c), since for $R \geq k$ we have that $\phi(x, y) = \Gamma(x, y)$ and

$$g(\bar{x}, \bar{y}) = \frac{1}{2}(2b\bar{x}\bar{y} - a\bar{x}^2 - b\bar{y}^2) + \frac{3}{4}(A\bar{x}^{4/3} + B\bar{y}^{4/3}) + N$$
$$\leq -K_1R^2 + K_2R^{4/3} + K_3 \to -\infty$$

as $R \to \infty$, it is clearly possible to choose r continuous and satisfying its condition. □

Corollary 7.7 *Assume that the conditions of Lemma 7.5 are satisfied, then ϕ has three critical points and so problem I has 3 solutions.*

Proof. Evident from (7.3) and (7.6). □

Bibliography

1. Ambrosetti, A., Rabinowitz, P.H. (1973): Dual variational methods in critical point theory and applications. J. Funct. Anal **14**, 349-381
2. Chang, K.-C. (1985): Infinite Dimensional Morse Theory and its Applications. S.M.S. les presses de l'Université de Montréal
3. Chang, K.-C., Long, Y., Zender, E.: Forced oscillations for the Triple Pendulum. preprint
4. Clark, D.C. (1972): A variant of the Lusternik-Schnirelman Theory. Indiana J. Math. **22**, 65-74
5. Fadell, E. (1986): Cohomological methods in non-free G-spaces with applications to general Borsuk-Ulam theorems and critical point theorems for invariant functionals. Nonlinear Functional Analysis and its Applications, S.P. Singh (ed.) D. Reidel Publishing Company, 1-45
6. Fournier, G., Willem, M. (1989): Multiple solutions of the forced double pendulum equation. Analyse Non Linéaire, Contribution in l'honneur de J.-J. Moreau, eds. H. Attouch, J.-P. Aubin, F. Clarke, I.Ekeland, CRM Gauthier-Villars, Paris, 259-281
7. Fournier, G., Willem, M.: Relative Category and the Calculus of Variations, preprint
8. Fournier, G., Willem, M. (1989): Simple variational methods for unbounded potentials. In Topological Fixed Point Theory and Applications, Boju Jiang Ed., Springer-Verlag, Berlin-Heidelberg-New York, 75-82

9. Mawhin, J., Willem, M. (1984): Multiple solutions of the periodic boundary value problem for some forced pendulum-type equations. J. Differential Equations **52**, 264-287
10. Ni, W.N. (1980): Some minimax principles and their applications in nonlinear elliptic equations. Journal d'Analyse Mathématiques **37**, 248-275
11. Palais, R. (1966): The Lusternik-Schnirelman theory on Banach manifolds. Topology **5**, 115-132
12. Palais, R. (1970): Critical point theory and the minimax principle. Global Analysis, Proc. Symp. Pure Math. **15** S.S. Chern (ed.), A.M.S., Providence, 185-202
13. Szulkin, A.: A relative category and applications to critical point theory for strongly indefinite functionals. Reports (1989-1) Dep of Math, U. of Stockholm, Sweden
14. Tian, G. (1984): On the mountain pass theorem. Kexue Tongbao **29**, 1150-1154

Extensions of an Algorithm for the Analysis of Nongeneric Hopf Bifurcations, with Applications to Delay-Difference Equations

Jeffery M. Franke[1] *and Harlan W. Stech*[2]

[1] McDonnell Douglas Corporation, St. Louis, Missouri 63082
[2] Department of Mathematics and Statistics, University of Minnesota, Duluth, Duluth, Minnesota 55812

Dedicated to Kenneth Cooke in Honor of his 65th *Birthday*

Abstract

A previously derived algorithm for the analysis of the Hopf bifurcation in functional differential equations is extended, allowing the elementary approximation of an existence and stability – determining scalar bifurcation function. With the assistance of the symbolic manipulation program MACSYMA [5], [9] this algorithm is used to implement the algorithm and to investigate the nature of nongeneric Hopf bifurcations in scalar delay – difference equations.

1 Introduction

The practical application of the now well – understood theory of Hopf bifurcations in functional differential equations still poses many significant computational issues. The thorough analysis of the bifurcation structures (including questions of stability and direction of bifurcation) for specific applications often requires a sizeable amount of computation. Even when a computer - assisted analysis is considered adequate, the selection of the appropriate technique is an important consideration.

Over the last 15 years, many techniques have been developed to treat such problems [10]. Among them, three have been most extensively discussed in the literature. Specifically, we refer to the method of averaging [3] [4], the use of the Poincaré normal form [8], and the method of Liapunov-Schmidt [13]. Of course, each of these methods must ultimately produce the same result when applied to a specific equation. However, the ease of application of each of these methods can vary significantly.

Our purpose in this paper is to report on the use of symbolic manipulation software in the implementation of the third of these methods. This method differs from the other two in that it does not require the approximation of the center manifold existing near criticality at the equilibrium point under consideration. Thus it appears to have an advantage when hand calculations are attempted and, as we shall see, lends itself to a computationally efficient symbolic implementation, as well.

The specific technique to be considered here was introduced in [13]. A generalized algorithm appeared in [14], and a FORTRAN - based implementation was developed in [1], [11], [2]. We consider here the use of symbolic - manipulation software in the extension of the algorithm of [14], and the application of this algorithm to a class of scalar delay - difference equations. The material presented on these two topics is based on the results of [6], where additionally a MACSYMA [5], [9] - based symbolic manipulation package (BIPACK) was designed for analyzing generic and third-order nongeneric scalar FDE.

In the section to follow, the specific class of functional differential equations under consideration, and the technical assumptions required will be presented. Theorems 2.2 and 2.3 represent extensions of the results in [13] to the case of fifth order nongeneric systems. The need for such results is illustrated in [12] where within the class of scalar integro-differential equations, elementary necessary and sufficient conditions are derived for third order degeneracy. A corollary addresses the important case of systems with odd nonlinearities. Section 3 is devoted to the application of these results to scalar delay - difference equations.

2 The bifurcation function

In this section, we begin by making assumptions which remain throughout this paper. We define $C = C([-1,0] : \mathbb{R}^n)$, $L(\alpha) : C \to \mathbb{R}^n$, and $H(\alpha) : C \to \mathbb{R}^n$ and consider the system of equations

$$\dot{y}(t) = L(\alpha)y_t + H(\alpha; y_t) \tag{1}$$

where L and H are continuous, and α is a parameter in some (Euclidean) space. For fixed α, we assume $H(\alpha;\psi)$ can be expressed in the following expansion

$$H(\alpha;\psi) = \sum_{j=2}^{7} H_j(\psi^j) + \mathcal{O}(\|\psi\|^8), \tag{2}$$

where the H_j's, $j = 2,\ldots,7$ are α-dependent, continuous, symmetric, j-linear forms taking values in $\mathbb{R}^n$. By the term symmetric, we mean that each H_j is invariant under a permutation of its j arguments. More precisely, we assume L and H are continuous in (α,ψ), and for fixed α, $H(\alpha;\psi)$ is at least 9 times continuously differentiable in ψ. As in [14], we assume that for $\psi \in C$ with derivatives $\psi^{(j)} \in C, j = 1,2,\ldots,7$, the functions $L(\alpha)\psi$, $H_j(\alpha;\psi)$, and $H(\alpha;\psi)$ are C^7 functions of α. Such assumptions are not uncommon to applications, where often derivatives of all orders are present.

Observe that $y \equiv 0$ defines a steady state for (1). The linearized equation

$$\dot{y}(t) = L(\alpha)y_t \tag{3}$$

has nontrivial solutions of the form $y(t) = \xi e^{\lambda t}$ with $\xi \in \mathbb{C}^n$ if and only there is a nontrivial ξ satisfying the characteristic system

$$0 = [\lambda I - L(\alpha)e^{\lambda \cdot}]\xi \equiv \Delta(\alpha;\lambda)\xi. \tag{4}$$

Assume for α near α_0 (4) possesses a nontrivial solution with $\lambda = \lambda(\alpha)$ such that $\lambda(\alpha_0) = i\omega, \omega \neq 0$. As usual, we assume that $\lambda = i\omega$ is a simple root of $det\Delta(\alpha_0;\lambda) = 0$ and all other roots (other than $\pm i\omega$) have negative real parts. Define $\xi^* = \xi^*(\alpha) \neq 0$ to be any solution of $\xi^*(\alpha)\Delta(\alpha;\lambda(\alpha)) = 0$ for α near α_0, and for λ near $\lambda(\alpha)$, let

$$\hat{\xi} = \hat{\xi}(\alpha;\lambda) \equiv \xi^*/[\xi^*\Delta'(\alpha;\lambda)\xi], \tag{5}$$

where $\Delta' = \partial\Delta/\partial\lambda$. See [13], [14] for details.

Our primary goal is to provide computational means of resolving the structure of Hopf bifurcations for (1) near criticality. The following proposition, proved in [13], asserts the existence of a scalar bifurcation function $g(\alpha, c)$ that facilitates such a study.

Proposition 2.1 *For ω in a neighborhood of ω_0 there exists a computable real-valued function g defined and C^8 in a neighborhood of $(\alpha_0, 0)$ whose zeros correspond in a 1-1 fashion with the small periodic solutions of (1) with period near $2\pi/\omega$. Under this correspondence, the periodic solution of (1) associated to a root c of $g(\alpha;\cdot)$ has the form*

$$y(t,\alpha;c,\nu) = 2\mathrm{Re}\{\xi(\alpha)e^{\nu it}\}c + \mathcal{O}(c^2), \tag{6}$$

(up to phase shift). Moreover, $y(t)$ is orbitally asymptotically stable (unstable) if and only if c is stable (unstable) when viewed as an equilibrium of the scalar equation $\dot{c} = g(\alpha;c)$.

Essential to the application of this result to specific equations is the effective approximation of the scalar bifurcation function g. This issue is considered in [14], where an inductive approximation algorithm is derived. It is shown in that reference that the small periodic solutions of (1) with periods $2\pi/\nu$ and α near α_0 coincide with those of the (complex) scalar bifurcation equation

$$0 = G(\alpha;\nu,c) \tag{7}$$

$$= [\lambda(\alpha) - i\nu]c + \frac{\nu}{2\pi}\int_0^{2\pi/\nu} e^{-\nu iu}\hat{\xi}\cdot H(\alpha;y_u)du \tag{8}$$

$$= (\lambda(\alpha) - i\nu)c + M_3(\alpha;\nu)c^3 + M_5(\omega;\nu)c^5 + M_7(\omega;\nu)c^7 + \cdots, \tag{9}$$

where $y(t) = 2\mathrm{Re}\{c\varphi(t)\} + \sum_{l=2}^{m} y^{(l)}(t)c^l + \ldots$ for $m < 8$, is defined inductively according to the following algorithm:

1. The expansion $y^{(l)}(t)$ has the form

$$y^{(l)}(t) = A_{l,l}e^{l\nu it} + A_{l,l-2}e^{(l-2)\nu it} + \cdots + A_{l,-l}e^{-l\nu it},$$

where $\overline{A_{l,j}} = A_{l,-j}$.

2. $y^{(1)}(t) = 2Re\{\varphi(t)\} = A_{1,1}e^{\nu it} + \bar{A}_{1,1}e^{-\nu it}$, with $A_{1,1} = \xi(\alpha)$ and $\varphi(s) \equiv \xi(\alpha)e^{\nu is}$,

3. Define $[\hat{\xi}\cdot\ ]$ to be the linear map from $\mathbb{C}^n$ to $\mathbb{C}$ given by

$$\hat{\xi}\cdot h = \sum_{j=1}^{n} \hat{\xi}_j \cdot h_j.$$

If, for $l \geq 2$, the coefficient of c^l in

$$\sum_{j=2}^{l} H_j(\alpha; [\sum_{m=1}^{l-1} y_i^{(m)} c^m]^j)$$

is $\sum_j B_{i,j}(\alpha;\nu)e^{j\nu it}$, then

$$A_{l,j}(\alpha;\nu) = \begin{cases} \Delta^{-1}(\alpha; j\nu i)B_{l,j}(\alpha;\nu) & \text{for } j \neq \pm 1, \\ (\Delta^{-1}(\alpha;\nu i) - \frac{1}{\nu i - \lambda(\alpha)}\xi[\hat{\xi}\cdot\ ])B_{l,1}(\alpha;\nu) & \text{for } j = 1. \end{cases}$$

The singularity at $\lambda = \lambda(\alpha)$, in

$$\Delta^{-1}(\alpha;\lambda) - \frac{1}{\lambda - \lambda(\alpha)}\xi[\hat{\xi}\cdot]$$

is removable. In particular, for $h \in \mathbb{C}^n$, and λ near $\lambda(\alpha)$, we have the expansion

$$\begin{aligned} &\Delta^{-1}(\alpha;\lambda)h - \frac{1}{\lambda-\lambda(\alpha)}\xi[\hat{\xi}\cdot h] = \\ &\quad d - [\hat{\xi}\Delta'(\alpha;\lambda(\alpha))d]\xi - \frac{1}{2}[\hat{\xi}\Delta''(\alpha;\lambda(\alpha))\xi][\hat{\xi}\cdot h]\xi \\ &\quad + \Big[e - [\hat{\xi}\Delta'(\alpha;\lambda(\alpha))e]\xi + \frac{1}{2}[\hat{\xi}\Delta''(\alpha;\lambda(\alpha))\xi]\ [\hat{\xi}\Delta'(\alpha;\lambda(\alpha))d]\xi \\ &\quad - \frac{1}{2}[\hat{\xi}\Delta''(\alpha;\lambda(\alpha))d]\xi + \Big\{(\frac{1}{2}[\hat{\xi}\Delta''(\alpha;\lambda(\alpha))\xi])^2 \\ &\quad - \frac{1}{6}[\hat{\xi}\Delta'''(\alpha;\lambda(\alpha))\xi]\Big\}[\hat{\xi}\cdot h]\xi\Big](\lambda-\lambda(\alpha)) + \mathcal{O}((\lambda-\lambda(\alpha))^2), \end{aligned}$$

where $d \in \mathbb{C}^n$ is any solution of

$$\Delta(\alpha;\lambda(\alpha))d = h - \Delta'(\alpha\lambda(\alpha))\xi[\hat{\xi}\cdot h],$$

and $e \in \mathbb{C}^n$ is any solution of

$$\Delta(\alpha;\lambda(\alpha))e = -\Delta'(\alpha;\lambda(\alpha))d + \Delta'(\alpha;\lambda(\alpha))\xi[\hat{\xi}\Delta'(\alpha;\lambda(\alpha))d] + \left\{-\frac{1}{2}\Delta''(\alpha;\lambda(\alpha))\xi + \frac{1}{2}[\hat{\xi}\Delta''(\alpha;\lambda(\alpha))\xi]\Delta'(\alpha;\lambda(\alpha))\xi\right\}[\hat{\xi}\cdot h]$$

For details, see [14], where the first term in this expansion is derived.

Implementation of this algorithm is obviously difficult to do by hand. We have choosen to perform the necessary details with the aid of the symbolic manipulation software MACSYMA [5], [9]; see [6] for complete details. As a result, we obtain the following theorem, which represents an extension of Theorem 2.1 [13], where the expansion through order 5 is presented.

Theorem 2.2 *Under the above hypotheses, there are $\varepsilon > 0$ and C^7 functions $G(\alpha;c,\nu)$ ($\mathbb{C}$ -valued), $y(t,\alpha;c,\nu)$ ($\mathbb{R}^n$-valued and $\frac{2\pi}{\nu}$ -periodic in t) defined for real c, $|c| < \varepsilon$, $|\nu-\omega| < \varepsilon$, $\|\alpha-\alpha_0\| < \varepsilon$, and $t \in \mathbb{R}$ such that (1) has a $2\pi/\nu$ -periodic solution $y(t)$ with $|y| < \varepsilon$, $|\nu-\omega| < \varepsilon$, and $\|\alpha-\alpha_0\| < \varepsilon$ if and only if $y(t) = y(t,\alpha;c,\nu)$ (up to phase shift) and (α,c,ν) solves the bifurcation equation: $G(\alpha;c,\nu) = 0$. Moreover, y satisfies (6), G is odd in c and*

$$G(\alpha;c,\nu) = [\lambda - i\nu]c + M_3(\alpha;\nu,\lambda)c^3 + M_5(\alpha;\nu,\lambda)c^5 + M_7(\alpha;\nu,\lambda)c^7 + \mathcal{O}(c^9), \quad (10)$$

where $\lambda = \lambda(\alpha)$, $M_3(\alpha;\nu,\lambda) = \hat{\xi}(\alpha;\lambda)\cdot N_3(\alpha;\nu)$,

$$N_3(\alpha;\nu) \equiv 3H_3(\varphi^2,\bar{\varphi}) + 2H_2(\bar{\varphi},A_{2,2}e^{2\nu i\cdot}) + 2H_2(\varphi,A_{2,0}), \quad (11)$$

with $\varphi(s) = \xi(\alpha)e^{i\nu s}$ for $s \le 0$ and $A_{2,2}, A_{2,0}$ the unique solutions of

$$\Delta(\alpha;2\nu i)A_{2,2} = H_2(\varphi^2),$$
$$\Delta(\alpha;0)A_{2,0} = 2H_2(\varphi,\bar{\varphi}),$$

respectively.

Similarly, $M_5(\alpha;\nu,\lambda) = \hat{\xi}(\alpha;\lambda)\cdot N_5(\alpha;\nu)$, where

$$\begin{aligned} N_5(\alpha;\nu) =& 2H_2(\varphi,A_{4,0}) + 2H_2(\bar{\varphi},A_{4,2}e^{2\nu i\cdot}) + 2H_2(A_{2,2}e^{2\nu i\cdot},\bar{A}_{3,1}e^{-\nu i\cdot}) \\ &+ 2H_2(\bar{A}_{2,2}e^{-2\nu i\cdot},A_{3,3}e^{3\nu i\cdot}) + 2H_2(A_{2,0},A_{3,1}e^{\nu i\cdot}) \\ &+ 3H_3(\varphi^2,\bar{A}_{3,1}e^{-\nu i\cdot}) + 6H_3(\varphi,\bar{\varphi},A_{3,1}e^{\nu i\cdot}) \\ &+ 3H_3(\bar{\varphi}^2,A_{3,3}e^{3\nu i\cdot}) + 6H_3(\bar{\varphi},A_{2,2}e^{2\nu i\cdot},A_{2,0}) \\ &+ 6H_3(\varphi,A_{2,2}e^{2\nu i\cdot},\bar{A}_{2,2}e^{-2\nu i\cdot}) + 3H_3(\varphi,(A_{2,0})^2) \\ &+ 12H_4(\varphi,\bar{\varphi}^2,A_{2,2}e^{2\nu i\cdot}) + 12H_4(\varphi^2,\bar{\varphi},A_{2,0}) \\ &+ 4H_4(\varphi^3,\bar{A}_{2,2}e^{-2\nu i\cdot}) + 10H_5(\varphi^3,\bar{\varphi}^2), \end{aligned}$$

with $A_{3,3}, A_{3,1}, A_{4,2}, A_{4,0}$ the unique solutions of

$$\Delta(\alpha;3\nu i)A_{3,3} = H_3(\varphi^3) + 2H_2(\varphi,A_{2,2}e^{2\nu i\cdot})$$

$$A_{3,1} = d - [\hat{\xi}\Delta'(\alpha;\lambda(\alpha))d]\xi - \frac{1}{2}[\hat{\xi}\Delta''(\alpha;\lambda(\alpha))\xi]M_3\xi$$
$$+ \Big[e - [\hat{\xi}\Delta'(\alpha;\lambda(\alpha))e]\xi + \frac{1}{2}[\hat{\xi}\Delta''(\alpha;\lambda(\alpha))\xi]\,[\hat{\xi}\Delta'(\alpha;\lambda(\alpha))d]\xi$$
$$- \frac{1}{2}[\hat{\xi}\Delta''(\alpha;\lambda(\alpha))d]\xi + \Big\{(\frac{1}{2}[\hat{\xi}\Delta''(\alpha;\lambda(\alpha))\xi])^2$$
$$- \frac{1}{6}[\hat{\xi}\Delta'''(\alpha;\lambda(\alpha))\xi]\Big\}M_3\xi\Big](i\nu - \lambda(\alpha)),$$

where d is any solution of $\Delta(\alpha;\lambda(\alpha))d = N_3 - (\Delta'\xi)M_3$, e is any solution of

$$\Delta(\alpha;\lambda(\alpha))e = -\Delta'(\alpha;\lambda(\alpha))d + \Delta'(\alpha;\lambda(\alpha))\xi[\hat{\xi}\Delta'(\alpha;\lambda(\alpha))d]$$
$$+ \Big\{-\frac{1}{2}\Delta''(\alpha;\lambda(\alpha))\xi + \frac{1}{2}[\hat{\xi}\Delta''(\alpha;\lambda(\alpha))\xi]\Delta'(\alpha;\lambda(\alpha))\xi\Big\}M_3,$$

and

$\Delta^i \equiv (\partial^i\Delta/\partial\lambda^i)(\alpha;\lambda(\alpha)); i = 1, 2, \ldots$

$$\Delta(\alpha;2\nu i)A_{4,2} = 2H_2(\varphi, A_{3,1}e^{\nu i\cdot}) + 2H_2(\bar{\varphi}, A_{3,3}e^{3\nu i\cdot}) + 2H_2(A_{2,2}e^{2\nu i\cdot}, A_{2,0})$$
$$+ 6H_3(\varphi,\bar{\varphi},A_{2,2}e^{2\nu i\cdot}) + 3H_3(\varphi^2, A_{2,0}) + 4H_4(\varphi^3,\bar{\varphi}),$$

$$\Delta(\alpha;0)A_{4,0} = 2H_2(\varphi,\bar{A}_{3,1}e^{-\nu i\cdot}) + 2H_2(\bar{\varphi}, A_{3,1}e^{\nu i\cdot}) + H_2((A_{2,0})^2)$$
$$+ 2H_2(A_{2,2}e^{2\nu i\cdot},\bar{A}_{2,2}e^{-2\nu i\cdot}) + 3H_3(\varphi^2,\bar{A}_{2,2}e^{-2\nu i\cdot})$$
$$+ 3H_3(\bar{\varphi}^2, A_{2,2}e^{2\nu i\cdot}) + 6H_3(\varphi,\bar{\varphi},A_{2,0}) + 6H_4(\varphi^2,\bar{\varphi}^2).$$

Finally, $M_7(\alpha;\nu,\lambda) = \hat{\xi}(\alpha;\lambda)\cdot N_7(\alpha;\nu)$, where at $\alpha = \alpha_0$ and $\nu = \omega$

$$N_7(\alpha;\nu) = 2H_2(\bar{\varphi}, A_{6,2}e^{2\nu i\cdot}) + 2H_2(\varphi, A_{6,0}) + 2H_2(\bar{A}_{2,2}e^{-2\nu i\cdot}, A_{5,3}e^{3\nu i\cdot})$$
$$+ 2H_2(A_{2,0}, A_{5,1}e^{\nu i\cdot}) + 2H_2(\bar{A}_{3,3}e^{-3\nu i\cdot}, A_{4,4}e^{4\nu i\cdot})$$
$$+ 2H_2(\bar{A}_{3,1}e^{-\nu i\cdot}, A_{4,2}e^{2\nu i\cdot}) + 2H_2(A_{3,1}e^{\nu i\cdot}, A_{4,0})$$
$$+ 2H_2(\bar{A}_{4,2}e^{-2\nu i\cdot}, A_{3,3}e^{3\nu i\cdot}) + 2H_2(\bar{A}_{5,1}e^{-\nu i\cdot}, A_{2,2}e^{2\nu i\cdot})$$
$$+ 3H_3(\bar{\varphi}^2, A_{5,3}e^{3\nu i\cdot}) + 6H_3(\bar{\varphi},\varphi, A_{5,1}e^{\nu i\cdot})$$
$$+ 6H_3(\bar{\varphi},\bar{A}_{2,2}e^{-2\nu i\cdot}, A_{4,4}e^{4\nu i\cdot}) + 6H_3(\bar{\varphi},\bar{A}_{2,0}, A_{4,2}e^{2\nu i\cdot})$$
$$+ 6H_3(\varphi,\bar{A}_{2,2}e^{-2\nu i\cdot}, A_{4,2}e^{2\nu i\cdot}) + 6H_3(\bar{\varphi}, A_{2,2}e^{2\nu i\cdot}, A_{4,0})$$
$$+ 6H_3(\varphi, A_{2,0}, A_{4,0}) + 6H_3(\bar{A}_{2,2}e^{-2\nu i\cdot}, A_{2,0}, A_{3,3}, e^{3\nu i\cdot})$$
$$+ 6H_3(\bar{A}_{3,3}e^{-3\nu i\cdot},\varphi, A_{3,3}, e^{3\nu i\cdot}) + 6H_3(\bar{\varphi},\bar{A}_{3,1}e^{-\nu i\cdot}, A_{3,3}, e^{3\nu i\cdot})$$
$$+ 3H_3(\bar{\varphi},(A_{3,1}e^{\nu i\cdot})^2) + 6H_3(\bar{A}_{2,2}e^{-2\nu i\cdot}, A_{2,2}e^{2\nu i\cdot}, A_{3,1}e^{\nu i\cdot})$$
$$+ 3H_3((A_{2,0})^2, A_{3,1}e^{\nu i\cdot}) + 6H_3(\bar{A}_{3,1}e^{-\nu i\cdot},\varphi, A_{3,1}e^{\nu i\cdot})$$
$$+ 3H_3(\bar{A}_{3,3}e^{-3\nu i\cdot},(A_{2,2}e^{2\nu i\cdot})^2) + 6H_3(\bar{A}_{3,1}e^{-\nu i\cdot}, A_{2,0}, A_{2,2}e^{2\nu i\cdot})$$
$$+ 6H_3(\bar{A}_{4,2}e^{-2\nu i\cdot},\varphi, A_{2,2}e^{2\nu i\cdot}) + 3H_3(\bar{A}_{5,1}e^{-\nu i\cdot},\varphi^2)$$
$$+ 4H_4((\bar{\varphi})^3, A_{4,4}e^{4\nu i\cdot}) + 12H_4((\bar{\varphi})^2,\varphi, A_{4,2}e^{2\nu i\cdot})$$
$$+ 12H_4(\bar{\varphi},\varphi^2, A_{4,0}) + 12H_4((\bar{\varphi})^2, A_{2,0}, A_{3,3}e^{3\nu i\cdot})$$

$$
\begin{aligned}
&+24H_4(\bar{\varphi},\varphi,\bar{A}_{2,2}e^{-2\nu i\cdot},A_{3,3}e^{3\nu i\cdot})+12H_4((\bar{\varphi})^2,A_{2,2}e^{2\nu i\cdot},A_{3,1}e^{\nu i\cdot})\\
&+24H_4(\bar{\varphi},\varphi,A_{2,0},A_{3,1}e^{\nu i\cdot})+12H_4(\bar{A}_{2,2}e^{2\nu i\cdot},\varphi^2,A_{3,1}e^{\nu i\cdot})\\
&+12H_4(\bar{\varphi},\bar{A}_{2,2}e^{-2\nu i\cdot},(A_{2,2}e^{2\nu i\cdot})^2)+12H_4(\bar{\varphi},(A_{2,0})^2,A_{2,2}e^{2\nu i\cdot})\\
&+24H_4(\bar{A}_{2,2}e^{2\nu i\cdot},\varphi,A_{2,0},A_{2,2}e^{2\nu i\cdot})+12H_4(\bar{A}_{3,3}e^{3\nu i\cdot},\varphi^2,A_{2,2}e^{2\nu i\cdot})\\
&+24H_4(\bar{\varphi},\bar{A}_{3,1}e^{\nu i\cdot},\varphi,A_{2,2}e^{2\nu i\cdot})+4H_4(\varphi,(A_{2,0})^3)\\
&+12H_4(\bar{A}_{3,1}e^{-\nu i\cdot},(\varphi)^2,A_{2,0})+4H_4(\bar{A}_{4,2}e^{-2\nu i\cdot},(\varphi)^3)\\
&+20H_5(\bar{\varphi}^3,\varphi,A_{3,3})+30H_5(\bar{\varphi}^2,\varphi^2,A_{3,1}e^{\nu i\cdot})\\
&+10H_5(\bar{\varphi}^3,(A_{2,2}e^{2\nu i\cdot})^2)+60H_5(\bar{\varphi}^2,\varphi,A_{2,0},A_{2,2}e^{2\nu i\cdot})\\
&+60H_5(\bar{\varphi},\bar{A}_{2,2}e^{-2\nu i\cdot},\varphi^2,A_{2,2}e^{2\nu i\cdot})+30H_5(\bar{\varphi},\varphi^2,(A_{2,0})^2)\\
&+20H_5(\bar{A}_{2,2}e^{-2\nu i\cdot},\varphi^3,A_{2,0})+5H_5(\bar{A}_{3,3}e^{-3\nu i\cdot},\varphi^4)\\
&+20H_5(\bar{\varphi},\bar{A}_{3,1}e^{-\nu i\cdot},\varphi^3)+60H_6(\bar{\varphi}^3,\varphi^2,A_{2,2}e^{2\nu i\cdot})\\
&+60H_6(\bar{\varphi}^2,\varphi^3,A_{2,0})+60H_6(\bar{\varphi},\varphi^4,\bar{A}_{2,2}e^{-2\nu i\cdot})+35H_7(\bar{\varphi}^3,\varphi^4).
\end{aligned}
$$

In addition, $A_{4,4}, A_{5,1}, A_{5,3}, A_{6,0}$, and $A_{6,2}$ are unique solutions of

$$
\begin{aligned}
\Delta(\alpha;4\nu i)A_{4,4} &= 2H_2(\varphi,A_{3,3}3\nu i\cdot)+H_2((A_{2,2}e^{2\nu i\cdot})^2)\\
&\quad +H_3(\varphi^2,A_{2,2}e^{2\nu i\cdot})+H_4(\varphi^4),\\
A_{5,1} &= f-[\hat{\xi}\Delta' f]\xi-\frac{1}{2}[\hat{\xi}\Delta''\xi]M_5\xi,
\end{aligned}
$$

where f is any solution of $\Delta(\alpha;\lambda(\alpha))f = N_5-(\Delta'\xi)M_5$,

$$
\begin{aligned}
\Delta(\alpha;3\nu i)A_{5,3} =&2H_2(\bar{\varphi},A_{4,4}e^{4\nu i\cdot})+2H_2(\varphi,A_{4,2}e^{2\nu i\cdot})\\
&+2H_2(A_{2,0},A_{3,3}e^{3\nu i\cdot})+2H_2(A_{2,2}e^{2\nu i\cdot},A_{3,1}e^{\nu i\cdot})\\
&+6H_3(\bar{\varphi},\varphi,A_{3,3}e^{3\nu i\cdot})+3H_3(\varphi^2,A_{3,1}e^{\nu i\cdot})\\
&+3H_3(\bar{\varphi},(A_{2,2}e^{2\nu i\cdot})^2)+6H_3(\varphi,A_{2,0},A_{2,2}e^{2\nu i\cdot})\\
&+12H_4(\bar{\varphi},\varphi^2,A_{2,2}e^{2\nu i\cdot})+4H_4(\varphi^3,A_{2,0})+5H_5(\bar{\varphi},\varphi^4),
\end{aligned}
$$

$$
\begin{aligned}
\Delta(\alpha;0)A_{6,0} =\ &2H_2(\bar{\varphi},A_{5,1}e^{\nu i\cdot})+2H_2(\bar{A}_{2,2}e^{-2\nu i\cdot},A_{4,2}e^{2\nu i\cdot})\\
&+2H_2(A_{2,0},A_{4,0})+2H_2(\bar{A}_{3,3}e^{-3\nu i\cdot},A_{3,3}e^{3\nu i\cdot})\\
&+2H_2(\bar{A}_{3,1}e^{-\nu i\cdot},A_{3,1}e^{\nu i\cdot})+2H_2(\bar{A}_{4,2}e^{-2\nu i\cdot},A_{2,2}e^{2\nu i\cdot})\\
&+2H_2(\bar{A}_{5,1}e^{-\nu i\cdot},\varphi)+3H_3(\bar{\varphi}^2,A_{4,2}e^{2\nu i\cdot})\\
&+6H_3(\bar{\varphi},\varphi,A_{4,0})+6H_3(\bar{\varphi},\bar{A}_{2,2}e^{-2\nu i\cdot},A_{3,3}e^{3\nu i\cdot})\\
&+6H_3(\bar{\varphi},A_{2,0},A_{3,1}e^{\nu i\cdot})+6H_3(\bar{A}_{2,2}e^{-2\nu i\cdot},\varphi,A_{3,1}e^{\nu i\cdot})\\
&+6H_3(\bar{A}_{2,2}e^{-2\nu i\cdot},A_{2,0},A_{2,2}e^{2\nu i\cdot})+6H_3(\bar{A}_{3,3}e^{-3\nu i\cdot},\varphi,A_{2,2}e^{2\nu i\cdot})\\
&+6H_3(\bar{\varphi},\bar{A}_{3,1}e^{-\nu i\cdot},A_{2,2}e^{2\nu i\cdot})+H_3((A_{2,0})^3)\\
&+6H_3(\bar{A}_{3,1}e^{-\nu i\cdot},\varphi,A_{2,0})+3H_3(\bar{A}_{4,2}e^{-2\nu i\cdot},\varphi^2)\\
&+4H_4(\bar{\varphi}^3,A_{3,3}e^{3\nu i\cdot})+12H_4(\bar{\varphi}^2,\varphi,A_{3,1}e^{\nu i\cdot})\\
&+12H_4(\bar{\varphi}^2,A_{2,0},A_{2,2}e^{2\nu i\cdot})+24H_4(\bar{\varphi},\bar{A}_{2,2}e^{-2\nu i\cdot},\varphi,A_{2,2}e^{2\nu i\cdot})
\end{aligned}
$$

$$
\begin{aligned}
&+ 12H_4(\bar{\varphi}, \varphi, (A_{2,0})^2) + 12H_4(\bar{A}_{2,2}e^{-2\nu i\cdot}, \varphi^2, A_{2,0}) \\
&+ 4H_4(\bar{A}_{3,3}e^{-3\nu i\cdot}, \varphi^3) + 12H_4(\bar{\varphi}, \bar{A}_{3,1}e^{-\nu i\cdot}, \varphi^2) \\
&+ 20H_5(\bar{\varphi}^3, \varphi, A_{2,2}e^{2\nu i\cdot}) + 30H_5(\bar{\varphi}^2, \varphi^2, A_{2,0}) \\
&+ 20H_5(\bar{\varphi}, \bar{A}_{2,2}e^{-2\nu i\cdot}, \varphi^3) + 20H_6(\bar{\varphi}^3, \varphi^3),
\end{aligned}
$$

$$
\begin{aligned}
\Delta(\alpha; 2\nu i)A_{6,2} = {} & 2H_2(\bar{\varphi}, A_{5,3}e^{3\nu i\cdot}) + 2H_2(\varphi, A_{5,1}e^{\nu i\cdot}) \\
&+ 2H_2(\bar{A}_{2,2}e^{-2\nu i\cdot}, A_{4,4}e^{4\nu i\cdot}) + 2H_2(A_{2,0}, A_{4,2}e^{2\nu i\cdot}) \\
&+ 2H_2(A_{2,2}e^{2\nu i\cdot}, A_{4,0}) + 2H_2(\bar{A}_{3,1}e^{-\nu i\cdot}, A_{3,3}e^{3\nu i\cdot}) \\
&+ H_2((A_{3,1}e^{\nu i\cdot})^2) + 3H_3(\bar{\varphi}^2, A_{4,4}e^{4\nu i\cdot}) \\
&+ 6H_3(\bar{\varphi}, \varphi, A_{4,2}e^{2\nu i\cdot},) + 3H_3(\varphi^2, \bar{A}_{4,0}) \\
&+ 6H_3(\bar{\varphi}, A_{2,0}, A_{3,3}e^{3\nu i\cdot}) + 6H_3(\bar{A}_{2,2}e^{-2\nu i\cdot}, \varphi, A_{3,3}e^{3\nu i\cdot}) \\
&+ 6H_3(\bar{\varphi}, A_{2,2}e^{2\nu i\cdot}, A_{3,1}e^{\nu i\cdot}) + 6H_3(\bar{A}_{2,0}, \varphi, A_{3,1}e^{\nu i\cdot}) \\
&+ 6H_3(\varphi, \bar{A}_{3,1}e^{-\nu i\cdot}, A_{2,2}e^{2\nu i\cdot}) + 3H_3((A_{2,0})^2, A_{2,2}e^{2\nu i\cdot}) \\
&+ 3H_3((A_{2,2}e^{2\nu i\cdot})^2, \bar{A}_{2,2}e^{-2\nu i\cdot}) + 12H_4(\bar{\varphi}^2, \varphi, A_{3,3}e^{3\nu i\cdot}) \\
&+ 12H_4(\bar{\varphi}, \varphi^2, A_{3,1}e^{\nu i\cdot}) + 6H_4(\bar{\varphi}^2, (A_{2,2}e^{2\nu i\cdot})^2) \\
&+ 24H_4(\bar{\varphi}, \varphi, A_{2,0}, A_{2,2}e^{2\nu i\cdot}) + 12H_4(\bar{A}_{2,2}e^{-2\nu i\cdot}, \varphi^2, A_{2,2}e^{2\nu i\cdot}) \\
&+ 6H_4(\varphi^2, (A_{2,0})^2) + 4H_4(\bar{A}_{3,1}e^{-\nu i\cdot}, \varphi^3) \\
&+ 20H_5(\bar{\varphi}, \varphi^3, A_{2,0}) + 30H_5(\bar{\varphi}^2, \varphi^2, A_{2,2}e^{2\nu i\cdot}) \\
&+ 5H_5(\bar{A}_{2,2}e^{-2\nu i\cdot}, \varphi^4) + 15H_6(\bar{\varphi}^2, \varphi^4).
\end{aligned}
$$

Assuming $\lambda(\alpha) = \mu(\alpha) + i\omega(\alpha)$, the real and imaginary parts of $G(\alpha; c, \nu) = 0$ become

$$
\begin{aligned}
0 &= \mu(\alpha)c + \mathrm{Re}\{M_3(\alpha; \nu, \lambda)\}c^3 + \mathrm{Re}\{M_5(\alpha; \nu, \lambda)\}c^5 \\
&\quad + \mathrm{Re}\{M_7(\alpha; \nu, \lambda)\}c^7 + \mathcal{O}(c^9), \qquad (12) \\
\nu &= \omega(\alpha) + \mathrm{Im}\{M_3(\alpha; \nu, \lambda)\}c^2 + \mathrm{Im}\{M_5(\alpha; \nu, \lambda)\}c^4 \\
&\quad + \mathrm{Im}\{M_7(\alpha; \nu, \lambda)\}c^6 + \mathcal{O}(c^8), \qquad (13)
\end{aligned}
$$

for $c \neq 0$.

The following theorem (proved by iteration on equation (13) and elimination of variable ν) relates the real bifurcation function g of Proposition 2.1 to the complex bifurcation function G of the previous theorem.

Theorem 2.3 *The reduced bifurcation equation for higher order bifurcations is given by*

$$
0 = g(\alpha; c) = \mu(\alpha)c + K_3(\alpha)c^3 + K_5(\alpha)c^5 + K_7(\alpha)c^7 + \mathcal{O}(c^9), \qquad (14)
$$

where

$$
\begin{aligned}
K_3 &= \mathrm{Re}\{M_3(\alpha;\omega(\alpha),\lambda(\alpha))\},\\
K_5 &= \mathrm{Re}\{M_5(\alpha;\omega(\alpha),\lambda(\alpha))\} + \mathrm{Re}\{\frac{\partial}{\partial\nu}(M_3(\alpha;\nu,\lambda(\alpha)))|_{\nu=\omega(\alpha)}\}\cdot w_2,\\
K_7 &= \mathrm{Re}\{M_7(\alpha;\omega(\alpha),\lambda(\alpha))\} + \mathrm{Re}\{\frac{\partial}{\partial\nu}(M_5(\alpha;\nu,\lambda(\alpha)))|_{\nu=\omega(\alpha)}\}\cdot w_2\\
&\quad + \mathrm{Re}\{\frac{\partial}{\partial\nu}(M_3(\alpha;\nu,\lambda(\alpha)))|_{\nu=\omega(\alpha)}\}\cdot w_4\\
&\quad + \frac{1}{2}\mathrm{Re}\{\frac{\partial^2}{\partial\nu^2}(M_3(\alpha;\nu,\lambda(\alpha)))|_{\nu=\omega(\alpha)}\}\cdot(w_2)^2,
\end{aligned}
$$

and

$$
\begin{aligned}
w_2 &= \mathrm{Im}\{M_3(\alpha;\omega(\alpha),\lambda(\alpha))\},\\
w_4 &= \mathrm{Im}\{M_5(\alpha;\omega(\alpha),\lambda(\alpha))\}\\
&\quad + \mathrm{Im}\{\frac{\partial}{\partial\nu}(M_3(\alpha;\nu,\lambda(\alpha)))|_{\nu=\omega(\alpha)}\}\cdot\mathrm{Im}\{M_3(\alpha;\omega(\alpha),\lambda(\alpha))\}.
\end{aligned}
$$

The analysis of a particular equation then rests on identifying the critical parameter α_0 and the associated characteristic values and vectors, computing the terms in the expansion of the bifurcation function G in Theorem 2.2, then the evaluation of the expansion of g from the previous theorem. See [6] for a MACSYMA – based implemetation of these formulas for scalar functional differential equations. A FORTRAN -based approach (numerical evaluation of K_3 and K_5) for systems is described in [2]. Only under very special circumstances can one hope to apply such a lengthy algorithm by hand calculation. However, in some important situations, many of the higher order terms H_j are identically zero causing significant simplifications. One such situation is that of equations with odd nonlinearities.

Corollary 2.4 *Under the above hypotheses, if H is odd there are $\varepsilon > 0$ and C^7 functions $G(\alpha;c,\nu)$ ($\mathbb{C}$ -valued), $y(t,\alpha;c,\nu)$ ($\mathbb{R}^n$-valued and $\frac{2\pi}{\nu}$ -periodic in t) defined for real c, $|c| < \varepsilon$, $|\nu-\omega| < \varepsilon$, $\|\alpha-\alpha_0\| < \varepsilon$, and $t \in \mathbb{R}$ such that (1) has a $2\pi/\nu$ -periodic solution $y(t)$ with $|y| < \varepsilon$, $|\nu-\omega| < \varepsilon$, and $\|\alpha-\alpha_0\| < \varepsilon$ if and only if $y(t) = y(t,\alpha;c,\nu)$ (up to phase shift) and (α,c,ν) solves the bifurcation equation: $G(\alpha;c,\nu) = 0$. Moreover, relation (6) holds, G is odd in c, and*

$$G(\alpha;c,\nu) = [\lambda-i\nu]c+M_3(\alpha;\nu,\lambda)c^3+M_5(\alpha;\nu,\lambda)c^5+M_7(\alpha;\nu,\lambda)c^7+\mathcal{O}(c^9), \quad (15)$$

where $\lambda = \lambda(\alpha), M_3(\alpha;\nu,\lambda) = \hat{\xi}(\alpha;\lambda)\cdot N_3(\alpha;\nu)$,

$$N_3(\alpha;\nu) \equiv 3H_3(\varphi^2,\bar{\varphi}), \quad (16)$$

with $\varphi(s) = \xi(\alpha)e^{i\nu s}$ *for* $s \le 0$. *Similarly,* $M_5(\alpha;\nu,\lambda) = \hat{\xi}(\alpha;\lambda)\cdot N_5(\alpha;\nu)$, *where*

$$
\begin{aligned}
N_5(\alpha;\nu) &= 3H_3(\varphi^2,\bar{A}_{3,1}e^{-\nu i\cdot}) + 6H_3(\varphi,\bar{\varphi},A_{3,1}e^{\nu i\cdot})\\
&\quad + 3H_3(\bar{\varphi}^2,A_{3,3}e^{3\nu i\cdot}) + 10H_5(\varphi^3,\bar{\varphi}^2),
\end{aligned}
$$

with $A_{3,3}, A_{3,1}$ the unique solutions of

$$\begin{aligned}
\Delta(\alpha;3\nu i)A_{3,3} =& H_3(\varphi^3) \\
A_{3,1} =& d - [\hat{\xi}\Delta'(\alpha;\lambda(\alpha))d]\xi - \frac{1}{2}[\hat{\xi}\Delta''(\alpha;\lambda(\alpha))\xi]M_3\xi \\
&+ \Big[e - [\hat{\xi}\Delta'(\alpha;\lambda(\alpha))e]\xi + \frac{1}{2}[\hat{\xi}\Delta''(\alpha;\lambda(\alpha))\xi]\ [\hat{\xi}\Delta'(\alpha;\lambda(\alpha))d]\xi \\
&- \frac{1}{2}[\hat{\xi}\Delta''(\alpha;\lambda(\alpha))d]\xi + \Big\{(\frac{1}{2}[\hat{\xi}\Delta''(\alpha;\lambda(\alpha))\xi])^2 \\
&- \frac{1}{6}[\hat{\xi}\Delta'''(\alpha;\lambda(\alpha))\xi]\Big\}M_3\xi\Big](i\nu - \lambda(\alpha)),
\end{aligned}$$

where d and e are any solutions of

$$\begin{aligned}
\Delta(\alpha;\lambda(\alpha))d &= N_3 - (\Delta'\xi)M_3, \\
\Delta(\alpha;\lambda(\alpha))e &= -\Delta'(\alpha;\lambda(\alpha))d + \Delta'(\alpha;\lambda(\alpha))\xi[\hat{\xi}\Delta'(\alpha;\lambda(\alpha))d] \\
&+ \Big\{-\frac{1}{2}\Delta''(\alpha;\lambda(\alpha))\xi + \frac{1}{2}[\hat{\xi}\Delta''(\alpha;\lambda(\alpha))\xi]\Delta'(\alpha;\lambda(\alpha))\xi\Big\}M_3
\end{aligned}$$

and $\Delta^i \equiv (\partial^i\Delta/\partial\lambda^i)(\alpha;\lambda(\alpha)); i = 1,2,3$. Likewise, $M_7(\alpha;\nu,\lambda) = \hat{\xi}(\alpha;\lambda)\cdot N_7(\alpha;\nu)$, where at $\alpha = \alpha_0$ and $\nu = \omega$

$$\begin{aligned}
N_7(\alpha;\nu) =& 3H_3(\bar{\varphi}^2, A_{5,3}e^{3\nu i\cdot}) + 3H_3(\bar{\varphi},\varphi,A_{5,1}e^{\nu i\cdot}) \\
&+ 6H_3(\bar{A}_{3,3}e^{-3\nu i\cdot},\varphi,A_{3,3},e^{3\nu i\cdot}) + 6H_3(\bar{\varphi},\bar{A}_{3,1}e^{-\nu i\cdot},A_{3,3},e^{3\nu i\cdot}) \\
&+ 3H_3(\bar{\varphi},(A_{3,1}e^{\nu i\cdot})^2) + 6H_3(\bar{A}_{3,1}e^{-\nu i\cdot},\varphi,A_{3,1}e^{\nu i\cdot}) \\
&+ 3H_3(\bar{A}_{5,1}e^{-\nu i\cdot},\varphi^2) + 20H_5(\bar{\varphi}^3,\varphi,A_{3,3}) \\
&+ 30H_5(\bar{\varphi}^2,\varphi^2,A_{3,1}e^{\nu i\cdot}) + 5H_5(\bar{A}_{3,3}e^{-3\nu i\cdot},\varphi^4) \\
&+ 20H_5(\bar{\varphi},\bar{A}_{3,1}e^{-\nu i\cdot},\varphi^3) + 35H_7(\bar{\varphi}^3,\varphi^4),
\end{aligned}$$

and

$$\begin{aligned}
\Delta(\alpha;3\nu i)A_{5,3} =& 6H_3(\bar{\varphi},\varphi,A_{3,3}e^{3\nu i\cdot}) \\
&+ 3H_3(\varphi^2,A_{3,1}e^{\nu i\cdot}) + 5H_5(\bar{\varphi},\varphi^4), \\
A_{5,1} =& f - [\hat{\xi}\Delta' f]\xi - \frac{1}{2}[\hat{\xi}\Delta''\xi]M_5\xi,
\end{aligned}$$

where f is any solution of $\Delta(\alpha;\lambda(\alpha))f = N_5 - (\Delta'\xi)M_5$.

Example 2.5 The case of integrodifferential equations

$$\dot{y} = \alpha_1 y(t) + \alpha_2\int_{-1}^{0} g(y_t(s))d\eta(s)$$

where $g(y) = y + h_2y^2 + h_3y^3 + \ldots$ illustrates the type of results obtainable, and their complexity. The previous results imply that $K_3(\alpha;\omega) = c_1(\alpha,\omega)h_3 + c_2(\alpha,\omega)h_2^2$, with c_1 and c_2 computable functions of the bifurcation parameters

$\alpha = (\alpha_1, \alpha_2)$ and frequency ω. (See [12] for an examination of the generic case in greater detail, and a derivation of conditions under which $K_3 \equiv 0$ for all choices of h_2 and h_3.) Similarly, one sees that K_5 will be a linear combination of the coefficient combinations h_5, h_2h_4, $h_3h_2^2$, h_3^2, and h_2^4, while K_7 will be a linear combination of the eleven terms h_7, h_5h_3, $h_5h_2^2$, $h_4h_2h_3$, $h_4h_2^3$, h_4^2, $h_3^2h_2^2$, h_3^3, $h_3h_2^4$, h_2^6 and h_2h_6. These reduce greatly in the case of odd nonlinearities since $h_2 = h_4 = h_6 = 0$.

3 Scalar delay–difference equations

In this final section we will consider the scalar delay difference equation

$$\begin{aligned} \dot{x}(t) &= f(x(t), x(t-1)) \\ &= \alpha x(t) + \beta x(t-1) + h(x(t), x(t-1)) \end{aligned} \tag{17}$$

where $h(x,y) = a_2x^2 + b_2xy + c_2y^2 + a_3x^3 + b_3x^2y + c_3xy^2 + d_3y^3 + \ldots$ is assumed to be smooth. Our goal is to illustrate the results of the previous section and provide insight into the nongeneric bifurcation structure for this important equation.

The analysis of the linearized equation $\dot{z}(t) = \alpha z(t) + \beta z(t-1)$ is found in [7]. With $\Delta(\alpha, \beta; \lambda) = \lambda - \alpha - \beta e^{-\lambda}$ one easily identifies the line $\alpha + \beta = 0$ to characterize those parameter values at which $\lambda = 0$ is a characteristic root. Similarly, substituting $\lambda = i\omega$ into the characteristic equation and separating the real and imaginary parts leads to the parametrization $\beta = \tilde{\beta}(\omega) \equiv -\omega/\sin(\omega)$; $\alpha = \tilde{\alpha}(\omega) \equiv -\tilde{\beta}(\omega)\cos(\omega)$ characterizing those parameter values along which there are (simple) imaginary root pairs $\lambda = \pm i\omega$; $\omega > 0$. The interval $0 < \omega < \pi$ generates the remaining boundary of the region Ω_- of parameter values at which all characteristic roots have negative real parts. This region contains the negative half-axis $\alpha < 0, \beta = 0$, and is pictured in Figure 5.1 (page 109) of [7] subject to the elementary change of variables $a = -\alpha, b = -\beta$, and $r = 1$.. See Section 2 of [12] for generalizations.

Along the imaginary root curve the usual transversality criteria are easy to verify, and at $(\tilde{\alpha}(\omega), \tilde{\beta}(\omega))$ all characteristic roots other that $\lambda = \pm i\omega$ have negative real parts. The representation of the higher order terms $h(x(t), x(t-1))$ in terms of symmetric, multilinear functionals is trivial, allowing one to apply Theorems 2.2 and 2.3 directly. The generic bifurcation constant $K_3 = K_3(\omega)$ with $\alpha = \tilde{\alpha}(\omega)$, $\beta = \tilde{\beta}(\omega)$ is seen to take the form

$$\begin{aligned} K_3(\omega) =& c_{a_2a_2}(\omega)a_2^2 + c_{a_2b_2}(\omega)a_2b_2 + c_{b_2b_2}(\omega)b_2^2 \\ &+ c_{a_2c_2}(\omega)a_2c_2 + c_{c_2c_2}(\omega)c_2^2 + c_{b_2c_2}(\omega)b_2c_2 \\ &+ c_{a_3}(\omega)a_3 + c_{b_3}(\omega)b_3 + c_{c_3}(\omega)c_3 + c_{d_3}(\omega)d_3 \end{aligned}$$

where by direct (but symbolically assisted) computation

$$
\begin{aligned}
c_{a_3}(\omega) &= 3\sin(\omega)(\sin(\omega) - \omega\cos(\omega))/D_1(\omega)\\
c_{b_3}(\omega) &= \sin(\omega)(3\cos(\omega)\sin(\omega) - 2\omega\cos(\omega)^2 - \omega)/D_1(\omega)\\
c_{c_3}(\omega) &= (-3\omega\cos(\omega)\sin(\omega) - 2\cos(\omega)^4 + \cos(\omega)^2 + 1)/D_1(\omega)\\
c_{d_3}(\omega) &= 3\sin(\omega)(\cos(\omega)\sin(\omega) - \omega)/D_1(\omega)\\
c_{a_2a_2}(\omega) &= 2(\cos(\omega)+1)[3(2\cos(\omega)+3)\sin(\omega)\\
&\quad - \omega(\cos(\omega)+2)(4\cos(\omega)+1)]/D_2(\omega)\\
c_{a_2b_2}(\omega) &= (\cos(\omega)+1)[3(2\cos(\omega)+3)(3\cos(\omega)+1)\sin(\omega)\\
&\quad - \omega(8\cos(\omega)^3 + 26\cos(\omega)^2 + 19\cos(\omega) + 7)]/D_2(\omega)\\
c_{b_2b_2}(\omega) &= (\cos(\omega)+1)^2[(4\cos(\omega)^2 + 10\cos(\omega) + 1)\sin(\omega)\\
&\quad - \omega(8\cos(\omega)^2 + 4\cos(\omega) + 3)]/D_2(\omega)\\
c_{b_2c_2}(\omega) &= -(\cos(\omega)+1)[(8\cos(\omega)^4 - 8\cos(\omega)^3\\
&\quad - 32\cos(\omega)^2 - 19\cos(\omega) - 9)\sin(\omega)\\
&\quad - \omega(4\cos(\omega)^3 - 20\cos(\omega)^2 - 37\cos(\omega) - 7)]/D_2(\omega)\\
c_{c_2c_2}(\omega) &= -2(\cos(\omega)+1)[(4\cos(\omega)^3 - 4\cos(\omega)^2 - 13\cos(\omega) - 2)\sin(\omega)\\
&\quad - \omega(2\cos(\omega)^2 - 6\cos(\omega) - 11)]/D_2(\omega)\\
c_{a_2c_2}(\omega) &= 2(\cos(\omega)+1)^2[3(2\cos(\omega)+3)\sin(\omega) - \omega(8\cos(\omega)+7)]/D_2(\omega)
\end{aligned}
$$

where

$$
\begin{aligned}
D_1(\omega) &= \sin(\omega)^2 - 2\omega\cos(\omega)\sin(\omega) + \omega^2\\
D_2(\omega) &= \omega(4\cos(\omega)+5)(\sin(\omega)^2 - 2\omega\cos(\omega)\sin(\omega) + \omega^2).
\end{aligned}
$$

Along the curve $0 < \omega < \pi$ the scalar equation $\dot{c} = \mu c + K_3(\omega)c^3$ completely characterizes the generic Hopf bifurcation structure of the equation (3.1). For example, as the coefficients $c_{a_3}(\omega) > 0$ and $c_{d_3}(\omega) < 0$ for ω in that interval, increases in the corresponding coefficients a_3 and d_3 are seen to have destabilizing and stabilizing effects, respectively, on the equilibrium $x = 0$ at criticality, as well as on nearby Hopf bifurcations.

The special case $\omega = \pi/2$ is of particular importance. With $\alpha = 0$ and $\beta = -\pi/2$ one computes

$$
\begin{aligned}
K_3(\pi/2) = {} & 2[2(c_3 + 3a_3) - \pi(b_3 + 3d_3)]/(\pi^2 + 4)\\
& + 4[4(9-\pi)a_2^2 + (2 - 3\pi)b_2^2 + 2(4 - 11\pi)c_2^2\\
& + (18 - 7\pi)(a_2b_2 + 2a_2c_2 + b_2c_2)]/(5\pi(\pi^2 + 4))
\end{aligned}
$$

This extends Example 4.1 of [13]. Again the effects of the coefficients $a_2, b_2, \ldots, d_3$ on the stability of Hopf bifurcations can be easily deduced. Where $K_3(\omega) = 0$ (a cone in (a_2, b_2, c_2) space), one must compute (at least) $K_5(\omega)$ to fully understand the bifurcation structure for (17). This can be accomplished symbolically/numerically without serious difficulty. We illustrate this point by considering the quadratic delay difference equation

$$
\dot{x}(t) = \alpha x(t) + \beta x(t-1) + a_2x^2(t) + b_2x(t)x(t-1) + c_2x^2(t-1). \tag{19}
$$

For such an equation, one might consider asking an analogue of Hilbert's 16th Problem: How many simultaneous periodic orbits can this equation support? While this question is clearly difficult, our results of Section 2 shed light on the number of *small* periodic orbits that can be created via Hopf bifurcation at $x = 0$.

Using the quadratic nature of (19), we can normalize the coefficients of the higher order terms as $a_2 = \cos(\phi)$, $b_2 = \sin(\phi)\sin(\theta)$ and $c_2 = \sin(\phi)\cos(\theta)$, with $0 \le \phi \le \frac{\pi}{2}; 0 \le \theta \le 2\pi$ now defining our parameter space (ω fixed). An examination of the results of the previous section shows that the K_5 and K_7 are homogeneous polynomials of degree 4 and 8, respectively, in the variables a_2, b_2, c_2, the coefficients of these polynomials again being functions of ω. As these polynomials and their coefficients are quite complicated we will restrict our attention to specific selections for ω, and identify the curves $K_5 = 0, K_7 = 0$ by numerical evaluation.

Figure 3.1 depicts the situation at $\omega = \pi/2$. Each of the coefficients K_3, K_5, K_7 are observed to be positive for $\phi = 0$ (corresponding to $a_2 = 1, b_2 = c_2 = 0$). A careful examination of these curves reveals that there are no simultaneous solutions of $K_3 = K_5 = K_7 = 0$ other than the trivial case $a_2 = b_2 = c_2 = 0$. (The apparent simultaneous zero near $\phi = .9, \theta = 1.5$ is an artifact of the low graphics resolution and large scale). Thus $K_3 = K_5 = 0$ *implies* $K_7 \neq 0$ and consequently at $\omega = \pi/2$ the complete Hopf bifurcation structure for (17) can be described by the normal equation $\dot{c} = \mu c + K_3 c^3 + K_5 c^5 + K_7 c^7$ with $\mu, K_3, K_5 \approx 0; K_7 \neq 0$. We conclude that the equation

$$\dot{x}(t) = \beta x(t-1) + a_2 x^2(t) + b_2 x(t) x(t-1) + c_2 x^2(t-1). \tag{20}$$

for $\beta \approx -\pi/2$ can support at most three small periodic solution families bifurcating from $x = 0$.

A similar numerical analysis at other selected values of ω suggests this behavior to be generic for (19). However, by an examination of the crossing orders of the curves $K_j = 0; j = 3, 5, 7$ and observing their apparent continuity in ω, we are lead to conclude the existence of at least one value of ω in the interval $(2\pi/3, 3\pi/4)$ at which $K_3 = K_5 = K_7 = 0$ nontrivially. At such a value, a complete resolution of the Hopf bifurcation structure for (19) would require (at least) the computation of K_9. Such a computation, while theoretically within the scope of the algorithm of [14], would be a nontrivial task likely requiring careful partitioning of the calculations and hundreds of hours of cpu time on a current SUN or VAX-like workstation.

See [6] where Corollary 2.4 is used to derive analogous computations for K_3, K_5 and K_7 for (17) when $h(x(t), x(t-1))$ is assumed to be odd.

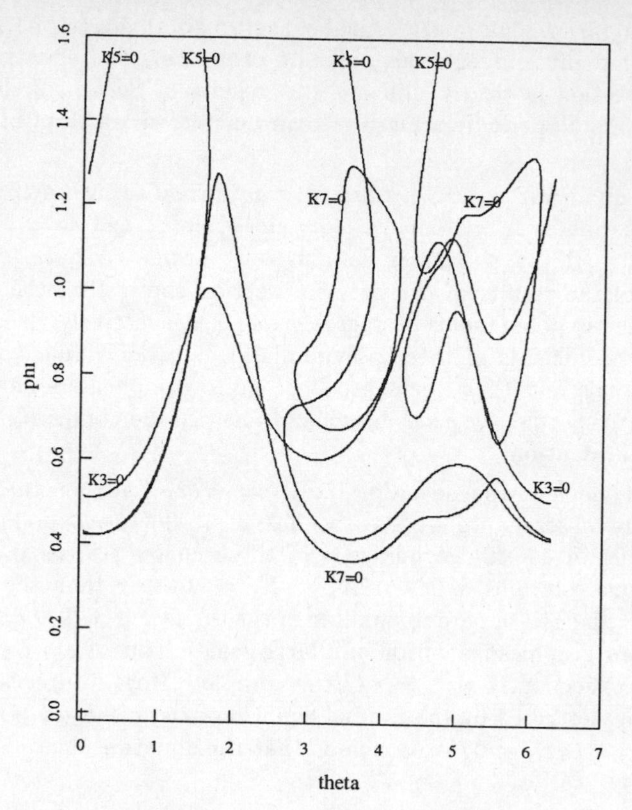

Fig. 3.1. Zero sets for Kj, j=3,5,7

Acknowledgments

The work of the authors was supported by grants AFOSR 87-0268, NSF DMS-8701456, NSF DMS-8901893, and the Department of the Commerce, National Bureau of Standards, through Georgia Tech. A grant of computing time from the Minnesota Supercomputer Institute and additional support from the Institute for Mathematics and its Applications are gratefully acknowledged.

References

1. Aboud, N. (1988): Contributions to the Computer-Aided Analysis of Functional Differential Equations. Master's Thesis, University of Minnesota
2. Aboud, N., Sathaye, A. Stech H. (1988): BIFDE: Software for the Investigation of the Hopf Bifurcation Problem in Functional Differential Equations. Proceedings of the 27th Conference on Decision and Control, IEEE, 821-824
3. Chow, S.-N., Mallet-Paret, J. (1977): Integral averaging and Hopf bifurcation. J. Differential Equations, **26**, 112-159
4. Claeyssen, J. R. (1980): The integral-averaging bifurcation method and the general one-delay equation. J. Math. Anal. Appl, **78**, 429-439

5. Drinkard, R. D., Sulinski N. (1981): MACSYMA: A Program For Computer Algebraic Manipulation (Demonstrations and Analysis). Naval Underwater Systems Center Technical Document 6401, Reprinted by SYMBOLICS
6. Franke, J. (1989): Symbolic Hopf Bifurcations for Functional Differential Equations. Master's Thesis, University of Minnesota
7. Hale, J. K. (1971): Functional Differential Equations. Applied Math. Sci., Vol. 3, Springer-Verlag, New York
8. Hassard, B., Kazarinoff, N., Wan, Y-H. (1981): Theory and Applications of Hopf Bifurcation. London Math. Soc. Lecture Notes, No. 41, Cambridge University Press, Cambridge
9. MACSYMA Reference Manual, Version 11 (1986), prepared by the MACSYMA group of SYMBOLICS. Inc. 11 Cambridge Center, Cambridge, MA 02142
10. Marsden, J. E., McCracken, M. (1976): The Hopf Bifurcation and its Applications. Applied Math. Sciences, Vol. 19, Springer-Verlag, New York
11. Sathaye, A. (1986): BIFDE: A Numerical Software Package for the Hopf Bifurcation Problem in Functional Differential Equations. Master's Thesis, Virginia Polytechic Institute and State University
12. Stech, H. W.: Generic Hopf bifurcations for a class of integro-differential equations. Submitted
13. Stech, H. W. (1985): Hopf bifurcation calculations for functional differential equations. Journal of Math Analysis and Applications, **109**, No. 2, 472-491
14. Stech, H. W. (1985): Nongeneric Hopf Bifurcations in Functional Differential Equations. SIAM J. Appl. Math., **16**, No. 6, 1134-1151

The Forced Spherical Pendulum Does Have Forced Oscillations

Massimo Furi[1] *and Maria Patrizia Pera*[2]

[1] Dipartimento di Matematica Applicata, Università di Firenze, Via S. Marta, 3 - 50139 Firenze
[2] Dipartimento di Matematica, Università di Siena, Via del Capitano, 15 - 53100 Siena

Consider the differential equation

$$\ddot{\theta}(t) = f(t, \theta(t)), \quad t \in \mathbf{R}, \tag{*}$$

where $f : \mathbf{R} \times \mathbf{R} \to \mathbf{R}$ is continuous, T periodic with respect to the first variable and 2π-periodic with respect to the second one. Because of the 2π-periodicity of f with respect to θ, (*) can be regarded as a second order differential equation on a circle S^1. It is in fact the motion equation of the forced planar pendulum (without friction). That is, a mass point (of mass one) constrained on a circle (by a weightless rigid rod, for example) and acted on by a T-periodic (normal to the rod) force f. Clearly, a solution $\theta(\cdot)$ of (*) is a forced oscillation (i.e., a periodic solution on S^1 of the same period as that of the forcing term) if and only if it satisfies the periodic boundary conditions;

$$\begin{cases} \theta(T) - \theta(0) = 2k\pi, & \text{for some } k \in \mathbf{Z} \\ \dot{\theta}(T) - \dot{\theta}(0) = 0. \end{cases}$$

It is known that (*) may not have forced oscillations, unless f satisfies some suitable assumptions. To see this observe that any non-vanishing autonomous tangent vector field on S^1 may be regarded as a periodic forcing term of any arbitrary period. Clearly, in this case, there are no periodic oscillations, since the energy of any solution of the motion equation is unbounded as $t \to +\infty$. For an extensive survey paper on the forced ordinary pendulum we recommend [6].

The above argument, however, does not work for the spherical pendulum. In fact, in this case the constraint is the 2-dimensional sphere S^2 and, as a consequence of the Poincaré-Hopf theorem, any tangent vector field on the sphere vanishes somewhere. So one gets equilibrium points, which can be regarded, of course, as periodic solutions (of arbitrary period).

In [4] we made the conjecture that the spherical pendulum (or, more generally, a constrained system, whose configuration space is a compact smooth manifold with nonzero Euler-Poincaré characteristic) admits forced oscillations even in the non-autonomous case, i.e., when it is excited on by a time periodic force. In [4], and independently in [2] with completely different methods, it has been proved that if the coefficient of friction is non-zero than the forced spherical pendulum (or more generally, a compact constrained system with nonzero Euler-Poincaré characteristic) does have forced oscillations. Moreover, in [2] V. Benci and M. Degiovanni proved that in the case when the exciting force is sufficiently small the system admits forced oscillations (even in the frictionless case). Another interesting result related to this argument has been obtained by V. Benci in [1], where he proved the existence of infinitely many forced oscillations for a system whose constraint is a smooth manifold with finite fundamental group (recall that $\pi^1(S^2)$ is trivial), provided that the force admits a time periodic Lagrangian satisfying certain physically reasonable assumptions.

Our aim here is to prove that the forced spherical pendulum has forced oscillations even in the frictionless case, in spite of the fact that, because of the presence of closed geodesics, *a priori* estimates for the speed of the forced oscillations cannot be established.

A crucial idea for our proof is to use the concept of the classical winding number in order to assign, in a continuous and canonical way, to any periodic orbit with sufficiently high speed, an integer, which, roughly speaking, counts the number of rotations the mass point makes in a subset of the sphere obtained by removing two appropriate antipodal points (which depend only on the given orbit). To get the proof, this idea is combined with a global (Rabinowitz type) bifurcation result for parametrized forced constrained systems recently obtained by the authors in [5].

We point out that the concept of winding number has been recently and successfully used in a joint interesting paper by A. Capietto, J. Mawhin and F. Zanolin in order to obtain existence results and information about the structure of the set of solutions of a parametrized forced first order system in $\mathbf{R}^n$ (see [3]). Their integer, however, depends on the choice of a two dimensional subspace of $\mathbf{R}^n$ and indubitably the strong geometric properties of the sphere cannot be used in that general case.

In what follows the inner product of two vectors v and w in $\mathbf{R}^3$ will be denoted by $\langle v, w\rangle$, the vector product by $v \times w$ and $|v|$ will stand for the euclidean norm of v (i.e. $|v| = \langle v, v\rangle^{1/2}$).

Consider a (frictionless) forced spherical pendulum, that is a point of mass m suspended by a rigid weightless rod of length R and acted on by a time-periodic force which we assume to include the force of gravity. The configuration space of this problem is the two-dimensional sphere

$$S = \{q \in \mathbf{R}^3 : |q| = R\}.$$

The motion of this constrained system is described by the following second order differential equation on S

$$m\ddot{x}(t) + m(|\dot{x}(t)|^2/R^2)x(t) = f(t, x(t)), \quad t \in \mathbf{R}, \quad x(t) \in S \tag{1}$$

where $f : \mathbf{R} \times S \to \mathbf{R}^3$ is a T-periodic ($T > 0$) continuous forcing term which may be assumed to be orthogonal to the rod (just replace $f(t,q)$ with $f(t,q)$ −

$(\langle q, f(t,q)\rangle / R^2)q$, if necessary). The term $r(q,v) = m(|v|^2/R^2)q$ is the *reactive force* (or *force of constraint*) at $q \in S$ corresponding to the velocity $v \in T_qS$, where

$$T_qS = \{v \in \mathbf{R}^3 : \langle q, v\rangle = 0\}$$

denotes the tangent space of S at q.

Theorem 1. *The equation (1) admits a forced oscillation.*

The proof will make use of Theorem 2 below, which is a special case of a global bifurcation result recently obtained by the authors in [5] in the more general context of a constrained system whose configuration space is a compact boundaryless manifold with nonzero Euler-Poincaré characteristic.

In order to state Theorem 2 we need some preliminaries.

Consider in S the parametrized motion equation

$$m\ddot{x}(t) + m(|\dot{x}(t)|^2/R^2)x(t) = \lambda f(t, x(t)), \quad t \in \mathbf{R}, \ \lambda \geq 0 \tag{2}$$

Let $C^1_T(S)$ denote the metric subspace of the Banach space $(C^1_T(\mathbf{R}^3), \|\cdot\|_1)$ of all the T-periodic C^1 maps $x : \mathbf{R} \to S$. An element $(\lambda, x) \in [0,\infty) \times C^1_T(S)$ will be called a solution pair of (2) provided that x is a solution of (2) corresponding to λ.

Let X denote the subset of $[0,\infty) \times C^1_T(S)$ of all the solution pairs of (2) and observe that, because of Ascoli's theorem, X is a locally compact closed set. Clearly, any element $q \in S$ is an equilibrium point of (2) corresponding to the value $\lambda = 0$ of the parameter. So, S can be considered as a subset of X just taking the embedding which assigns to any $q \in S$ the trivial solution pair $(0, q)$. We shall call S the trivial-solution's manifold of (2). A nontrivial solution pair will be an element of $X \backslash S$.

We observe that not all the solution pairs of the form $(0, x)$ are trivial. In fact, there are infinitely many T-periodic solutions of the inertial equation

$$m\ddot{x}(t) + m(|\dot{x}(t)|^2/R^2)x(t) = 0$$

spinning along any given maximal circle of S (necessarily, with constant speed $u = 2k\pi R/T, k = 1, 2, \cdots$). Consequently, the forced oscillations of (2) are not *a priori* bounded in the C^1 norm when λ ranges in the interval $[0, 1]$. This makes the classical continuation methods hard to be applied in this situation.

Theorem 2 *The parametrized equation (2) admits an unbounded connected branch of nontrivial solution pairs whose closure meets the trivial solution's manifold $S \subset [0,\infty) \times C^1_T(S)$.*

The following two lemmas provides some inequalities directly involving the mechanics of the considered motion and will be crucial to prove our result.

Lemma 1 *Let $x : \mathbf{R} \to S$ be a T-periodic solution of (2) corresponding to a given $\lambda \geq 0$. Let $u(t) = |\dot{x}(t)|$ and $F = \max\{|f(t,q)| : t \in \mathbf{R}, q \in S\}$. Then the* *norm of the momentum vector* $p(t) = m\dot{x}(t)$ *is a Lipschitz function with constant* λF. *So, in particular, for any* $t_1, t_2 \in \mathbf{R}$, *one has*

$$m|u(t_2) - u(t_1)| \leq \lambda FT. \tag{3}$$

Proof. Since $x(\cdot)$ satisfies the differential equation (2), if $u(t) \neq 0$, one has

$$m\dot{u}(t) = m\langle \dot{x}(t), \ddot{x}(t)\rangle / u(t) = \langle \dot{x}(t), \lambda f(t, x))\rangle / u(t).$$

Thus, $|m\dot{u}(t)| \leq \lambda F$ for all $t \in \mathbf{R}$ such that $u(t) \neq 0$.

Now, let $t_1, t_2 \in \mathbf{R}$ with $t_1 < t_2$. If $u(t) \neq 0$ for all $t \in (t_1, t_2)$, then the inequality (3) is obvious. Otherwise, without loss of generality, we may assume $u(t_1) \leq u(t_2)$ and $u(t_2) > 0$. Let $c = \max\{t \in [t_1, t_2] : u(t) = 0\}$. Then, since the function u is nonnegative, one obtains

$$m|u(t_2) - u(t_1)| \leq mu(t_2) = m(u(t_2) - u(c)) \leq \lambda F|t_2 - c| \leq \lambda F|t_2 - t_1|.$$

□

Lemma 2 *Let* $x(\cdot), u(\cdot)$, F *be as in Lemma 1. Assume that* $mu(t) > \lambda FT$ *for each* $t \in \mathbf{R}$. *Take* $\tau \in \mathbf{R}$ *and let* α *be the straight line through the origin spanned and oriented by the vector product* $x(\tau) \times \dot{x}(\tau)$. *Denote by* $\rho(t)$ *the distance of* $x(t)$ *from the* α*-axis. Then, for any* $t \in \mathbf{R}$ *the angular momentum* $M_\alpha(t)$ *with respect to the* α*-axis is such that*

$$(mu(\tau) - \lambda FT)R \leq M_\alpha(t) \leq (mu(\tau) + \lambda FT)R \tag{4}$$

Moreover, the distance $\rho(t)$ *of* $x(t)$ *from the* α*-axis satisfies the inequality*

$$\rho(t) \geq \frac{mu(\tau) - \lambda FT}{mu(\tau) + \lambda FT} R. \tag{5}$$

So, in particular $x(\cdot)$ *lies in* $S \backslash \alpha$.

Proof. Since $M_\alpha(t)$ is the projection of the angular momentum $x(t) \times m\dot{x}(t)$ onto the α-axis, one has

$$M_\alpha(t) \leq m\rho(t)u(t), \quad \text{for all } t \in \mathbf{R},$$

and

$$M_\alpha(\tau) = mRu(\tau).$$

Moreover,

$$|\dot{M}_\alpha(t)| \leq \lambda\rho(t)|f(t, x(t))| \leq \lambda FR.$$

Therefore, for any $t \in \mathbf{R}$,

$$|M_\alpha(t) - M_\alpha(\tau)| = |\int_\tau^t \dot{M}_\alpha(s) ds| \leq \lambda FRT,$$

so that

$$(mu(\tau) - \lambda FT)R \leq M_\alpha(t) \leq (mu(\tau) + \lambda FT)R.$$

Now, by applying the inequality (3) to t and τ, one obtains

$$(mu(\tau) - \lambda FT)R \leq M_\alpha(t) \leq m\rho(t)u(t) \leq (mu(\tau) + \lambda FT)\rho(t)$$

which implies

$$\rho(t) \geq \frac{mu(\tau) - \lambda FT}{mu(\tau) + \lambda FT} R.$$

□

The following lemma turns out to be useful in the proof of our existence result (see [5]).

Lemma 3 *Let C be an unbounded closed connected subset of a metric space Y. Assume that any bounded subset of C is relatively compact. If Y_0 is a bounded closed subset of Y which intersects C, then there exists an unbounded connected subset $\tilde{C}$ of $C \setminus Y_0$ whose closure intersects Y_0.*

We are now in a position to give the

Proof (of Theorem 1). Let us associate to equation (1) the parametrized forced equation on S

$$m\ddot{x}(t) + m(|\dot{x}(t)|^2/R^2)x(t) = \lambda f(t, x(t)), \quad t \in \mathbf{R}, \quad \lambda \geq 0. \tag{2}$$

By Theorem 2, the equation (2) admits an unbounded branch $\Sigma \subset [0, \infty) \times C_T^1(S)$ of nontrivial solution pairs (λ, x) of (2) whose closure intersects S. We will prove Theorem 1 by showing that Σ must contain a solution of the form $(1, x)$. Suppose not. Thus Σ is contained in $[0, 1) \times C_T^1(S)$. So, necessarily, its projection onto $C_T^1(S)$ is unbounded. Let us prove that this leads us to a contradiction.

We say that a curve $x \in C_T^1(S)$ is *admissible* if, for any $\tau, t \in \mathbf{R}$ one has $\dot{x}(\tau) \neq 0$ and $\rho(t) > 0$, where $\rho(t)$ denotes the distance of $x(t)$ from the axis through the origin spanned by the vector product $x(\tau) \times \dot{x}(\tau)$. It is evident that the set of all the admissible curves is an open subset of $C_T^1(S)$. We will assign, in a continuous manner, an integer to any admissible curve and we will show that forced oscillations with sufficiently high energy are admissible.

Observe first that, because of the T-periodicity, any curve $x \in C_T^1(S)$ can be considered as defined on the unit circle S^1 of the complex plane by identifying t with $\exp(2\pi i t/T)$. Moreover, if x is admissible and $\tau \in S^1$, we have

$$\langle y(\tau), x(t)\rangle^2 + \langle z(\tau), x(t)\rangle^2 = \rho(t)^2 > 0, \quad \text{for all } \ t \in S^1.$$

where $y(\tau)/|x(\tau)|$ and $z(\tau) = \dot{x}(\tau)/|\dot{x}(\tau)|$. So, we may consider the map $x_\tau : S^1 \to S^1$ given by

$$x_\tau(t) = \frac{\langle y(\tau), x(t)\rangle + i(\langle z(\tau), x(t)\rangle}{\sqrt{\langle y(\tau), x(t)\rangle^2 + \langle z(\tau), x(t)\rangle^2}}.$$

Clearly, the Brouwer degree $\deg(x_\tau)$ of x_τ, because of the homotopy property, is independent of τ) (see e.g., [7]). Hence the integer $i(x) = \deg(x_\tau)$ is well defined and will be called the *index* of the admissible curve x. Again for the homotopy property of the degree, the index is a locally constant function defined on the open set of all the admissible curves. Moreover, it is well known that the degree of a map $\sigma : S^1 \to S^1$ coincides with the winding number of the closed curve σ around the origin. So, since x_τ is C^1, one has

$$i(x) = \frac{1}{2\pi}\int_0^T \dot{\phi}(t)dt,$$

where $\phi(t) = \arg(x_\tau(t))$.

Now, since S is bounded, Lemma 1 implies the existence of a constant $M > 0$ with the property that if $x(\cdot)$ is a forced oscillation of (2), with $||x||_1 \geq M$ and corresponding to some $\lambda \in [0,1)$, then $m|\dot{x}(t)| > FT$ for all $t \in S^1$. So, by Lemma 2, the curve $x(\cdot)$ is admissible. To evaluate $i(x)$, observe that, given $\tau \in S^1$, the rate of change of the angle $\phi(t) = \arg(x_\tau(t))$ is given, with the notation of Lemma 2, by $\dot{\phi}(t) = M_\alpha(t)/m\rho^2(t)$. Hence, from (4) one obtains

$$\dot{\phi}(t) \geq \frac{(m|\dot{x}(\tau)| - \lambda FT)R}{m\rho^2(t)} \geq \frac{m|\dot{x}(\tau)| - FT}{mR},$$

so that

$$i(x) \geq (T/2\pi)\frac{m|\dot{x}(\tau)| - FT}{mR}.$$

Consequently, since S is bounded and τ arbitrary, one can find two constants $a, b > 0$ such that the inequality

$$i(x) \geq a||x||_1 - b \tag{6}$$

holds for any forced oscillation x of (2) corresponding to some $\lambda < 1$, provided that $||x||_1 > M$.

Let us now go back to our branch Σ. Denote by C the closure of Σ in the space $Y = [0,1] \times C^1_T(S)$ and set

$$Y_0 = \{(\lambda, x) \in Y : ||x||_1 \leq M\}.$$

Ascoli's theorem shows that any element $(\lambda, x) \in C$ is still a solution pair and any bounded subset of C is relatively compact in Y. Moreover, the intersection $C \cap Y_0$ is nonempty, since, by Theorem 2, C meets the subset S of Y_0. Therefore Lemma 3 applies to get an unbounded connected subset $\tilde{C}$ of $C \backslash Y_0$. It is clear, by the definition of Y_0, that any forced oscillation x in the projection Γ of $\tilde{C}$ onto $C^1_T(S)$ is such that $||x||_1 > M$ and, consquently, admissible. Now, since Γ is connected, the index, which is a continuous integer valued function, must be constant on Γ. Therefore, from the inequality (6), one obtains that Γ cannot be unbounded. This contradiction shows that there exists $x_0 \in C^1_T(S)$ such that $(1, x_0) \in \Sigma$, as claimed. □

References

1. Benci V. (1986): Periodic solutions of Lagrangian systems on a compact manifold. J. Differential Equations, **63**, 135-161
2. Benci V., Degiovanni M.: Periodic solutions of dissipative dynamical systems. Preprint
3. Capietto A., Mawhin J., Zanolin F.: A continuation approach to superlinear periodic boundary value problems. To appear in J. Differential Equations
4. Furi M., Pera M.P. (1990): On the existence of forced oscillations for the spherical pendulum. Boll. Un. Mat. Ital. **4**-B, 381-390
5. Furi M., Pera M.P. (1989): A continuation principle for the forced spherical pendulum. To appear in Proc. Intern. Conf. on Fixed Point Theory and Applications, (CIRM), Marseille-Luminy), Pitman Research Series
6. Mawhin J. (1988): The forced pendulum: A paradigm for nonlinear analysis and dynamical systems. Expo. Math. **6**, 271-287
7. Milnor J.W. (1965): Topology from the Differentiable Viewpoint. Univ. Press of Virginia, Charlottesville

Radial Bounds for Schrödinger Operators in Euclidean Domains

David Gurarie[1], *Gerhard Kalisch*[2], *Mark Kon*[3], *and Edward Landesman*[4]

[1] Case Western Reserve University
[2] University of California, Irvine
[3] Boston University and Columbia University
[4] University of California, Santa Cruz

Abstract

We extend study of regularity properties of elliptic operators (c.f. [2]) to second order operators on domains bounded by finite numbers of hyperplanes. Previous results for Euclidean space and the symmetry of the domains are exploited to obtain resolvent bounds. Corollaries include semigroup generation, essential self-adjointness, and regularity of eigenfunction expansions for such operators. The present work provides basic results aimed at extending regularity information for partial differential operators (especially with singular coefficients) to a general class of operators in domains with boundary. In one dimension these results encompass a body of work in Sturm-Liouville theory on the half-line.

Introduction

In this note we study regularity properties of second order elliptic operators on certain domains of $\mathbf{R}^n$. This work is part of a program directed at ascertaining behavior of elliptic operators (with possibly singular coefficients) on domains, through study of resolvent kernels and other functions of such operators. In [2] and [3] regularity properties are derived by bounding resolvent kernels with L^1 convolution kernels.

Here we consider analogous results for Schrödinger operators $A = -\Delta + b(x)$ on domains $\Gamma \subset \mathbf{R}^n$ of the form $\Gamma = (\mathbf{R}^+)^k \times \mathbf{R}^{n-k}$, i.e. regions bounded by a finite number of hyperplanes $x_j = 0 \quad (j = 1, \ldots, k)$. We study the resolvent kernel (Green function) $(\zeta - A)^{-1}$ for operators A under Dirichlet and Neumann boundary conditions. A consequence is a variety of results concerning A, including L^p-spectral bounds, closedness, semigroup generation, and summability.

Our results (in particular their consequences for eigenfunction expansions) specialize in one dimension to results in Sturm-Liouville theory (see for instance [5], [4]). In particular, a body of work in spectral theory and summability for Sturm-Liouville operators on the half line is subsumed in the summability, self-adjointness, and semigroup results here (note that a half line is an example of the domains under consideration).

Results similar to ours have been previously obtained in various situations without boundary, e.g., for Schrödinger operators on $\mathbf{R}^n$, elliptic operators on compact manifolds, and perturbations of elliptic operators on $\mathbf{R}^n$ ([2],[3]) to list some. Little has been previously known (in the direction of the present results) regarding operators arising from boundary value problems. The limitations of what might be called standard approaches correspond to difficulties in construction of the resolvent kernel even for the unperturbed operator $A_0 = -\Delta$. We circumvent this by exploiting symmetries which allow explicit construction of the free resolvent kernel $(\zeta + \Delta)^{-1}$, and then the full kernel as a perturbation of the free resolvent. Indications are that our results extend qualitatively to the general situation involving an elliptic operator in a (not necessarily compact) domain with boundary.

1 The unperturbed resolvent kernel

Let A_0 denote $(-\Delta)$ on Γ, with boundary condition

$$\alpha_j u + \beta_j \partial_n u = 0 \tag{M}$$

on the hyperplane $x_j = 0$, $j = 1, 2, \ldots, k$. Here $(\alpha_j, \beta_j) = (0,1)$ or $(1,0)$ for each j, and ∂_n denotes the normal derivative on $\partial\Gamma$. Clearly, (M) includes both Dirichlet $(u|_{\partial\Gamma} = 0)$, and Neumann $(\partial_n u|_{\partial\Gamma} = 0)$ boundary conditions, which may differ on different parts of the boundary.

We denote by $G^0 = G^0_\zeta(x, y)$ the resolvent kernel of A_0 and by $R = R_\zeta(x, y)$ that of $-\Delta$ on all of $\mathbf{R}^n$. The kernel R_ζ is well known, being for each complex ζ, $\operatorname{Re}\zeta < 0$ a convolution kernel $R_\zeta(|x - y|)$, with

$$R_\zeta(|x|) = \int_0^\infty e^{t\zeta - \frac{x^2}{4t}} (4\pi t)^{-\frac{n}{2}} dt, \tag{1}$$

a radial function. It is shown, e.g., in [3] that, uniformly for $|\arg\zeta| \geq \theta_0$ (for any given $\theta_0 > 0$), R_ζ can be written as an L^1-dilation

$$R_\zeta(|x|) = \rho^{\frac{n}{2}-1} R_\theta(\sqrt{\rho}|x|), \quad \rho = |\zeta| \tag{2a}$$

of the radial function R_θ $(\theta = \arg\zeta)$, which is estimated for $n \geq 2$ by

$$|R_\theta(|x|)| \le h_\theta(|x|) = \frac{C}{|\sin\frac{\theta}{2}|^s}\begin{cases} |x|^{2-n}, \ (1-\log|x| \text{ if } n=2); & |x| \le 1 \\ \frac{e^{-\gamma|\sin\frac{\theta}{2}||x|}}{|x|^t}; & |x|>1 \end{cases}, \tag{2b}$$

with $s > \frac{n+1}{2}$, $t = \frac{n-1}{2} + s > n$, any $\gamma < 1$, and a constant $C = C(\gamma, s)$. For our purposes it is significant that h_θ is in L^1, and is radial and monotonically decreasing.

To express the resolvent kernel G^0 for the domain Γ in terms of R we can use standard reflection principle arguments. For convenience we use the following notation: let $\epsilon = (\epsilon_1, \ldots, \epsilon_k)$ be a k-tuple with entries ± 1; $\operatorname{sgn} \epsilon = \prod_1^k \epsilon_j$. Each vector $x \in \Gamma$ will be written as a pair (x', x'') with $x' \in (\mathbf{R}^+)^k$, $x'' \in \mathbf{R}^{n-k}$, and $\epsilon(x)$ will denote the point $(\epsilon_1 x_1, \ldots, \epsilon_k x_k;\ x'') \in \mathbf{R}^n$.

We introduce $b_j = \beta_j - \alpha_j$, $(j = 1, \ldots, k)$ and denote by $\{c_\epsilon\}_\epsilon$ the coefficients in the formal expansion of the product

$$\prod_{j=1}^k (e^{x_j} + b_j e^{-x_j}) = \sum_\epsilon c_\epsilon e^{\epsilon \cdot x}, \qquad x = (x_1, \ldots, x_k);$$

the summation is over all ϵ, and $\epsilon \cdot x$ denotes a dot product. Then standard reflection principle arguments yield the representation

$$G^0_\zeta(x,y) = \sum_\epsilon c_\epsilon R_\zeta(x - \epsilon(y)). \tag{3}$$

Indeed, it is easy to check that $G^0_\zeta(x,y)$ solves the Green function equation, and satisfies the boundary condition (M). Thus G^0 is a linear combination of "convolution-type" kernels.

2 Resolvent of the perturbed operator

We now construct the resolvent $G = G_\zeta(x,y)$ of the operator $A = -\Delta + b(x)$, with the perturbation series identity (cf. [5], [2])

$$G = G^0 \sum_{k=0}^\infty (bG^0)^k. \tag{4}$$

Above, b denotes the operator consisting of multiplication by $b(x)$.

The kth term of (4) is a composition of convolution-type kernels G^0 and multiplications with $b(x)$. The problem of analyzing (4) reduces to the Euclidean case $\mathbf{R}^n$, by the following extension procedure. We extend each function f on Γ to all of $\mathbf{R}^n$ by setting

$$\tilde{f}(\epsilon(x)) = c_\epsilon f(x), \qquad \forall \epsilon,\ x \in \Gamma.$$

Obviously, the subsets $\{\epsilon(\Gamma)\}_\epsilon$ disjointly partition $\mathbf{R}^n$. Hence $\|\tilde{f}\|_{L^p(\mathbf{R}^n)} \sim \|f\|_{L^p(\Gamma)}$ and by (3)

$$G^0(f) = R * \tilde{f},$$

with convolution on all of $\mathbf{R}^n$, where above $R \equiv R_\zeta$. Now we proceed to establish the following:

Theorem 1 *Let $b(x) \in L^r + L^\infty$ on $\mathbf{R}^n$ or on the quotient Γ/L, where L is a subspace of $\mathbf{R}^n$, $\dim \mathbf{R}^n/L = m$. Let $\frac{m}{r} < 2$. Then*

(I) For ζ outside $\Omega_1 = \left\{\zeta = \rho e^{i\theta} : \frac{C_1 \rho^{\frac{m}{2r}-1}}{|\sin\frac{\theta}{2}|^{\frac{n-1}{2}}} < 1\right\}$, a parabolic domain about $\mathbf{R}^+$, the operator $b(\zeta - A_0)^{-1}$ is bounded in all $L^p(\Gamma)$, $1 \le p \le r$, and its norm is estimated by

$$\|bG^0_\zeta(f)\|_p \le \frac{C_1}{|\sin\frac{\theta}{2}|^s}\rho^{d-1}\|f\|_p,$$

where $d = \frac{m}{2r} < 1$, and $s > \frac{n+1}{2}$, $\zeta = \rho e^{i\theta}$.

(II) The geometric series (4) converges absolutely for all ζ in the complement of $\Omega_2 = \left\{\zeta = \rho e^{i\theta} : \frac{C_2 \rho^{\frac{m}{2r}-1}}{|\sin\frac{\theta}{2}|^{\frac{n-1}{2}}} < 1\right\}$, and defines an integral kernel $G_\zeta(x,y)$ which admits the bound

$$|G_\zeta(x,y)| \le C(\rho,\theta)\rho^{\frac{n}{2}-1}\, h_\theta(\sqrt{\rho}|x-y|),$$

with h_θ given by (2a) and

$$C(\rho,\theta) = \frac{C_2}{|\sin\frac{\theta}{2}|^s}\left(1 - C_2\frac{\rho^{\frac{m}{2r}-1}}{|sin\frac{\theta}{2}|^s}\right)^{-1}.$$

(III) If $\zeta \notin \Omega_2$, then for all L^p-spaces, $1 \le p \le r$, G_ζ is the resolvent of $A = -\Delta + b(x)$ in L^p.

Proof. We first prove II by considering the perturbation series (4). By writing out the kernels of the terms in (4) we conclude that the jth term $L_j = G^0(BG^0)^j$ has kernel

$$L_j(x,y) = \sum_{\bar{\epsilon}} c_{\bar{\epsilon}} \int_\Gamma \cdots \int_\Gamma R(x - \epsilon_1(z_1))b(z_1)R(z_1 - \epsilon_2(z_2)) \\ \ldots b(z_j)R(z_j - \epsilon_{j+1}(y))dz_1 \ldots dz_j. \tag{5}$$

The summation above is over all tuples of reflections $\bar{\epsilon} = (\epsilon_1, \ldots, \epsilon_{m+1})$ (that is, it is a sum over the variables $\epsilon_1, \epsilon_2, \ldots, \epsilon_{m+1}$), $c_{\bar{\epsilon}} = \prod_{i=1}^{m+1} c_{\epsilon_i}$, and $x, y, z_i \in \Gamma$.

We now assume that $b \in L^r$ on Γ; the same argument will hold in the case $b \in L^\infty$, and thus more generally in the case $b \in L^r + L^\infty$. Applying a multiple Hölder inequality to (5), we have

$$|L_j(x,y)| \le \|b\|_r^j \sum_{\bar{\epsilon}} |c_{\bar{\epsilon}}| \|R(x - \epsilon_1(z_1)) \ldots R(z_j - \epsilon_{j+1}(y))\|_{r',\ldots,r'}, \tag{6}$$

with $r' = \frac{r}{r-1}$, where the subscript $r', r', ..., r'$; denotes an $L^{r'}$ norm with respect to all of the variables $z_1, z_2, \ldots, z_j$. We remark that the norm $||b||_r$ is taken in the quotient space $\mathbf{R}^n/L$, as Hölder's inequality is applied here to the integration only with respect to the variables on which b depends (b does not depend on the variables complementary to L). In the case where L is nontrivial, the last term in (6) involves a mixed $L^{r'}$ norm; see the discussion after (9).

We now define a modification of the standard convolution, which we denote as the p, ϵ convolution. For two functions f, g on $\mathbf{R}^n$, this is defined by

$$(f *_{p,\epsilon} g)(x) = \left(\int_\Gamma |f(x - \epsilon(y))g(y)|^p dy\right)^{\frac{1}{p}}. \tag{7}$$

If we define $g^\epsilon(x) = g(\epsilon^{-1}(x))$, (7) becomes the "$\Gamma$ part" of a p-convolution on all of $\mathbf{R}^n$:

$$\begin{aligned}(f *_{p,\epsilon} g)(x) &= \left(\int_\Gamma |f(x - y)g^\epsilon(y)|^p dy\right)^{\frac{1}{p}} \\ &\leq \left(\int_{\mathbf{R}^n} |f(x - y)\, g^\epsilon(y)|^p \, dy\right)^{\frac{1}{p}} \equiv (f *_p g^\epsilon)(x),\end{aligned}$$

where the last equality is the definition of the p-convolution $*_p$. It is immediate from the definition of the $*_{p,\epsilon}$-convolution that the last term of (6) can be written

$$R *_{r',\epsilon} R *_{r',\epsilon} \ldots *_{r',\epsilon} R.$$

We then bound each kernel $R = R_\zeta$ using the radial function (2a) with the help of the following Lemma, proved in [3]:

Lemma *Let*

$$h_{s,t,\gamma}(|z|) \equiv \begin{cases} |z|^{-s} \ \ (1 - \log |z|); & |z| \leq 1 \\ |z|^{-t} e^{-\gamma|z|}; & |z| \geq 1 \end{cases} \qquad (s < n < t, \ \gamma > 0);$$

then there is a constant C (which may depend on s_i, t_i, and γ) such that

$$h_{s_1,t_1,\gamma} *_{p,\epsilon} h_{s_2,t_2,\gamma} \leq C h_{s,t,\gamma}; \tag{8}$$

with $s = \min\{s_1, s_2\}$, $t = \min\{t_1, t_2\}$.

By induction (8) extends to an arbitrary number of (p, ϵ)-convolutions.

Considering the dilating factor $\sqrt{\rho}$ $(\rho = |\zeta|)$, we estimate the term L_j by observing that the $*_p$ convolution has the following scaling property:

$$f(\sqrt{\rho}x) *_p g(\sqrt{\rho}x) = \sqrt{\rho}^{-n/p}(f *_p g)(\sqrt{\rho}x).$$

Combining this with the Lemma and the bounds (2), we have

$$|L_j(x,y)| \leq C' \left(\sum |c_\epsilon|\right) \rho^{\frac{n}{2}-1} \left(\frac{C'||b||_r \rho^{\frac{m}{2r}-1}}{|\sin\frac{\theta}{2}|^s}\right)^j h_\theta(\sqrt{\rho}|x-y|), \tag{9}$$

with h_θ given by (2), where $d = \frac{m}{2r}$. Note that the exponent of ρ is $\frac{m}{2r} - 1$, where $m = \dim \mathbf{R}^n/L$. If $m \neq n$ (i.e., if L is nontrivial), then the convolutions in (7) as they are used to bound the last term in (6) must be slightly modified, whence the modified scaling in this situation yields the power $\frac{m}{2r} - 1$ for $m \neq n$. Summing up a geometric series of bounds (9), (II) follows.

To prove (I) is straightforward now, since to estimate $||bG_\zeta^0||$, we need only replace G_ζ^0 by its convolution kernel upper bound above, and then apply first Young's inequality, followed by Hölder's inequality. Similarly, (III) now follows, since it is clear that the kernel defined by the series (4) must define the L^p-resolvent of the full operator A if it converges in L^p. The constraint $1 \leq p \leq r$ is necessary because even though the series (4) defines a right inverse of $\zeta - A$ for all p, it is clear that if the potential $b \notin L^p$, then the operator $\zeta - A$ may not even have a dense domain (e.g., if b has a dense set of singularities). This completes the proof of the theorem. □

Remark. The above argument shows that the main contribution to the kernel $G(x,y)$ comes, as in the Euclidean case, from the sum of "true convolution" terms, in (5), i.e. $\epsilon_1 = \ldots = \epsilon_{m+1} = I$. The influence of the remaining (reflected) terms ($\epsilon_i \neq I$) decays as $\zeta \to \infty$; this can be observed from (9), when one studies the behavior of the integral in reflected domains $\epsilon(\Gamma)$, ($\epsilon \neq I$). It is expected that the corresponding observation for an analogous reflection procedure involving higher order operators (on more general domains) will imply the same conclusions for resolvent kernels (and thus for related analytic properties of operators, see, e.g., below).

Theorem 1 has a number of functional analytic applications.

Corollary 1 *The operator A is closeable in all L^p, $1 \leq p \leq r$ and its spectrum is included in Ω_1.*

This follows immediately from the theorem, since $(\zeta - A)^{-1}$ is bounded for $\zeta \notin \Omega_2$.

Corollary 2 *The domain of A in L^p $(1 \leq p \leq r)$ is equal to the domain of $A_0 = -\Delta$, the latter being the Sobolev space $\mathcal{L}_2^p(\Gamma)$.*

This corollary follows from the bounds on the resolvents of A_0 and A, which in turn immediately lead to a priori inequalities between A_0 and A. The standard (Kato-Rellich) perturbation theorems then apply to show that A_0 and A have the same domain. The following corollary also follows from the standard theory:

Corollary 3 *If $r \geq 2$ and $b(x)$ is real, then A is semibounded from below, and hence essentially self-adjoint on $C_0^\infty(\Gamma)$.*

We also have

Theorem 2 *(resolvent summability)*

For each $f \in L^p$ $(1 \le p < \infty)$ or C^0, the operator

$$\zeta G_\zeta(f) \to f(x), \tag{10}$$

as $\zeta \to \infty$, uniformly in each sector $\Omega_\theta = \{\zeta |\ |\arg \zeta| \ge \theta > 0\}$, in L^p-norm and pointwise on the Lebesgue set of $f \in L^p$.

Proof. The convolution bounds on G_ζ established in the Theorem reduce the problem to showing that a scaled convolution kernel behaves as an approximate identity, which reduces the problem to elementary harmonic analysis (see [2, Theorem 1]). □

Summability theory for eigenfunction expansions has been studied extensively in certain contexts. Previous classes of results, however, have been restricted to one dimensional domains, compact manifolds (with or without boundary), or general domains without boundary.

Note the theorem implies that the convergence in (10) holds a.e.

We briefly indicate some other consequences. With Theorem 2 one can study other multipliers $\phi(A)$ and analytic summation families, $\{\phi_\delta(A)\}_\delta$, using the Dunford functional calculus, i.e. Cauchy integration along a suitable contour γ in $\mathbf{C}$,

$$\phi(A) = \frac{1}{2\pi i} \int_\gamma \phi(\zeta) G_\zeta d\zeta. \tag{11}$$

Using $\phi(\zeta) = e^{-t\zeta}$ gives the semigroup. Specifically, it is easy to show (using the convolution bounds on the integrand obtained from Theorem 1) that in this case (11) represents a holomorhpic semigroup. Corresponding bounds on the kernel of (11), obtained by substituting the bounds of Theorem 1 into (11), then show in the same way as we did in Theorem 2 for $\zeta \to \infty$ that the behavior of the semigroup as $t \to 0$ is correct. Indeed, the statements are equivalent, since $t \to 0$ corresponds by a change of variables in (11) to the limit $\zeta \to \infty$ in the integrand. Thus we have:

Theorem 3 *The operator A is the generator of a holomorphic semigroup $T_t = e^{-At}$ in the right half plane $Re\, t > 0$, in all L^p-spaces $(1 \le p \le r)$. The semigroup is continuous at $\{0\}$, i.e.*

$$T_t f(x) \underset{t \to 0}{\longrightarrow} f(x)$$

in L^p and pointwise on the Lebesgue set of f, uniformly in any sector $\Omega'_\theta = \{|\arg t| \le \theta < \frac{\pi}{2}\}$.

Remark. The above analysis applies to higher order constant-coefficient operators with boundary conditions analogous to (M), and a general class of lower order perturbations of these, (see [2]).

References

1. Copson, E.T. (1965): Asymptotic Expansions. Cambridge University Press, London
2. Gurarie, D., Kon, M. (1984): Radial bounds for perturbations of elliptic operators. J. Functional Analysis **56**, 99-123
3. Gurarie, D. (1984): Kernels of elliptic operators: bounds and summability. J. Differential Equations **55**, 1-29
4. Levitan, B.M., Sargsjan, I.S. (1975): Introduction to Spectral Theory: Self-Adjoint Ordinary Differential Operators. American Mathematical Society, Providence
5. Titchmarsh, E. (1946): Eigenfunction Expansions Associated with Second Order Differential Equations. Vol. I, II. Oxford University Press, Oxford

Jumping Nonlinearity for 2nd Order ODE with Positive Forcing

P. Habets,[1] *M. Ramos*[2*] *and L. Sanchez*[2]

[1] Mathem. Institute, Université de Louvain, Chemin du Cyclotron 2, Louvain-la-Neuve, Belgium
[2] INIC/CMAF, Av. Professor Gama Pinto, 2, P 1699 Lisboa, Portugal

1 Introduction

In this paper we study the Neumann boundary value problem (BVP)

$$\begin{aligned} &u'' + \mu u^+ - \nu u^- = p(t) \\ &u'(0) = u'(\pi) = 0, \end{aligned} \tag{1.1}$$

where $p(t)$ is a positive function, $\mu > 0$, $\nu > 0$, $u^+(t) = \max(u(t), 0)$ and $u^-(t) = max(-u(t), 0)$.

Our main motivation goes back to the theory of suspension bridges. If one looks for solutions of the form

$$u(t)\sin(\pi x/L)$$

(see [5]) it can be shown that u must satisfy a BVP

$$\begin{aligned} &u'' + \mu u^+ - \nu u^- = p(t), \\ &u(0) = u(2\pi),\ u'(0) = u'(2\pi). \end{aligned} \tag{1.2}$$

If further we assume the forcing satisfies the symmetry condition

$$p(\pi + t) = p(\pi - t), 0 \le t \le \pi,$$

one obtains solutions of (1.2) from solutions of (1.1).

The periodic problem (1.2) has been investigated by A. C. Lazer and P. J. McKenna [5], [6]. Some results on the Neumann problem (1.1) can be found in D. C. Hart, A. C. Lazer and P. J. McKenna [4]. In all these papers, the forcing is a small perturbation of a positive constant.

In the present work, we investigate some general positive forcing p for some regions of the (μ, ν) plane.

In case $0 < \mu \le 1/4$, we prove (1.1) has a unique positive solution. In the region R_1

$$1/4 < \mu < 1,\ \nu > 0,\ \frac{1}{2\sqrt{\mu}} + \frac{1}{2\sqrt{\nu}} > 1$$

* On leave from Faculdade de Ciências de Lisboa with a scholarship from I.N.I.C.

we give necessary and sufficient condition for existence of a positive solution, of a solution u with a simple zero such that $u(0) > 0$ and of a solution u with a simple zero such that $u(0) < 0$. Putting these results together we obtain existence and uniqueness of solutions in this region and the positive cone of forcings can be divided into three cones, each of them corresponding to solutions of a given type. Such necessary and sufficient conditions were obtained for Dirichlet problem and in a very specific case by L. Aguinaldo and K. Schmitt [1]. We also obtain a similar description of solutions in the region R_2

$$1/4 < \mu < 1,\ \nu > 0,\ 1 > \frac{1}{2\sqrt{\mu}} + \frac{1}{2\sqrt{\nu}}$$

where three solutions may exist simultaneously. For nonlinear problems such as

$$\begin{aligned} u'' + \mu u^+ - \nu u^- &= p(t,u) \\ u'(0) = u'(\pi) &= 0 \end{aligned}$$

similar results can be obtained and will be published elsewhere.

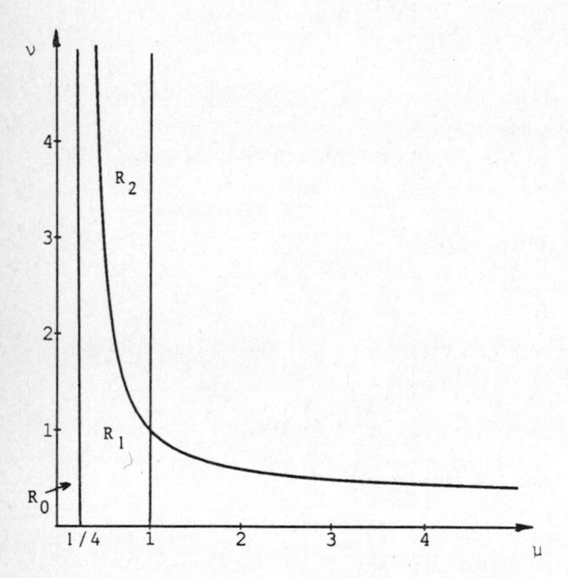

Fig. 1.

2 Auxiliary Lemmas

In this section, we collect some elementary results which we use in the sequel. We denote by $\mathcal{C}[a,b]$ the space of continuous functions $u : [a,b] \to \mathbb{R}$ and write $\mathcal{C}$ for $\mathcal{C}[0,\pi]$.

Lemma 2.1 *Let $p \in C[a,b]$ be such that $p(t) > 0$, let $\gamma > 0$ and $u(t)$ be a solution of*

$$\begin{aligned} u'' + \gamma u = p(t) \\ u'(a) = u(b) = 0. \end{aligned} \tag{2.1}$$

Then

(a) $u(t) > 0$ *in* (a,b) *implies* $b - a \geq \pi/2\sqrt{\gamma}$ *;*
(b) $u(t) < 0$ *in* $[a,b)$ *if and only if* $b - a < \pi/2\sqrt{\gamma}$.

If further $\gamma \neq (2n+1)^2\pi^2/4(b-a)^2$, $n \in \mathbf{N}$ *one has :*

(c) $u'(b) = \frac{1}{\cos(\sqrt{\gamma}(b-a))} \int_a^b p(t)\cos(\sqrt{\gamma}(t-a))\,dt.$

Lemma 2.2 *Let $p \in C[a,b]$ be such that $p(t) > 0$, let $\gamma > 0$ and $u(t)$ be a solution of*

$$\begin{aligned} u'' + \gamma u = p(t) \\ u(a) = u'(b) = 0. \end{aligned} \tag{2.2}$$

Then

(a) $u(t) > 0$ *in* (a,b) *implies* $b - a \geq \pi/2\sqrt{\gamma}$ *;*
(b) $u(t) < 0$ *in* $(a,b]$ *if and only if* $b - a < \pi/2\sqrt{\gamma}$.

If further $\gamma \neq (2n+1)^2\pi^2/4(b-a)^2$, $n \in \mathbf{N}$ *one has :*

(c) $u'(a) = \frac{-1}{\cos(\sqrt{\gamma}(b-a))} \int_a^b p(t)\cos(\sqrt{\gamma}(b-t))\,dt.$

Lemma 2.3 *Let $p \in C[a,b]$ be such that $p(t) > 0$, $\gamma > 0$ and let $u(t)$ be a solution of*

$$\begin{aligned} u'' + \gamma u = p(t) \\ u(a) = u(b) = 0. \end{aligned} \tag{2.3}$$

Then

(a) $u(t) < 0$ *in* (a,b) *if and only if* $b - a < \pi/\sqrt{\gamma}$ *;*
(b) $u(t) < 0$ *in* (a,b) *implies* $u'(a) < 0$ *and* $u'(b) > 0$.

Lemma 2.3 is contained in Theorem 1.14 and Propositions 1.15 and 1.16 in D. De Figueiredo [2]. The other lemmas can be proved in a similar way.

3 Positive solutions

Our first result is a necessary and sufficient condition for existence of positive solution u, i.e. such that $\forall t \in [0,\pi]$, $u(t) \geq 0$, in case $\mu \in (0,1)$.

Proposition 3.1. *Let $0 < \mu < 1$, $\nu \in \mathbf{R}$ and $p \in C$ be such that $p(t) > 0$. Then the BVP*

$$\begin{aligned} u'' + \mu u_+ - \nu u_- &= p(t) \\ u'(0) = u'(\pi) &= 0 \end{aligned} \tag{3.1}$$

has a positive solution if and only if

$$A := \int_0^\pi p(s)\cos(\sqrt{\mu}s)\,ds \geq 0$$

and (3.2)

$$B := \int_0^\pi p(s)\cos(\sqrt{\mu}(\pi - s))ds \geq 0$$

Proof. If u is a positive solution of (3.1), one computes

$$u(t) = \int_0^\pi G(t,s)p(s)ds \tag{3.3}$$

where

$$\begin{aligned} G(t,s) &= \cos(\sqrt{\mu}s)\cos(\sqrt{\mu}(\pi - t))/\sqrt{\mu}\sin(\sqrt{\mu}\pi), \text{ if } s \leq t, \\ G(t,s) &= \cos(\sqrt{\mu}t)\cos(\sqrt{\mu}(\pi - s))/\sqrt{\mu}\sin(\sqrt{\mu}\pi), \text{ if } s > t. \end{aligned}$$

Hence one has

$$A = u(\pi)\sqrt{\mu}\sin(\sqrt{\mu}\pi) \geq 0 \text{ and } B = u(0)\sqrt{\mu}\sin(\sqrt{\mu}\pi) \geq 0.$$

Reciprocally, let u be defined by (3.3). The conditions (3.2) imply $u(0) \geq 0$ and $u(\pi) \geq 0$. Let $t_0 \in (0,\pi)$ be such that $u(t_0)$ is negative. One deduces then from Lemmas 2.1 and 2.2

$$\pi = t_0 + (\pi - t_0) \geq \frac{\pi}{2\sqrt{\mu}} + \frac{\pi}{2\sqrt{\mu}} > \pi$$

which is a contradiction. Hence $u(t)$ is positive. □

Remark 3.1. The conditions $A \geq 0$, $B \geq 0$ can be replaced by $u(0) \geq 0$ and $u(\pi) \geq 0$. Also, we proved that the positive solution is such that $u(t) > 0$ on $(0,\pi)$.

The next result proves the uniqueness of the positive solution in one region of the μ, ν plane.

Proposition 3.2. *Let $0 < \mu < 1$ and $\nu > 0$ be such that*

$$\frac{1}{2\sqrt{\mu}} + \frac{1}{2\sqrt{\nu}} > 1.$$

Then, if u is a positive solution of (3.1), it is the only solution of (3.1).

Proof. Let v be a solution of (3.1) and assume $w = u - v \neq 0$. The function w is such that

$$\begin{aligned} w'' + Mw &= (M - \alpha(t))w \\ w'(0) &= w'(\pi) = 0, \end{aligned} \tag{3.4}$$

where $m = \min(\mu, \nu) \leq \alpha(t) \leq \max(\mu, \nu) < M$.

If w has constant sign, direct integration of (3.4) gives

$$0 = \int_0^\pi \alpha(t) w(t)\, dt = \int_0^\pi \alpha(t) \mid w(t) \mid dt$$

and since $\alpha(t) \geq m > 0$, one deduces $w(t) \equiv 0$, which is a contradiction.

If w changes sign at $t = b$, by uniqueness of solutions of (3.4), one has $w'(b) \neq 0$. Hence, we can find $0 \leq a < b < c \leq \pi$ such that $w'(a) = w(b) = w'(c) = 0$ and either

$$w(t) > 0 \text{ on } [a, b) \text{ and } w(t) < 0 \text{ on } (b, c]$$

or

$$w(t) < 0 \text{ on } [a, b) \text{ and } w(t) > 0 \text{ on } (b, c].$$

In the first case, Lemma 2.1 implies $b - a \geq \pi/2\sqrt{M}$, and since $c - b = \frac{\pi}{2\sqrt{\mu}}$ one has

$$\pi \geq (b - a) + (c - b) \geq \frac{\pi}{2\sqrt{M}} + \frac{\pi}{2\sqrt{\mu}}.$$

We can choose M small enough so that $\frac{\pi}{2\sqrt{M}} + \frac{\pi}{2\sqrt{\mu}} > \pi$, which is a contradiction. A similar argument holds in the second case. □

Corollary 3.1. *Let $0 < \mu < 1/4$, $\nu > 0$ and $p \in C$ be such that $p(t) > 0$. Then the BVP (3.1) has exactly one solution which is positive.*

Proof. If $\mu \leq 1/4$, one has $A \geq 0$ and $B \geq 0$. □

In case $\mu = 1$, we can still give a necessary and sufficient condition for existence of positive solution but in that case there exist a continuum of solutions.

Proposition 3.3. *Let $\mu = 1$, $\nu \in \mathbf{R}$ and $p \in C$ be such that $p(t) > 0$. Then the BVP (3.1) has a positive solution if and only if*

$$\int_0^\pi p(s) \cos s\, ds = 0. \tag{3.5}$$

Further, all positive solutions are

$$u(t) = u_0 \cos t + \int_0^t p(s) \sin(t - s)\, ds, \tag{3.6}$$

where $u_0 \in [0, V]$ and $V := \int_0^\pi p(s) \sin s\, ds > 0$.

Proof. To prove (3.5) is necessary, multiply the equation by $\cos t$ and integrate.

Notice that all possible positive solutions u of (3.1) are given by (3.6). Further we must have

$$u(0) = u_0 \geq 0 \text{ and } u(\pi) = -u_0 + \int_0^\pi p(s) \sin s \, ds \geq 0,$$

i.e. $u_0 \in [0, V]$.

At last, if (3.5) is satisfied and $u_0 \in [0, V]$, the function u defined by (3.6) is a positive function. Indeed, one has $u(0) \geq 0$ and $u(\pi) \geq 0$. Let then t_0 be such that $u(t_0) \leq 0$. One has from Lemmas 2.1 and 2.2 that

$$t_0 \geq \frac{\pi}{2} \text{ and } \pi - t_0 \geq \frac{\pi}{2},$$

i.e. $t_0 = \pi/2$, and one computes

$$u(t_0) = u\left(\frac{\pi}{2}\right) = \int_0^{\pi/2} p(s) \sin\left(\frac{\pi}{2} - s\right) ds > 0.$$

□

If $\mu > 1$, necessary and sufficient conditions for existence of positive solutions do not seem to exist. However one can write sufficient conditions.

Proposition 3.4. *Let $1 < \mu < 4$ and $p \in C$ be such that $p(t) > 0$. Then the BVP (3.1) has a positive solution if*

$$A = \int_0^\pi p(s) \cos(\sqrt{\mu} s) \, ds \leq 0, \; B = \int_0^\pi p(s) \cos(\sqrt{\mu}(\pi - s)) \, ds \leq 0$$

and

$$\int_0^{\pi/2} p(s) \cos(\sqrt{\mu} s) \, ds \geq 0, \; \int_{\pi/2}^\pi p(s) \cos(\sqrt{\mu}(\pi - s)) \, ds \geq 0. \tag{3.7}$$

Proof. Let u be defined by (3.3). As in Proposition 3.1 we prove $u(0) \geq 0$, $u(\pi) \geq 0$. Let then $t_0 \in (0, \pi)$ be a minimum of u. Multiply the equation

$$u'' + \mu u = p(t)$$

by $\cos(\sqrt{\mu} s)$ or $\cos(\sqrt{\mu}(\pi - s))$ and integrate to obtain

$$u(t_0) = \frac{1}{\sqrt{\mu} \sin(\sqrt{\mu} t_0)} \int_0^{t_0} p(s) \cos \sqrt{\mu} s \, ds$$

or

$$u(t_0) = \frac{1}{\sqrt{\mu} \sin(\sqrt{\mu}(\pi - t_0))} \int_{t_0}^\pi p(s) \cos \sqrt{\mu}(\pi - s) \, ds.$$

From (3.7), one can see that $u(t_0) \geq 0$. Hence, u is a positive solution of (3.1).

□

4 A uniqueness result

In this section, we prove the uniqueness of solutions of (3.1) in case $(\mu, \gamma) \in R_0 \cup R_1$, i.e.,

$$0 < \mu < 1,\ \nu > 0 \text{ and } \frac{1}{2\sqrt{\mu}} + \frac{1}{2\sqrt{\nu}} > 1.$$

We shall need the following lemmas, the proof of which are consequences of Lemmas 2.1 and 2.2.

Lemma 4.1. *Let $0 < \mu < 1$, $\nu > 0$, $p \in C$ be such that $p(t) > 0$ and let v be any solution of (3.1). Then v has at most one zero.*

Proof. If $u(0) > 0$ and $u(\pi) > 0$ we know from Remark 3.1 that $u(t) > 0$ on $(0, \pi)$. Hence, we can assume $u(0) \leq 0$ or $u(\pi) \leq 0$. In case $u(0) \leq 0$, let $a \geq 0$ be the first zero of u. We know from Lemma 2.3 that $u > 0$ on $\left(a, a + \frac{\pi}{\sqrt{\mu}}\right)$ and we are done. The same argument holds true if $u(\pi) \leq 0$. □

Lemma 4.2. *Let $0 < \mu < 1$, $\nu > 0$ be such that*

$$\frac{1}{2\sqrt{\mu}} + \frac{1}{2\sqrt{\nu}} > 1$$

and let $p \in C$ be such that $p(t) > 0$. Let $u(t)$ be the solution of

$$\begin{aligned} u'' + \mu u &= p(t) \\ u'(0) = u'(\pi) &= 0 \end{aligned} \tag{4.1}$$

and $v(t)$ be any solution of (3.1). Then

$$sign\ u(0) = sign\ v(0).$$

Proof. By direct integration of the equations (3.1) or (4.1), it is easy to see that u and v take positive values. If u or v is positive, then by proposition 3.2, $u(t) \equiv v(t)$. If u and v are not positive, we know from Lemma 4.1 that these functions have exactly one zero and

$$u(0)u(\pi) < 0, \quad v(0)v(\pi) < 0. \tag{4.2}$$

Let us assume first

$$u(0) > 0 > v(0). \tag{4.3}$$

From Lemmas 2.1 and 2.2, one has

$$u(t) > 0 \text{ on } \left[0, \frac{\pi}{2\sqrt{\mu}}\right),\ v(t) > 0 \text{ on } \left(\pi - \frac{\pi}{2\sqrt{\mu}}, \pi\right].$$

Further on $\left[\pi - \frac{\pi}{2\sqrt{\mu}}, \pi\right]$,

$$w(t) := u(t) - v(t) = (u(\pi) - v(\pi)) \cos \sqrt{\mu}(\pi - t).$$

From (4.2) and (4.3) we have $v(\pi) > 0 > u(\pi)$ so that

$$w'\left(\pi - \frac{\pi}{2\sqrt{\mu}}\right) = (u(\pi) - v(\pi))\sqrt{\mu} < 0.$$

Define

$$a = \sup\{t < \pi - \frac{\pi}{2\sqrt{\mu}} \mid w'(t) = 0\} < \pi - \frac{\pi}{2\sqrt{\mu}}.$$

On $\left[a, \pi - \frac{\pi}{2\sqrt{\mu}}\right)$, we have $w(t) > 0$ and

$$w'' + Mw = (M - \alpha(t))w > 0$$

$$w'(a) = w\left(\pi - \frac{\pi}{2\sqrt{\mu}}\right) = 0$$

where $M > \max(\mu, \nu) \geq \alpha(t)$. From Lemma 2.1

$$\pi - \frac{\pi}{2\sqrt{\mu}} - a \geq \frac{\pi}{2\sqrt{M}}$$

and we can choose M such that

$$\pi \geq \pi - a \geq \frac{\pi}{2\sqrt{M}} + \frac{\pi}{2\sqrt{\mu}} > \pi$$

which is a contradiction.

A similar argument proves $u(0) < 0 < v(0)$ is impossible and the lemma in proved. □

Proposition 4.1. *Let* $0 < \mu < 1, \nu > 0$ *be such that*

$$\frac{1}{2\sqrt{\mu}} + \frac{1}{2\sqrt{\nu}} > 1$$

and let $p \in C$ *be such that* $p(t) > 0$*. Then the BVP (3.1) has at most one solution.*

Proof. Let u and v be two solutions of (3.1). From Lemma 4.2 we know sign $u(0) =$ sign $v(0)$. Assume $u(0) > v(0) > 0$. We deduce then from Lemma 2.1 that $\forall t \in [0, \pi/2\sqrt{\mu})$, $u(t) > 0$, $v(t) > 0$ and

$$w(t) := v(t) - u(t) = (v(0) - u(0)) \cos \sqrt{\mu} t.$$

Hence

$$w'(\pi/2\sqrt{\mu}) = \sqrt{\mu}(u(0) - v(0)) > 0.$$

If we define

$$b = \inf\{t > \pi/2\sqrt{\mu} : w'(t) = 0\} > \pi/2\sqrt{\mu}$$

we have $w(t) > 0$ on $(\pi/2\sqrt{\mu}, b]$ and

$$w'' + Mw = (M - \alpha(t))w > 0$$

$$w(\pi/2\sqrt{\mu}) = w'(b) = 0,$$

where $\alpha(t) \leq \max(\mu,\nu) < M$. Next we deduce from Lemma 2.2 that $b - \pi/2\sqrt{\mu} \geq \pi/2\sqrt{M}$ and we can choose M such that

$$\pi \geq b \geq \frac{\pi}{2\sqrt{\mu}} + \frac{\pi}{2\sqrt{M}} > \pi,$$

which is a contradiction.

If $u(0) > v(0) = 0$, Lemma 2.3 implies v is positive and uniqueness follows from Proposition 3.2.

A similar conclusion holds if $u(0) < v(0) \leq 0$. Hence $u(0) = v(0)$ and $u = v$. □

5 One zero solutions below the first Fučik curve

We have seen (Corollary 3.1) that if $\mu \leq 1/4$, there exists a unique solution of (3.1) which is positive. If $\mu > 0$, $\nu > 0$ and

$$\frac{1}{2\sqrt{\mu}} + \frac{1}{2\sqrt{\nu}} > 1$$

it is easy to show using degree argument (see e.g. [3]) that there exists at least one solution and from section 4, if $\mu < 1$, this solution is unique. In this section we shall characterize this solution from the forcing p. Let $\mathcal{A}$ be the set of functions $u \in C^1[0,\pi]$ such that there exists $a \in (0,\pi)$ with the property that $u(t) > 0$ if $t \in [0,a)$, $u(t) < 0$ if $t \in (a,\pi]$ and a is a simple zero of u. Let also $\mathcal{B} = -\mathcal{A}$.

Proposition 5.1. *Let $\frac{1}{4} < \mu < 1$, $\nu > 0$ be such that*

$$\frac{1}{2\sqrt{\mu}} + \frac{1}{2\sqrt{\nu}} > 1,$$

let $p \in C$ be such that $p(t) > 0$. Then

(a) the BVP (3.1) has a solution $u \in \mathcal{A}$ if and only if

$$A = \int_0^\pi p(s)\cos(\sqrt{\mu}s)ds < 0,$$

(b) the BVP (3.1) has a solution $u \in \mathcal{B}$ if and only if

$$B = \int_0^\pi p(s)\cos(\sqrt{\mu}(\pi - s))ds < 0.$$

Proof. Assume first $A < 0$ and let a be such that

$$\frac{\pi}{2\sqrt{\mu}} < a < \pi.$$

According to Lemma 2.1, the BVP

$$\begin{aligned} v'' + \mu v &= p(t) \\ v'(0) = v(a) &= 0 \end{aligned} \tag{5.1}$$

has a unique solution v such that

$$v'(a) = \frac{1}{\cos(\sqrt{\mu}a)} \int_0^a p(t)\cos(\sqrt{\mu}t)\, dt.$$

As $\pi - a < \pi - \frac{\pi}{2\sqrt{\mu}} < \frac{\pi}{2\sqrt{\nu}}$, we deduce from Lemma 2.2 that the BVP

$$\begin{aligned} w'' + \nu w &= p(t) \\ w(a) = w'(\pi) &= 0 \end{aligned} \tag{5.2}$$

has a unique negative solution w such that

$$w'(a) = \frac{-1}{\cos(\sqrt{\nu}(\pi - a))} \int_a^\pi p(t)\cos(\sqrt{\nu}(\pi - t))\, dt < 0.$$

Let us now choose a such that

$$\begin{aligned} F(a) := \cos(\sqrt{\mu}a) \int_a^\pi p(t)\cos(\sqrt{\nu}(\pi - t))\, dt \,+ \\ + \cos(\sqrt{\nu}(\pi - a)) \int_0^a p(t)\cos(\sqrt{\mu}t) dt = 0. \end{aligned}$$

Such an a is easy to obtain from the intermediate value theorem since

$$F\left(\frac{\pi}{2\sqrt{\mu}}\right) = \cos\left(\sqrt{\nu}\left(\pi - \frac{\pi}{2\sqrt{\mu}}\right)\right) \int_0^{\pi/2\sqrt{\mu}} p(t)\cos(\sqrt{\mu}t)\, dt > 0$$

and

$$F(\pi) = \int_0^\pi p(t)\cos(\sqrt{\mu}t)\, dt < 0.$$

For such an a, one has $v'(a) = w'(a) < 0$ and since $a < \pi < \pi/\sqrt{\mu}$, it follows from Lemma 2.3 that $v(t) > 0$ on $[0, a)$. The solution u of (3.1) can now be obtained placing side by side v and w.

If $B < 0$, a similar argument proves there exists a solution $u \in \mathcal{B}$.

At last, we notice that $A + B > 0$ so that only three cases are possible : (i) $A \geq 0$, $B \geq 0$; (ii) $A < 0$, $B > 0$; (iii) $A > 0$, $B < 0$.

In each of them we proved existence of a solution of a specific type. As these solutions are unique, we also have that these conditions are necessary for existence of a solution of the given type. □

Remark 5.1. If existence is already proved, Proposition 5.1 can be deduced from Lemma 4.2, Proposition 4.1 and Remark 3.1.

Sufficient conditions for existence of solutions $u \in \mathcal{A}$ or $u \in \mathcal{B}$ can be given if $\mu > 1$. For example one has the following proposition.

Proposition 5.2. *Let $\mu > 1$, $\nu > 0$ be such that*

$$\frac{1}{2\sqrt{\mu}} + \frac{1}{2\sqrt{\nu}} > 1$$

and let $p \in \mathcal{C}$ be such that $p(t) > 0$. Then :

(a) *The BVP (3.1) has a solution $u \in \mathcal{A}$ if*

$$\cos\left(\sqrt{\nu}\left(\pi - \frac{\pi}{\sqrt{\mu}}\right)\right)\int_0^{\pi/\sqrt{\mu}} p(t)\cos(\sqrt{\mu}t)\,dt - \int_{\pi/\sqrt{\mu}}^{\pi} p(t)\cos(\sqrt{\nu}(\pi - t))\,dt < 0;$$

(b) *The BVP (3.1) has a solution $u \in \mathcal{B}$ if*

$$\cos\left(\sqrt{\nu}\left(\pi - \frac{\pi}{\sqrt{\mu}}\right)\right)\int_{\pi-\pi/\sqrt{\mu}}^{\pi} p(t)\cos(\sqrt{\mu}(\pi - t))\,dt - \int_0^{\pi-\pi/\sqrt{\mu}} p(t)\cos(\sqrt{\nu}t)\,dt < 0.$$

6 One zero-solutions above the first Fučik curve

In this section, we study the existence of solutions of (3.1) in the region R_2, i.e.

$$\nu > 0, \frac{1}{4} < \mu < 1 \text{ and } \frac{1}{2\sqrt{\mu}} + \frac{1}{2\sqrt{\nu}} < 1 \tag{6.1}$$

Proposition 6.1. *Let μ, ν be such that (6.1) holds and $p \in \mathcal{C}$ such that $p(t) > 0$. Then*

(a) *if*

$$A = \int_0^{\pi} p(s)\cos(\sqrt{\mu}s)\,ds > 0$$

the BVP (3.1) has exactly one solution $u \in \mathcal{A}$;

(b) *if*

$$A \leq 0$$

the BVP (3.1) has at most two solutions $u \in \mathcal{A}$.

Proof. Let $u \in \mathcal{A}$ be a solution of (3.1) and a be its unique zero. From Lemma 2.2, we have

$$\pi - a < \frac{\pi}{2\sqrt{\nu}} \quad \text{i.e.} \quad a > \pi\left(1 - \frac{1}{2\sqrt{\nu}}\right).$$

Further, if we proceed as in Proposition 5.1, we have

$$F(a) = \cos(\sqrt{\mu}a)\int_a^{\pi} p(t)\cos(\sqrt{\nu}(\pi - t))\,dt + \cos(\sqrt{\nu}(\pi - a))\int_0^a p(t)\cos(\sqrt{\mu}t)\,dt = 0.$$

Reciprocally, if $a \in \left(\pi\left(1-\frac{1}{2\sqrt{\nu}}\right), \pi\right)$ is a zero of F, we can build a solution $u \in \mathcal{A}$ placing side by side the solutions v and w of (5.1) and (5.2). Hence solutions $u \in \mathcal{A}$ correspond to the zeros of F in $\left(\pi\left(1-\frac{1}{2\sqrt{\nu}}\right), \pi\right)$.

One computes

$$F\left(\pi\left(1-\frac{1}{2\sqrt{\nu}}\right)\right) = \cos\left(\sqrt{\mu}\left(1-\frac{1}{2\sqrt{\nu}}\right)\pi\right) \int_{\pi-\frac{\pi}{2\sqrt{\nu}}}^{\pi} p(s)\cos(\sqrt{\nu}(\pi-s))\,ds < 0,$$

$$F(\pi) = \int_0^\pi p(s)\cos(\sqrt{\mu}s)\,ds = A,$$

$$F'(a) = -\sqrt{\mu}\sin(\sqrt{\mu}a)\int_a^\pi p(s)\cos(\sqrt{\nu}(\pi-s))\,ds + \sqrt{\nu}\sin(\sqrt{\nu}(\pi-a))\int_0^a p(s)\cos(\sqrt{\mu}s)\,ds$$

$$\begin{aligned} F''(a) = &-\mu\cos(\sqrt{\mu}a)\int_a^\pi p(s)\cos(\sqrt{\nu}(\pi-s))\,ds \\ &-\nu\cos(\sqrt{\nu}(\pi-a))\int_0^a p(s)\cos(\sqrt{\mu}s)\,ds \\ &+\left[\sqrt{\mu}\sin(\sqrt{\mu}a)\cos(\sqrt{\nu}(\pi-a))\right. \\ &\qquad \left.+\sqrt{\nu}\sin(\sqrt{\nu}(\pi-a))\cos(\sqrt{\mu}a)\right] p(a). \end{aligned}$$

Therefore, if $F(a) = 0$, one has

$$F'(a) = \left[\sqrt{\mu}\ \mathrm{tg}\ (\sqrt{\mu}a) + \sqrt{\nu}\ \mathrm{tg}\ (\sqrt{\nu}(\pi-a)\right] \cos(\sqrt{\nu}(\pi-a))\int_0^a p(s)\cos(\sqrt{\mu}s)\,ds$$

and if $F(a) = F'(a) = 0$

$$F''(a) = (\mu-\nu)\cos(\sqrt{\nu}(\pi-a))\int_0^a p(s)\cos(\sqrt{\mu}s)\,ds.$$

If $A > 0$, we deduce from the intermediate value theorem, there exists a zero of F. Further this zero is unique since :

$$\begin{aligned} &F(a) = 0\ ,\ a < a_0 \text{ implies } F'(a) > 0; \\ &F(a) = 0\ ,\ a > a_0 \text{ implies } F'(a) < 0; \\ &F(a) = 0\ ,\ a = a_0 \text{ implies } F'(a_0) = 0 \text{ and } F''(a_0) < 0. \end{aligned}$$

Here, a_0 is the only zero in $\left(\pi\left(1-\frac{1}{2\sqrt{\nu}}\right), \pi\right)$ of

$$\sqrt{\mu}\ \mathrm{tg}\ (\sqrt{\mu}a) + \sqrt{\nu}\ \mathrm{tg}\ (\sqrt{\nu}(\pi-a)).$$

If $A \leq 0$, the same type of argument proves there exist at most two zeros of F. □

Using the same type of argument, one proves

Proposition 6.2. *Let μ, ν be such that (6.1) holds and $p \in C$ such that $p(t) > 0$. Then*

(a) if

$$B = \int_0^\pi p(s)\cos(\sqrt{\mu}(\pi - s))\,ds > 0$$

the BVP (3.1) has exactly one solution $u \in \mathcal{B}$,

(b) if

$$B \leq 0$$

the BVP (3.1) has at most two solutions $u \in \mathcal{B}$.

Remark 6.1. From proposition 3.1, 6.1 and 6.2 we can give a global description of solutions in the region (6.1) :

(a) if $A > 0$, $B > 0$, there is exactly one positive solution, one solution in $\mathcal{A}$ and one in $\mathcal{B}$;
(b) if $A < 0$, $B > 0$, there is no positive solution, at most two in $\mathcal{A}$ and exactly one in $\mathcal{B}$;
(c) if $A > 0$, $B < 0$, there is no positive solution, exactly one in $\mathcal{A}$ and at most two in $\mathcal{B}$.

Remark 6.2. In case $A < 0$, one can find forcings p such that there is exactly 0, 1 or 2 solutions in $\mathcal{A}$.

References

1. Aguinaldo, L., Schmitt, K. (1978): On the boundary value problem $u'' + u = \alpha u^- + p(t)$, $u(0) = 0 = u(\pi)$. Proc. AMS **68**, 64-68
2. De Figueiredo, D. (1982): Positive solutions of semilinear elliptic problems. Lect. Notes **957**, Springer Verlag, Berlin-Heidelberg-New York, 34-87
3. Habets, P., Metzen, G., (1989): Existence of periodic solutions of Duffing equations. J. Diff Eq. **78**, 1-32
4. Hart, D.C., Lazer, A.C., McKenna, P.J. (1986): Multiplicity of solutions of nonlinear boundary value problems. SIAM J. Math. Anal. **17**, 1332-1338
5. Lazer, A.C., McKenna, P.J. (1987): Large scale oscillatory behaviour in loaded asymmetric systems. Ann. Inst. Henri Poincaré **4**, 243-274
6. Lazer, A.C., McKenna, P.J. (1989): Existence, uniqueness and stability of oscillations in differential equations with asymmetric nonlinearities. Trans. AMS **315**, 721-739

Moment Conditions for a Volterra Integral Equation in a Banach Space

Kenneth B. Hannsgen and Robert L. Wheeler

Department of Mathematics and Interdisciplinary Center for Applied Mathematics, Virginia Polytechnic Institute and State University, Blacksburg, Virginia 24061-0123.

Abstract

For a linear Volterra equation of scalar type in a Banach space, sufficient conditions are given for three related resolvent kernels to be integrable with respect to certain weights on the positive half-line. The problem arises in the study of energy decay in viscoelastic solids, and the results lead to integral estimates for the rate of this decay.

1 Introduction

Let $\mathbf{X}$ be a Banach space and $\mathbf{L}$ a closed, densely defined, linear operator in $\mathbf{X}$. Assume that

$$\begin{array}{l}\mathbf{L} \text{ generates a strongly continuous cosine family} \\ \mathbf{C}(t) \text{ in } \mathbf{X} \text{ with } \|\mathbf{C}(t)\| \leq Me^{\omega_0|t|} \quad (t \in \mathbb{R})\end{array} \tag{1.1}$$

(see [3, 11]. $\|\cdot\|$ denotes both the norm in $\mathbf{X}$ and the operator norm in $\mathcal{L}(\mathbf{X})$, the space of bounded linear operators on $\mathbf{X}$).

Set $\omega_0(\mathbf{L} = \inf\{\,\omega_0 \mid (1.1) \text{ holds }\}$. Let

$$\mathbf{X}_1 = \{\mathbf{x} \in \mathbf{X} : \mathbf{C}(t)\,\mathbf{x} \in C^1(\mathbf{R}^+, \mathbf{X})\}$$

($\mathbb{R}^+ = [0, \infty)$) with norm

$$\|\mathbf{x}\|_1 = \|\mathbf{x}\| + \sup_{0 \leq t \leq 1} \|\dot{\mathbf{C}}(t)\mathbf{x}\| \ .$$

Thus in the important special case where

$$\mathbf{X} \text{ is a Hilbert space and } \mathbf{L} \text{ is a negative definite selfadjoint operator}, \tag{1.2}$$

$\omega_0(\mathbf{L}) = 0$ and $\mathbf{X}_1$ is the domain of $\mathbf{M} \equiv (-\mathbf{L})^{1/2}$.

We consider the problem

$$\ddot{\mathbf{u}}(t) = E\mathbf{L}\mathbf{u}(t)+\frac{d}{dt}\int_0^t a(t-\tau)\mathbf{L}\mathbf{u}(\tau)\,d\tau+\mathbf{f}(t), \qquad \mathbf{u}(0)=\mathbf{u}_0, \qquad \dot{\mathbf{u}}(0)=\mathbf{u}_1, \tag{P}$$

where $E > 0$

$$\mathbf{u}_0 \in \mathbf{X}_1, \quad \mathbf{u}_0 \in \mathbf{X}, \quad \mathbf{f} \in L^1(\mathbb{R}^+, \mathbf{X}) \ . \tag{1.3}$$

The kernel $a(t)$ satisfies either

$$\begin{aligned}&a \in C(0,\infty)\cap L^1(0,1)\cap AC^2_{loc}(0,\infty) \text{ with } a \text{ positive,}\\ &\text{non}-\text{increasing and log}-\text{convex on } (0,\infty) \text{ and}\\ &0 = a(\infty) < a(0^+) \leq \infty, \text{ or}\end{aligned} \tag{1.4a}$$

$$a \in AC^2_{loc}(\mathbb{R}^+) \text{ with } a(0) > 0,\ \dot{a}(0) < 0 \text{ and } a(t) \to 0 \text{ as } t \to \infty. \tag{1.4b}$$

Solutions of (1.3) can be studied by means of resolvent formulas such as

$$\begin{aligned}\mathbf{u}(t) &= \mathbf{U}(t)\mathbf{u}_0 + \mathbf{W}(t)\mathbf{u}_1 + \int_0^t \mathbf{W}(t-\tau)\mathbf{f}(\tau)\,d\tau,\\ \dot{\mathbf{u}}(t) &= \mathbf{V}(t)\mathbf{u}_0 + \mathbf{U}(t)\mathbf{u}_1 + \int_0^t \mathbf{U}(t-\tau)\mathbf{f}(\tau)\,d\tau,\end{aligned} \tag{1.5}$$

where the resolvent kernels can be defined formally either via the homogeneous version of (1.5) or via the Laplace transform formulas

$$\begin{aligned}\hat{\mathbf{U}}(s) = (s-\hat{A}(s)\mathbf{L})^{-1}, \qquad \hat{\mathbf{V}}(s) = \hat{A}(s)\mathbf{L}\hat{\mathbf{U}}(s),\\ s\hat{A}(s)\mathbf{L}\hat{\mathbf{W}}(s) = \hat{\mathbf{V}}(s),\end{aligned} \tag{1.6}$$

where $\hat{A}(s) = \hat{a}(s) + Es^{-1}$. Note that

$$\mathbf{U}(t) = \mathbf{I} + \int_0^t \mathbf{V}(\tau)\,d\tau, \qquad \mathbf{W}(t) = \int_0^t \mathbf{U}(\tau)\,d\tau. \tag{1.7}$$

J. Prüss [9, 10] has obtained detailed results on the existence, norm continuity and integrability of $\mathbf{U}(t)$ and $\mathbf{V}(t)$ under assumptions (1.1) and (1.4a) and with $E \geq 0$. Here we use Prüss's methods in the special situation where $E > 0$ (appropriate to the equations of motion for a viscoelastic solid) to deduce estimates on the integrability of these resolvents with respect to a weight function. As a consequence, in the case (1.2), we obtain integrability conditions on the energy of solutions of (P), defined by

$$\mathcal{E}(t) = \frac{1}{2}A(t)\|\mathbf{M}\mathbf{u}(t)\|^2 + \frac{1}{2}\|\dot{\mathbf{u}}(t)\|^2 - \frac{1}{2}\int_0^t \dot{a}(t-\tau)\|\mathbf{M}(\mathbf{u}(t)-\mathbf{u}(\tau))\|^2\,d\tau. \tag{1.8}$$

A more detailed presentation of some of these results, together with a discussion of an example in viscoelasticity, will be found in [5]. In particular, it is shown that boundary feedback mechanisms are ineffective in improving energy decay rates when viscoelastic creep dominates the asymptotic behavior. These applications were suggested by recent work of J. Lagnese [7] and G. Leugering [8].

2 Integrability of resolvents

A function $\rho(t)$ is called a (regular) weight on $\mathbb{R}^+$ if ρ is positive, continuous and nondecreasing on $\mathbb{R}^+$, $\rho(0) = 1$, $\rho(t+s) \leq \rho(t)\rho(s)$ $(t, s \in \mathbb{R}^+)$ and $\lim_{t\to\infty} t^{-1} \log \rho(t) = 0$. We have in mind such weights as

$$\begin{aligned} \rho_1(t) &= (1+t)^r, \qquad r \geq 0, \\ \rho_2(t) &= (1+\log(1+t))^\gamma \rho_1(t), \qquad \gamma \geq 0, \\ \rho_3(t) &= \exp(t^\alpha)\rho_2(t), \qquad 0 \leq \alpha < 1. \end{aligned}$$

Theorem 2.1 *Let $A(t) = E + a(t)$ where $E > 0$ and (1.4a) holds and $\dot{a}(0^+) = -\infty$. Let* $\mathbf{L}$ *satisfy (1.1) and the stability conditions*

$$\mathbf{L} \text{ has a bounded inverse,} \tag{2.1}$$

$$s/\hat{A}(s) \in \mathcal{R} \equiv \text{resolvent set of } \mathbf{L} \quad (\operatorname{Re} s \geq 0,\ s \neq 0). \tag{2.2}$$

Let $\rho(t)$ be a regular weight, and assume that

$$\int_1^\infty (|\dot{a}(t)| + |t\ddot{a}(t)| + |t^2 \dddot{a}(t)|)\rho(t)\,dt < \infty. \tag{2.3}$$

The transform relations (1.6) *determine strongly continuous functions* $\mathbf{U}(t)$: $\mathbb{R}^+ \to \mathcal{L}(\mathbf{X})$ *and* $\mathbf{V}(t), \mathbf{LW}(t) : \mathbb{R}^+ \to \mathcal{L}(\mathbf{X}_1, \mathbf{X})$, *and* $(1+t)\mathbf{U}(t) \in L^1(\mathbb{R}^+, \mathcal{L}(\mathbf{X}); \rho)$ *while both* $(1+t)^2\mathbf{V}(t)$ *and* $\mathbf{LW}(t)$ *belong to* $L^1(\mathbb{R}^+, \mathcal{L}(\mathbf{X}_1, \mathbf{X}); \rho)$.

As shown in [9], $\mathbf{U}(t)$ is a resolvent in the sense of [2], and one can justify (1.5). When $\mu \equiv \sqrt{A(0^+)} < \infty$ and $\kappa \equiv -\dot{A}(0^+)/2\mu$ are both finite, the resolvents cannot be measurable in the norm [9] and so cannot belong to the L^1 spaces of Theorem 2.1. The appropriate conclusion in this case is that the resolvents are ρ-integrable:

$$\begin{aligned} &\{\text{there exists } \varphi \in L^1(\mathbb{R}^+; \rho) \text{ such that} \\ &\{(1+t)\|\mathbf{U}(t)\| + (1+t)^2\|\mathbf{V}(t)\|_{\mathcal{L}(\mathbf{X}_1,\mathbf{X})} + \|\mathbf{LW}(t)\|_{\mathcal{L}(\mathbf{X}_1,\mathbf{X})}\} \leq \varphi(t) \\ &\text{a.e. on } \mathbb{R}^+. \end{aligned} \tag{2.4}$$

Theorem 2.2 *Let $A(t) = E + a(t)$ with $E > 0$, and suppose* (1.1) *and* (2.1) *hold and that $\mu + \kappa < \infty$ with $\omega_0(\mathbf{L}) < \kappa/\mu^2$.*
(i) If (1.4a), (2.2) *and* (2.3) *hold, then the conclusions of Theorem* 2.1 *hold with the L^1 inclusions replaced by* (2.4).
(ii) If (1.4b), (2.2), (2.3) *and*

$$q(s) \equiv 1 + \kappa\hat{A}(s) \neq 0, \ \operatorname{Re} s \geq 0, \ s \neq 0, \tag{2.5}$$

hold, then the relations (1.6) *determine strongly measurable functions* $\mathbf{U}(t)$: $\mathbb{R}^+ \to \mathcal{L}(\mathbf{X})$ *and* $\mathbf{V}(t), \mathbf{LW}(t) : \mathbb{R}^+ \to \mathcal{L}(\mathbf{X}_1, \mathbf{X})$ *and (2.4) holds.*

Remarks

(See [5] for further details.) In (i) above, the resolvents are again strongly continuous as in Theorem 2.1, and (1.5) holds. The same can be said in (ii) in the special case (1.2), provided

$$a(t) \text{ is positive, decreasing and convex.} \tag{2.6}$$

In addition, (2.6) implies (2.5) and, if $\omega_0(\mathbf{L}) = 0$ and $a \in AC^1_{loc}$, then (2.6) implies (2.2). If $t\dot{\rho}(t) \leq M\rho(t)$ $(t \geq 1)$, and if (2.6) holds and $-\dot{a}$ is convex, then (2.3) follows from

$$\int_1^\infty |\dot{a}(t)|\rho(t)\,dt < \infty; \tag{2.7}$$

this would include the kernels generally used for viscoelastic models with, say, $\rho = \rho_1$.

The following sketch of the proof of Theorem 2.2 (ii) for $\mathbf{U}$ will indicate the perturbation technique that is involved. We write

$$\mathbf{U}_1(t) = \mathbf{C}(\mu t)\exp(-\kappa t/\mu)$$

and

$$\mathbf{U}_0 = \mathbf{U} - \mathbf{U}_1.$$

Then $\mathbf{U}_1(t)$ decays exponentially and so is ρ-integrable with respect to any regular weight. Rearranging terms (as in [9, Theorem 11]) and working with Laplace transforms one obtains a Volterra equation

$$\mathbf{U}_0(t) = \mathbf{R}_1(t) + \mathbf{R}_2 * \mathbf{U}_0(t) \tag{2.8}$$

for $\mathbf{U}_0$, where $\hat{\mathbf{R}}_j(s)$ $(j = 1, 2)$ is a sum of terms that are either scalar or scalars multiplied by $\hat{\mathbf{U}}_1(s)$. Using local analyticity [6], one establishes that the scalar factors are transforms of functions in $L^1(\mathbb{R}^+; \rho_+)$, where $\rho_+(t) = (1+t)\rho(t)$; the scalar factors involved are $\hat{A}(s)/q(s)$, $\hat{\dot{a}}(s)/q(s)$ and $\hat{\ddot{a}}(s)/q(s)$. As a consequence, and by means of Gripenberg's version of the Paley-Wiener lemma [4], (2.8) can be solved for $\mathbf{U}_0$ in the Banach algebra $L^1(\mathbb{R}^+, \mathcal{L}(\mathbf{X}); \rho_+)$, and the proof is complete. □

Theorem 2.2 (i) is proved in much the same manner. For Theorem 2.1, Prüss's representation [9]

$$\mathbf{U}(t) = \int_0^\infty w_t(t,\tau)\mathbf{C}(\tau)\,d\tau,$$

where w solves a certain Rayleigh problem, plays the central role, and there are additional technical complications. Once again the model is [9, Theorem 11]. The proofs for $\mathbf{V}(t)$ are analogous (cf [10]), and they quickly yield the results for $\mathbf{W}(t)$, via (1.6).

3 Energy decay

In this section we assume (1.2), (1.3), $\mathbf{f} \in L^1(\mathbb{R}^+, \mathbf{X}; \rho)$ and the hypotheses of Theorem 1.2 with $a(0^+) < \infty$, or of Theorem 2.2 (either part) with (2.6) in case (ii). Then by (1.5) and the conclusions of Theorem 2.1 or 2.2 we get

$$\int_0^\infty (\|\mathbf{Mu}(t)\| + \|\dot{\mathbf{u}}(t)\|)\rho(t)\,dt \le K(\|\mathbf{Mu}_0\| + \|\mathbf{u}_1\| + \|\mathbf{f}\|_\rho) \tag{3.1}$$

with $\|\cdot\|_\rho$ =the norm in $L^1(\mathbb{R}^+, \mathbf{X}; \rho)$ and K depending only on the L^1 bounds of the various resolvents.

The energy computation of [1, Theorem 3.1] (together with an approximation argument) yields boundedness of $\|\mathbf{Mu}(t)\| + \|\dot{\mathbf{u}}(t)\|$ on $\mathbb{R}^+$. Combining this with (3.1), we obtain the following.

Theorem 3.1 *Under the assumptions of this section, the solution* $\mathbf{u}$ *of* (P) *satisfies*

$$\int_0^\infty (\|\mathbf{Mu}(t)\|^2 + \|\dot{\mathbf{u}}(t)\|^2)\rho(t)\,dt < \infty.$$

As a consequence, we deduce that $\mathcal{E}(t)$ is integrable with respect to ρ; the estimate for the viscoelastic stored energy term of (1.8) is

$$\begin{aligned}
&\int_0^\infty \rho(t) \int_0^t |\dot{a}(\tau)|\, \|\mathbf{M}(\mathbf{u}(t) - \mathbf{u}(t-\tau))\|^2\, d\tau dt \\
&\quad \le 2\int_0^\infty |\dot{a}(\tau)| \int_\tau^\infty \rho(t)(\|\mathbf{Mu}(t)\|^2 + \|\mathbf{Mu}(t-\tau)\|^2)\,dt d\tau \\
&\quad \le 2\int_0^\infty |\dot{a}(\tau)|\,d\tau \int_0^\infty \|\mathbf{Mu}(t)\|^2 \rho(t)\,dt \\
&\qquad + 2\int_0^\infty |\dot{a}(\tau)| \int_0^\infty \rho(t+\tau)\|\mathbf{Mu}(t)\|^2\,dt d\tau \\
&\quad \le 2a(0)\int_0^\infty \|\mathbf{Mu}(t)\|^2 \rho(t)\,dt \\
&\qquad + 2\int_0^\infty |\dot{a}(\tau)|\rho(\tau)\,d\tau \int_0^\infty \|\mathbf{Mu}(t)\|^2 \rho(t)\,dt < \infty.
\end{aligned}$$

Acknowledgment

This research was partly supported by the Air Force Office of Scientific Research under grant AFOSR-89-0268.

References

1. Dafermos, C.M. (1970): An Abstract Volterra Equation with Applications to Linear Viscoelasticity. J. Differential Equations **7**, 554–569
2. Da Prato, G., Iannelli, M. (1980): Linear Integro-differential Equations in Banach Spaces. Rend. Sem. Mat. Univ. Padova **62**, 207–219
3. Fattorini, H.O. (1985): Second Order Linear Differential Equations in Banach Spaces. Notas de Matemática **99**, Elsevier, Amsterdam, Netherlands
4. Gripenberg, G. (1987): Asymptotic Behaviour of Resolvents of Abstract Volterra Equations. J. Math. Anal. Appl. **122**, 427–438
5. Hannsgen, K.B., Wheeler, R.L.: Viscoelastic and Boundary Feedback Damping: Precise Energy Decay Rates When Creep Modes Are Dominant. J. Integral Equations Appl., to appear
6. Jordan, G.S., Staffans, O.J., Wheeler, R.L. (1982): Local Analyticity in Weighted L^1-spaces and Applications to Stability Problems for Volterra Equations. Trans. Amer. Math. Soc. **274**, 749–782
7. Lagnese, J.E. (1989): Boundary Stabilization of Thin Plates. SIAM Studies in Applied Math. **10**, SIAM Publications, Philadelphia
8. Leugering, G. (1990): Boundary Feedback Stabilization of a Viscoelastic Beam. Proc. Royal Soc. Edinburgh **114A**, 57–69
9. Prüss, J. (1987): Positivity and Regularity of Hyperbolic Volterra Equations in Banach Spaces. Math. Ann. **279**, 317–344
10. Prüss, J. (1989): Regularity and Integrability of Resolvents of Linear Volterra Equations. in Volterra Integrodifferential Equations in Banach Spaces and Applications, G. Da Prato and M. Iannelli, eds., Pitman Research Notes in Mathematics Series **190**, 339–367
11. Travis, C.C., Webb, G.F. (1978): Second Order Differential Equations in Banach Spaces. in Nonlinear Equations in Abstract Spaces, V. Lakshmikantham ed., Academic Press, 331–361

Asymptotic Behaviors of Solutions of a System of Linear Ordinary Differential Equations as $t \to \infty$

William A.Harris, Jr.[1] *and Yasutaka Sibuya*[2]

[1] Department of Mathematics, the University of Southern California, Los Angeles, California 90089-1113
[2] School of Mathematics, the University of Minnesota, Minneapolis, Minnesota 55455

1 Introduction.

A fundametal result of N. Levinson [2] describes the asymptotic behavior of solutions of a linear differential system

$$\frac{dx}{dt} = A(t)x \tag{1}$$

as $t \to +\infty$, where x is an n-dimensional vector and $A(t)$ is an $n \times n$ matrix. According to this theorem, if

$$A(t) = A_0 + V(t) + R(t),$$

where A_0 is an $n \times n$ constant matrix, $V(t)$ is an $n \times n$ matrix such that $|V(t)'|$ is integrable over the interval $[0, +\infty)$ and $R(t)$ is an $n \times n$ matrix such that $|R(t)|$ is integrable over the interval $[0, +\infty)$, and if eigenvalues $\lambda_k(t)(k = 1, \ldots, n)$ of the matrix $A_0 + V(t)$ satisfy some additional requirements, then system (1) can be reduced to

$$\frac{dy}{dt} = \Lambda(t)y, \qquad \Lambda(t) = \mathrm{diag}[\lambda_1(t), \ldots, \lambda_n(t)] \tag{2}$$

by a linear transformation

$$x = P(t)y \tag{3}$$

with an $n \times n$ matrix $P(t)$ such that $\lim_{t \to +\infty} P(t)$ exits and is nonsingular.

The basic idea of the proof of this result is
(i) regarding $A_0 + V(t)$ as a *constant* matrix, diagonalize $A_0 + V(t)$, i.e.

$$\Lambda(t) = Q(t)^{-1}\{A_0 + V(t)\}Q(t),$$

(ii) change system (1) to

$$\frac{du}{dt} = \{\Lambda(t) - Q(t)^{-1}Q'(t) + Q(t)^{-1}R(t)Q(t)\}u \tag{4}$$

by

$$x = Q(t)u,$$

(iii) show that $|Q(t)^{-1}Q'(t)|$ is integrable over an interval $[t_o, +\infty)$,
(iv) remove $-Q(t)^{-1}Q'(t) + Q(t)^{-1}R(t)Q(t)$ from the right-hand side of (4) by another linear transformation.

Step (iv) is now known as "Levinson's theorem" and has been shown (cf. Harris-Lutz [1]) to be the basis for many asymptotic integration results.

The method indicated in steps (i) and (iii) does not apply directly to the scalar equation

$$\frac{d^2\eta}{dt^2} + \{1 + h(t)\sin(\alpha t)\}\eta = 0, \tag{5}$$

when $h(t)$ is a small function such as $\frac{1}{t}, \frac{1}{t^{\frac{1}{2}}}, \frac{1}{ln\ t}, \frac{1}{ln\ t}\sin(t^{\frac{1}{2}})$, because derivatives of $h(t)\sin(\alpha t)$ do not become small enought, i.e. integrable. In this paper we will use Floquet theory with parameters to recast this type of problem so that the procedure outlined above can be applied directly.

Given an $n \times n$ matrix $A(t,\varepsilon)$ whose entries are periodic of period 1 in a real variable t and smooth in (t,ε), where ε is a vector-valued parameter, let us consider a system:

$$\frac{dx}{dt} = A(t, h(t))x, \tag{6}$$

where x is an n-dimensional vector, and $h(t)$ is a vector-valued function of t. To study the asymptotic behaviors of solutions of system (6) as $t \to +\infty$, we first look at the system:

$$\frac{du}{dt} = A(t,\varepsilon)u. \tag{7}$$

Utilizing the Floquent theorem, we change system (7) to

$$\frac{dv}{dt} = B(\varepsilon)v \tag{8}$$

by a linear transformation

$$u = Q(t,\varepsilon)v, \tag{9}$$

where Q is periodic of period 1 in t and smooth in (t,ε) (cf. Y. Sibuya [3]). Note that

$$B = Q^{-1}AQ - Q^{-1}\frac{\partial Q}{\partial t}.$$

Next, we change system (6) to

$$\frac{dw}{dt} = \left\{B(h(t)) - h'(t)Q(t,h(t))^{-1}\frac{\partial Q}{\partial \varepsilon}(t,h(t))\right\}w, \tag{10}$$

by the linear transformation

$$x = Q(t,h(t))w. \tag{11}$$

Then under the assumption that $|h'(t)|$ is integrable over a t-interval $[t_o, +\infty)$, we can apply the theorem of Levinson given above to system (10). Note that $B(h(t))$ depends only on $h(t)$, i.e. the periodic part is completely eliminated.

We may also condsider repeated applcations of this procedure when successive derivatives of h are smaller, e.g. $h(t) = \frac{\sin(t^{\frac{1}{2}})}{t^{\frac{1}{2}}}$.

2 An equation with periodic coefficients.

Let $A(t,\varepsilon)$ be an $n \times n$ matrix whose entries are continuous in $(t,\varepsilon) \in R \times \Delta(r)$ and holomorphic in $\varepsilon \in \Delta(r)$ for each fixed $t \in R$, where

$$\begin{cases} R = \{t; -\infty < t < +\infty\}, \\ \varepsilon = (\varepsilon_1, \ldots, \varepsilon_m) \in R^m, \\ |\varepsilon| = max_{1 \le j \le m}|\varepsilon_j|, \\ \Delta(r) = \{\varepsilon; |\varepsilon| < r\} \end{cases}$$

and r is a positive number. The following result was obtained by Y. Sibuya [4].

Theorem 2.1 *If A is periodic in t of period λ, i.e.*

$$A(t+\lambda,\varepsilon) = A(t,\varepsilon) \qquad for\ (t,\varepsilon) \in R \times \Delta(r), \tag{12}$$

where λ is a positive number, there exist $n \times n$ matrices $P(t,\varepsilon)$ and $H(\varepsilon)$ such that

(i) the entries of $P(t,\varepsilon)$ are continuous in $(t,\varepsilon) \in R \times \Delta(\hat{r})$ and holomorphic in $\varepsilon \in \Delta(\hat{r})$ for each fixed $t \in R$, where $\hat{r}$ is a suitable positive number,

(ii) $P(t+\lambda,\varepsilon) = P(t,\varepsilon)$ for $(t,\varepsilon) \in R \times \Delta(\hat{r})$,

(iii) $P(t,\varepsilon)$ is invertible for every $(t,\varepsilon) \in R \times \Delta(\hat{r})$,

(iv) the entries of $H(\varepsilon)$ are holomorphic in $\varepsilon \in \Delta(\hat{r})$,

(v) any two distinct eigenvalues of $H(0)$ do not differ by integral multiples of $\frac{2\pi i}{\lambda}$,

(vi) $\frac{\partial}{\partial t}P(t,\varepsilon)$ exists for $(t,\varepsilon) \in R \times \Delta(\hat{r})$ and given by

$$\frac{\partial}{\partial t}P(t,\varepsilon) = A(t,\varepsilon)P(t,\varepsilon) - P(t,\varepsilon)H(\varepsilon) \tag{13}$$

for $(t,\varepsilon) \in R \times \Delta(\hat{r})$.

Let $B_0(\varepsilon)$ be an $n \times n$ matrix whose entries are holomorphic in $\varepsilon \in \Delta(r)$, and let $B_1(t,\varepsilon,\mu)$ be an $n \times n$ matrix whose entries are continuous in $(t,\varepsilon,\mu) \in R \times \mathcal{D}(r)$ and holomorphic in $(\varepsilon,\mu) \in \mathcal{D}(r)$ for each fixed $t \in R$, where

$$\begin{cases} \varepsilon = (\varepsilon_1, \ldots, \varepsilon_m) \in R^m, & \mu = (\mu_1, \ldots, \mu_p) \in R^p, \\ |\varepsilon| = \max_{1 \le j \le m} |\varepsilon_j|, & |\mu| = \max_{1 \le h \le p} |\mu_h|, \\ \mathcal{D}(r) = \{(\varepsilon, \mu); |\varepsilon| + |\mu| < r\} \end{cases}$$

and r is a positive number. Our first fundamental result is the theorem below.

Theorem 2.2 *Assume that*

$$B_1(t + \lambda, \varepsilon, \mu) = B_1(t, \varepsilon, \mu) \qquad \text{for } (t, \varepsilon, \mu) \in R \times \mathcal{D}(r), \tag{14}$$

where λ is a positive number, and that

$$B_1(t, \varepsilon, 0) = 0 \qquad \text{for } (t, \varepsilon) \in R \times \Delta(r). \tag{15}$$

Assume also that any two distinct eigenvalues of $B_0(0)$ do not differ by integral multiples of $\frac{2\pi i}{\lambda}$. Then there exist $n \times n$ matrices $P_1(t, \varepsilon, \mu)$ and $H_1(\varepsilon, \mu)$ such that

(i) the entries of $P_1(t, \varepsilon, \mu)$ are continuous in $(t, \varepsilon, \mu) \in R \times \mathcal{D}(\hat{r})$ and holomorphic in $(\varepsilon, \mu) \in \mathcal{D}(\hat{r})$ for each fixed $t \in R$, where $\hat{r}$ is a suitable positive number,

(ii) $P_1(t + \lambda, \varepsilon, \mu) = P_1(t, \varepsilon, \mu)$ for $(t, \varepsilon, \mu) \in R \times \mathcal{D}(\hat{r})$,

(iii) $P_1(t, \varepsilon, 0) = 0$ for $(t, \varepsilon) \in R \times \Delta(\hat{r})$,

(iv) the entries of $H_1(\varepsilon, \mu)$ are holomorphic in $(\varepsilon, \mu) \in \mathcal{D}(\hat{r})$ and $H_1(\varepsilon, 0) = 0$ for $\varepsilon \in \Delta(\hat{r})$,

(v) $\frac{\partial}{\partial t} P_1(t, \varepsilon, \mu)$ exists for $(t, \varepsilon, \mu) \in R \times \mathcal{D}(\hat{r})$ and is given by

$$\begin{aligned} \frac{\partial}{\partial t} P_1(t, \varepsilon, \mu) &= \{B_0(\varepsilon) + B_1(t, \varepsilon, \mu)\}\{I + P_1(t, \varepsilon, \mu)\} \\ &\quad - \{I + P_1(t, \varepsilon, \mu)\}\{B_0(\varepsilon) + H_1(\varepsilon, \mu)\} \end{aligned} \tag{16}$$

for $(t, \varepsilon, \mu) \in R \times \mathcal{D}(\hat{r})$, where I is the $n \times n$ identity matrix.

Remark 2.3 Equation (16) can be simplified as

$$\begin{aligned} \frac{\partial}{\partial t} P_1(t, \varepsilon, \mu) &= B_0(\varepsilon) P_1(t, \varepsilon, \mu) - P_1(t, \varepsilon, \mu) B_0(\varepsilon) + \{B_1(t, \varepsilon, \mu) P_1(t, \varepsilon, \mu) \\ &\quad - P_1(t, \varepsilon, \mu) H_1(\varepsilon, \mu) + B_1(t, \varepsilon, \mu)\} - H_1(\varepsilon, \mu). \end{aligned} \tag{17}$$

Proof of Theorem 2.2. Given $m = (m_1, \ldots, m_p)$, where the m_j are non-negative integers, let us denote $\sum_{1 \le j \le p} |m_j|$ and $\mu_1^{m_1} \ldots \mu_p^{m_p}$ by $|m|$ and μ^m respectively. Set

$$\begin{cases} P_1(t, \varepsilon, \mu) = \sum_{|m| \ge 1} \mu^m P_{1,m}(t, \varepsilon), \\ B_1(t, \varepsilon, \mu) = \sum_{|m| \ge 1} \mu^m B_{1,m}(t, \varepsilon), \\ H_1(\varepsilon, \mu) = \sum_{|m| \ge 1} \mu^m H_{1,m}(\varepsilon). \end{cases}$$

Then equation (17) is equivalent to

$$\frac{\partial P_{1,m}}{\partial t} = B_0(\varepsilon)P_{1,m} - P_{1,m}B_0(\varepsilon) + Q_{1,m}(t,\varepsilon) - H_{1,m}(\varepsilon), \tag{18}$$

where

$$Q_{1,m}(t,\varepsilon) = \sum_{h+k=m,(|h|,|k|\geq 1)} (B_{1,h}P_{1,k} - P_{1,k}H_{1,h}) + B_{1,m}(\varepsilon). \tag{19}$$

Hence

$$\begin{aligned} P_{1,m}(t,\varepsilon) = \exp[t\ B_0(\varepsilon)]\Big\{ C(\varepsilon) + \int_0^1 \exp[-s\ B_0(\varepsilon)]\Big(Q_{1,m}(s,\varepsilon) \\ - H_{1,m}(s,\varepsilon)\Big) \exp[s\ B_0(\varepsilon)]ds \Big\} \exp[-t\ B_0(\varepsilon)], \end{aligned} \tag{20}$$

where $C(\varepsilon)$ and $H_{1,m}(\varepsilon)$ are $n \times n$ matrices to be determined by the condition that $P_{1,m}(t,\varepsilon)$ is periodic in t of period λ, i.e.

$$\begin{aligned} &\exp[\lambda\ B_0(\varepsilon)]C(\varepsilon) - C(\varepsilon)\exp[\lambda\ B_0(\varepsilon)] \\ &\quad - \exp[\lambda\ B_0(\varepsilon)] \int_0^\lambda \exp[-s\ B_0(\varepsilon)]H_{1,m}(\varepsilon)\exp[s\ B_0(\varepsilon)]ds \\ &\quad = \exp[\lambda\ B_0(\varepsilon)] \int_0^\lambda \exp[-s\ B_0(\varepsilon)]Q_{1,m}(s,\varepsilon)\exp[s\ B_0(\varepsilon)]ds. \end{aligned} \tag{21}$$

It is not difficult to see that condition (21) determines the matrices $C(\varepsilon)$ and $H_{1,m}(\varepsilon)$. Then the matrix $P_{1,m}(t,\varepsilon)$ is determined by (20). Convergence of power series P_1 and H_1 can be shown by utilizing suitable majorant series. □
For further details as well as questions relating to the sufficiency of condition (21) see Sibuya [3].

Remark 2.4 If $B_0(0)$ has ν distinct eigenvalues, we can assume without any loss of generality that $B_0(\varepsilon)$ has a block-diagonal form:

$$B_0(\varepsilon) = \begin{bmatrix} B_{0,1}(\varepsilon) & 0 & 0 & \dots & 0 & 0 \\ 0 & B_{0,2}(\varepsilon) & 0 & \dots & 0 & 0 \\ \vdots & \vdots & \vdots & \ddots & \vdots & \vdots \\ 0 & 0 & 0 & \dots & 0 & B_{0,\nu}(\varepsilon) \end{bmatrix}, \tag{22}$$

such that each block $B_{0,j}(\varepsilon)$ corresponds to an eigenvalue of $B_0(0)$. Then, we can determine $C(\varepsilon)$ and $H_{1,m}(\varepsilon)$ so that $H_{1,m}(\varepsilon)$ has the same block-form as $B_0(\varepsilon)$, i.e.

$$H_{1,m}(\varepsilon) = \begin{bmatrix} H_{1,m,1}(\varepsilon) & 0 & 0 & \dots & 0 & 0 \\ 0 & H_{1,m,2}(\varepsilon) & 0 & \dots & 0 & 0 \\ \vdots & \vdots & \vdots & \ddots & \vdots & \vdots \\ 0 & 0 & 0 & \dots & 0 & H_{1,m,\nu}(\varepsilon) \end{bmatrix}. \tag{23}$$

Applications of theorems 2.1 and 2.2 Let $A(t,\varepsilon), P(t,\varepsilon)$ and $H(\varepsilon)$ be the same as in theorem 2.1. Consider a differential equation

$$\frac{dy}{dt} = A(t, h(t))y, \tag{24}$$

where y is and unknown vector in R^n and $h(t) = (h_1(t), \dots, h_m(t))$ is an R^m-valued differentiable function in t on the interval $[0, +\infty)$.

Case I. Let us first prove the theorem below.

Theorem 3.1 *The transformation*

$$y = P(t, h(t))u \tag{25}$$

changes differential equation (24) to

$$\frac{du}{dt} = \left\{ H(h(t)) - P(t, h(t))^{-1} \sum_{1\leq j\leq m} h_j'(t) \frac{\partial P}{\partial \varepsilon_j}(t, h(t)) \right\} u. \tag{26}$$

Proof. In fact

$$\begin{aligned}\frac{du}{dt} &= P(t,h(t))^{-1}\left\{A(t,h(t))P(t,h(t)) - \frac{d}{dt}[P(t,h(t))]\right\}u \\ &= P(t,h(t))^{-1}\left\{A(t,h(t))P(t,h(t)) - \frac{\partial P}{\partial t}(t,h(t))\right. \\ &\qquad \left. - \sum_{1\leq j\leq m} h_j'(t)\frac{\partial P}{\partial \varepsilon_j}(t,h(t))\right\}u. \end{aligned} \tag{27}$$

Since $H(\varepsilon)$ is given by (13), we can derive (26) from (27). □

Remark 3.2
(i) $H(h(t))$ does not contain any periodic terms,
(ii) any two distinct eigenvalues of $H(0)$ do not differ by integral multiples of $\frac{2\pi i}{\lambda}$,
(iii) if we assume that

$$\begin{cases} \int_0^{+\infty} |h'(t)|dt < +\infty, \\ \lim_{t\to+\infty} h(t) = 0, \end{cases} \tag{28}$$

then

$$\int_0^{+\infty} \left| P(t,h(t))^{-1} \sum_{1\leq j\leq m} h_j'(t) \frac{\partial P}{\partial \varepsilon_j}(t,h(t)) \right| dt < +\infty. \tag{29}$$

Case II. If we set

$$\begin{cases} B_0(\varepsilon) = H(\varepsilon), p = m, \\ B_1(t,\varepsilon,\mu) = -P(t,\varepsilon)^{-1} \sum_{1 \le j \le m} \mu_j \frac{\partial P}{\partial \varepsilon_j}(t,\varepsilon), \end{cases} \tag{30}$$

we can apply theorem 2.2 to the two matrices $B_0(\varepsilon)$ and $B_1(t,\varepsilon,\mu)$. Note that $B_0(0) = H(0)$. Furthermore differential equation (26) can be written as

$$\frac{du}{dt} = \{B_0(h(t)) + B_1(t, h(t), h'(t))\}u.$$

In the same way as the proof of theorem 3.1, we can prove the theorem below.

Theorem 3.3 *The transformation*

$$u = \{I + P_1(t, h(t), h'(t))\}v \tag{31}$$

changes differential equation (26) to

$$\begin{aligned} \frac{dv}{dt} = \Big\{ & H(h(t)) + H_1(h(t), h'(t)) \\ & - \Big[I + P_1(t, h(t), h'(t))\Big]^{-1} \sum_{1 \le j \le m} \Big[h_j'(t) \frac{\partial P_1}{\partial \varepsilon_j}(t, h(t), h'(t)) \\ & + h_j''(t) \frac{\partial P_1}{\partial \mu_j}(t, h(t), h'(t))\Big]\Big\} v, \end{aligned} \tag{32}$$

where $P_1(t,\varepsilon,\mu)$ and $H_1(\varepsilon,\mu)$ are the two matrices given in Theorem 2.2.

Remark 3.4

(i) $H(h(t)) + H_1(h(t), h'(t))$ does not contain any periodic terms,
(ii) $H(0) + H_1(0,0) = H(0)$,
(iii) if we assume that

$$\begin{cases} \int_0^{+\infty} |h_j''(t)| dt < +\infty, & \int_0^{+\infty} |h_j'(t) h_k'(t)| dt < +\infty, \\ \lim_{t \to +\infty} h_j(t) = 0, & \lim_{t \to +\infty} h_j'(t) = 0, \end{cases} \tag{33}$$

then

$$\begin{aligned} \int_0^{+\infty} \Big| \Big[I + P_1(t, h(t), h'(t))\Big]^{-1} & \sum_{1 \le j \le m} \Big[h_j'(t) \frac{\partial P_1}{\partial \varepsilon_j}(t, h(t), h'(t)) \\ & + h_j''(t) \frac{\partial P_1}{\partial \mu_j}(t, h(t), h'(t))\Big]\Big| dt < +\infty, \end{aligned} \tag{34}$$

(iv) condition (33) is in general weaker than condition (28),

(v) since $P_1(t, \varepsilon, 0) = 0$, we can write

$$-\left[I + P_1(t, h(t), h'(t))\right]^{-1} \sum_{1 \le j \le m} \left[h_j'(t) \frac{\partial P_1}{\partial \varepsilon_j}(t, h(t), h'(t)) + h_j''(t) \frac{\partial P_1}{\partial \mu_j}(t, h(t), h'(t))\right]$$

$$= \sum_{1 \le j \le m} h_j'(t) h_k'(t) \mathcal{T}_{j,k}(t, h(t), h'(t)) + \sum_{1 \le j \le m} h_j''(t) \mathcal{T}_j(t, h(t), h'(t)), \tag{35}$$

where the matrices $\mathcal{T}_{j,k}(t, \varepsilon, \mu)$ and $\mathcal{T}_j(t, \varepsilon, \mu)$ satisfy the conditions below:

(a) the entries of $\mathcal{T}_{j,k}(t, \varepsilon, \mu)$ and $\mathcal{T}_j(t, \varepsilon, \mu)$ are continuous in $(t, \varepsilon, \mu) \in R \times \mathcal{D}(\rho)$ and holomorphic in $(\varepsilon, \mu) \in \mathcal{D}(\rho)$ for each fixed $t \in R$ if $\rho > 0$ is sufficiently small,

(b) $\mathcal{T}_{j,k}(t + \lambda, \varepsilon, \mu) = \mathcal{T}_{j,k}(t, \varepsilon, \mu)$ and $\mathcal{T}_j(t + \lambda, \varepsilon, \mu) = \mathcal{T}_j(t, \varepsilon, \mu)$ for $(t, \varepsilon, \mu) \in R \times \mathcal{D}(\rho)$.

Acknowledgment

The work of Y.Sibuya is partially supported by the National Science Foundation. The main part of this paper was written at the University of Southern California during this author's sabbatical leave (1988/89) from the University of Minnesota.

References

1. Harris, Jr., H.A., Lutz, D.A. (1977): A unified theory of asymptotic integration. J. Math. Anal. Appl. **57**, 571-586
2. Levinson, N. (1948): The asymptotic nature of solutions of linear differential equations. Duke Math J. **15**, 111-126
3. Sibuya, Y. (1958): Nonlinear ordinary differential equations with periodic coefficients. Funk. Ekva. **1**, 77-132
4. Sibuya, Y. (1954): Sur un système d'équations différentielles ordinaires linéaires à coefficents périodiques et cotenants des paramètres. Jour. Fac. Sci., Univ. Tokyo, I, 7, 229-241

Implicit Differential Equations which are not Solvable for the Highest Derivative

Tomasz Kaczynski

Département de Mathématiques et d'Informatique, Université de Sherbrooke, Sherbrooke, (Québec), Canada J1R 2R1

1 Introduction

This paper is motivated by the following example given by W.V. Petryshyn in [11]:

$$(P_1)\begin{cases} y'' = h(t) + y^3 + (y')^2 + ksin(y''), & t \in [0,1], \\ y(0) = y(1) = 0 \end{cases}$$

Petryshyn proved (with the use of A-proper mapping theory) that (P_1) has a solution $y \in C^2([0,1])$ provided $0 \leq k < 1$. We will show here, by entirely different methods, that (P_1) has solutions $y \in W_0^{2,1}([0,1])$ for any real constant k, and it has C^2 solutions if $|k| \leq 1$.

In general, we study the solvability of two-point boundary value problems for implicit differential equations

$$F(t, y, y', y'') = 0, \quad t \in [0,1]. \tag{1}$$

Ideally, we would like to solve the equation $F(t, y, p, x) = 0$ for x in order to reduce (1) to

$$y'' = f(t, y, y'), \quad t \in [0,1], \tag{2}$$

where a variety of existence results based on the Leray-Schauder theory is available. Let us remark that in the case of an initial value problem $\{(1)$ subject to $y(t_0) = y_0, y'(t_0) = p_0\}$ the local solvability of $F(t, y, p, x) = 0$ in x about a point (t_0, y_0, p_0, x_0) is entirely sufficient. In the case of a boundary value problem, however, we need a global continuous solution $x = f(t, y, p)$ defined for all $(t, y, p) \in [0,1] \times R^2$, and that is very rarely available. The situation changes if we set the problem in the space $W^{2,1}$ of functions $y \in C^1$, with absolutely continuous derivative. New results of Bielawski and Gorniewicz [1] are helpfull in reducing (1) to a differential inclusion

$$y'' \in \psi(t, y, y'), \ t \in [0,1], \tag{3}$$

where ψ is a lower semicontinuous selection of the multifunction

$$\varphi(t,y,p) = \{x \in R| \quad F(t,y,p,x) = 0\}. \tag{4}$$

Then the Fryszkowski selection theorem on mappings with decomposable values in L^1 (improved by Bressan and Colombo [2] in the completion with [10]) and a priori estimate techniques of Frigon [7] and Granas, Guennoun [9], are used to acomplish proofs of existence.

For the simplicity of the presentation, we restricted the study to equations involving functions $y : [0,1] \to R$ and to the homogeneous Dirichlet boundary value problem. Our results, however, have straightforward extensions to systems of equations (i.e. $y(t) \in R^n$) and to various classes of boundary conditions. We will briefly discuss those extensions at the end of this note.

2 Solvability of differential inclusions

We first study the homogeneous Dirichlet boundary value problem for

$$y'' \in \psi(t,y,y'), \ a.e. \ t \in [0,1], \tag{5}$$

where $\psi : [0,1] \times R^2 \to 2^R$ is a multifunction with closed values satisfying the following conditions:

(a) ψ is $\mathcal{L}\otimes\mathcal{B}$ measurable (Lebesgue on $[0,1]$, Borel on R^2) and $\psi(t,\cdot,\cdot)$ is lower semicontinuous for a.e. t;

(b) $\psi(t,y,p) \subset [f_1(t,y,p), f_2(t,y,p)]$ for all y,p and a.e. t, where f_1 and f_2 are two Carathéodory functions satisfying the growth conditions (H_1) and (H_2) stated below.

We recall from [9] that $f(t,y,p)$ is a Carathéodory function if it is measurable in t, continuous in (y,p) and, for (y,p) in a bounded $B \subset R^2$, dominated by an integrable funciton $g_B(t)$. We also restate the following conditions from [9]:

(H_1) There is a constant $M > 0$ such that $yf(t,y,0) > 0$ for all y with $|y| > M$ and a.e. t;

(H_2) $|f(t,y,p)| \leq \alpha(|p|)$, for all y with $|y| \leq M$, $p \in R$, and a.e. t, where α is a positive locally bounded function with $\int_0^\infty \frac{s}{\alpha(s)} ds > 2M$.

It is easily verified that (H_2) holds for any f satisfying the classical Bernstein growth restriction:

(H_2') There are constants $A, B > 0$ such that $|f(t,y,p)| \leq A + Bp^2$, for all y with $|y| \leq M$, $p \in R$ and a.e. t.

Theorem 1 *If ψ satisfies (a) and (b) then the inclusion (5) has a solution $u \in W_0^{2.1}$ such that $\|u\|_0 \leq M$ (i.e. $|u(t)| \leq M$ for a.e. t).*

Proof. Let f_1 and f_2 be as in (b), M a constant from (H_1) common to f_1 and f_2, α_i a function defined for f_i as in (H_2), $i = 1,2$, and let $M_1 > 0$ be defined by

$$\int_0^{M_1} \frac{s}{\alpha_i(s)} ds > 2M, \quad i = 1, 2.$$

Next, let $g \in L^1([0,1])$ be a positive function with $|f_i(t,y,p)| \leq g(t)$ for all $|y| \leq M, |p| \leq M_1$ and a.e. $t \in [0,1], i = 1,2$. We define

$$M_2 = \|g\|_1 = \int_0^1 |g(t)| dt$$

and let

$$U = \{y \in W_0^{2,1} \mid \|y\|_0 < M+1, \ \|y'\|_0 < M_1 + 1, \ \|y''\|_1 < M_2 + 1\}.$$

Consider the diagram

$$\begin{array}{ccc} C^1([0,1],R) & \xrightarrow{i} & C([0,1],R^2) \\ \uparrow j & & \downarrow \psi^* \\ W_0^{2,1}([0,1],R) & \xrightarrow{L} & L^1([0,1],R) \end{array}$$

where the operators are defined as follows: $Ly = y''$, j is the inclusion mapping, i is the imbedding defined by $i(y) = (y, y')$, and ψ^* is the multivalued mapping defined for ψ by

$$\psi^*(y,z)(t) = \{u \in L^1 | \quad u(t) \in \psi(t, y(t), z(t)) \quad a.e. \ t\}.$$

The inclusion (5) subject to $y(0) = y(1) = 0$ is equivalent to $Ly \in (\psi^* ij)(y)$, $y \in W_0^{2,1}$ and, next, to $y \in (L^{-1}\psi^* ij)(y)$, since L is bijective. Corollary 3.3 in [8] shows that the restrictions of ψ^* to compact subsets of $C([0,1], R^2)$ have continuous single-valued selections. Since j is completely continuous, the mapping $L^{-1}\psi^* ij$ restricted to $\bar{U}$ has a continuous selection $F : \bar{U} \to W_0^{2,1}$. Clearly, any fixed point of F also is a solution of (5) satisfying the boundary conditions. In order to complete the proof, we must show that all possible solutions of the parametrised family of equations

$$y = \lambda F(y), \quad 0 < \lambda < 1, \tag{6}$$

belong to U. If so, then the condition follows from the Theorem 5.1, Ch. II, §4 of [3]. Let us note that (6) and (b) imply

$$\lambda f_1(t,y,y') \leq y'' \leq \lambda f_2(t,y,y'), \quad a.e. \ t, \tag{7}$$

where the functions on both sides satisfy the growth conditions (H_1) and (H_2). Proving that $|y(t)| \leq M$ and $|y'(t)| \leq M_1$ for a.e. t is based on the same arguments as those in the proof of the assertion (ii), Lemma 3.1 in [9]. Next, $\|y''\|_1 \leq M_2$ by the definition of M_2 and by (7). This shows that $y \in U$.

3 Reduction of implicit equations to differential inclusions

We may now study the homogeneous Dirichlet problem for the differential equation (1). Let F be a continuous function from $[0,1] \times R^3$ to R and let $\varphi : [0,1] \times R^2 \to 2^R$ be the multifunction corresponding to F, defined by (4). The following assumptions are posed:

(*i*) For any $(t,y,p) \in [0,1] \times R^2$ there is an open $U \subset R$ such that $\varphi(t,y,p) \cap \partial U = \emptyset$ and $deg(F(t,y,p,\cdot),U) \neq 0$ (by deg we mean the Leray-Schauder degree).

(*ii*) For any $(t,y,p) \in [0,1] \times R^2$, the covering dimension of $\varphi(t,y,p)$ is zero (c.f. [4]).

(*iii*) $\varphi(t,y,p)$ satisfies the condition (b) of the previous section.

Note that φ is an u.s.c. mapping but not necessarily a l.s.c. mapping. The multivalued mapping $\varphi^* : C([0,1],R^2) \to cl\ L^1([0,1],R)$, corresponding to φ, is not l.s.c. in general, so Theorem 1 cannot be directly applied. However, our assumptions and Theorem 2.5 in [1] imply that φ has a lower semicontinous selection $\psi(t,y,p) \subset \varphi(t,y,p)$ with nonempty closed values. Geometrically, ψ is obtained from φ by deleting from the graph of $F(t,y,p,x) = 0$ all points at which that equation is not locally solvable for x. That and Theorem 1 imply the following:

Theorem 2 *If F satisfies the conditions (i), (ii), and (iii) then the equation (1) has at least one solution $y \in W_0^{2,1}$.*

Since the formulation of Theorem 2 is very general and the assumptions might seem not easily verified, we now give two more explicit examples. First let us consider the problem mentioned in the introduction:

$$(P_1) \begin{cases} y'' = h(t) + y^3 + (y')^2 + ksin(y''), & a.e. \quad t \in [0,1], h \in C([0,1]), \\ y(0) = y(1) = 0 \end{cases}$$

Corollary 1 *The problem (P_1) has solutions $u \in W_0^{2,1}$ for all real constants k. Moreover a C^2 solution exists if $|k| \leq 1$.*

Proof. Define $\varphi_1(z) = \{x \in R| \quad x = z + ksinx\}$. It is clear that the mapping $x \to z + ksinx - x$ has a nonzero degree on sufficiently large intervals $(-M_z, M_z)$ for all $z \in R$, in particular, $\varphi_1(z)$ is nonempty. The set $\varphi_1(z)$ is finite so, in particular, its covering dimension is zero. Consequently, the multifunction

$$\varphi(t,y,p) = \varphi_1(h(t) + y^3 + p^2)$$

satisfies the conditions (i) and (ii). Let us observe that $\varphi_1(z) \subset z + [-k,k]$, so that (iii) is verified with

$$f_i(t,y,p) = h(t) + y^3 + p^2 + (-1)^i|k|, \quad i = 1,2.$$

It is easily proved that f_1 and f_2 satisfy (H_1) and (H_2'), and we may now refer to the Theorem 1. Finally, if $|k| \leq 1$ then the equation $x = z + ksinx$ has a unique solution $x = \varphi_1(z)$ for all z, and φ_1 is a continuous function of z. Hence (1) is reduced to $y'' = \varphi_1(h(t) + y^3 + (y')^2)$ with the continuous right-hand side and the second conclusion follows.

The second example is

$$(P_2)\begin{cases} P_{2n+1}(y'') = f(t,y,y'), & a.e. \quad t \in [0,1], \\ y(0) = y(1) = 0 \end{cases}$$

where P_{2n+1} is a polynomial of odd degree $2n+1$ with constant coefficients.

Corollary 2 *Suppose that $f : [0,1] \times R^2 \to R$ is a continuous function satisfying (H_1) and the following conditions:*

(H_1'') *There exists $\varepsilon > 0$ such that*

$$\liminf_{|y|\to\infty} y^{-1} f(t,y,0) > \varepsilon \quad fora.e.t,$$

(H_2'') *For any $M > 0$, there are constants $A, B > 0$ such that, for all y with $|y| \leq M, p \in R$, and a.e. t, $|f(t,y,p)| \leq A + Bp^{4p+2}$,*

Then (P_2) has at least one solution $y \in W_0^{2,1}$.

Proof. We define $\varphi_1(z) = \{x| \quad P_{2n+1}(x) = z\}$ and repeat the arguments of the previous proof. For verifying that (iii) holds, let us note that the roots of the equation $P_{2n+1}(x) = z$ can be estimated by the roots of $x^{2n+1} = z$ so that

$$\varphi_1(z) \subset k_1 z^{1/2n+1} + [-k_2, k_2]$$

for some constants $k_1, k_2 > 0$. We define

$$f_i(t,y,p) = k_1 + (f(t,y,p))^{1/2n+1} + (-1)^i k_2.$$

The conditions (H_1'') and (H_2'') on f imply that f_1 and f_2 satisfy (H_1) and (H_2'), and the conclusion follows.

4 Remarks on possible generalisations

1) Our results have extensions to non-homogeneous Dirichlet, periodic, Neuman, or Sturm-Liouville boundary value problems considered in [9], with the analogue modification of the growth condition (H_2) in the case of the Sturm-Liouville boundary conditions.
2) Theorems 1 and 2 can be extended to systems of equations (or inclusions). That will be fullfilled upon replacing our growth conditions (H_1) and (H_2) by the growth conditions (H_1, H_2, H_3) of [5].

3) It seems that the continuity of $F(t,y,p,x)$ in t is not essential if we seek solutions in $W^{2,1}$, although it simplifies the arguments.
4) The assumption (ii) in section 3 is essential. However, if we go to another extreme and assume that $\varphi(t,y,p)$ is convex, then the results of [6] based on the theory of convex-valued u.s.c. mappings can be used instead. As an example, consider the equation analogous to (P_1):

$$y'' = h(t) + y^3 + (y')^2 + S(y''),$$

where the sinusoide in (P_1) is replaced by the "piece-wise linear sinusoide" of slopes ± 1 defined by

$$S(x) = \begin{cases} x \quad if \quad x \in [-1,1] \\ 2-x \quad if \quad x \in (1,3] \\ S(x-4n) \quad if \quad x \in [4n-1, 4n+3), \quad n = \pm 1, \pm 2, \ldots \end{cases}$$

Then the multifunction $\varphi(z) = \{x| \quad x = z + S(x)\}$ is u.s.c. with nonempty compact convex values and the results of [6] apply.

Acknowledgment

This research was supported by the NSERC grant No. OG1PN 016.

References

1. Bielawski, R., Górniewicz, L.: A fixed point index approach to some differential equations. To appear in Proc. of Conf. on Fixed Point Theory, ed. A. Dold, Lecture Notes in Math., Springer Verlag, Berlin-Heidelberg-New York
2. Bressan, A., Colombo, G. (1988): Extensions and selections of maps with decomposable values. Studia Mathematica **90**, 69-86
3. Dugundji, J., Granas, A. (1982): Fixed Point Theory. Vol. I, PWN, Warszawa
4. Engelking, E. (1979): Dimension Theory. PWN, Warszawa
5. Erbe, L.H., Krawcewicz, W.: Nonlinear boundary value problems for differential inclusion $y'' \in F(t,y,y')$. To appear in Annales Polonici Mat.
6. Erbe, L.H., Krawcewicz, W., Kaczynski, T.: Solvability of two-point boundary value problems for systems of nonlinear differential equations of the form $y'' = g(t,y,y',y'')$. To appear in Rocky Mt. J. of Math.
7. Frigon, M. (1983): Applications de la transversalité topologique à des problèmes nonlinéaires pour certaines classes d'équations différentielles ordinaires. To appear in Dissentationes Mathematicae
8. Fryszkowski, A. (1983): Continuous selections for a class of non-convex multivalued maps. Studia Mathematica **76**, 163-174
9. Granas, A., Guennoun, Z. (1988): Quelques résultats dans la théorie de Bernstein-Carathéodory de l'équation $y'' = f(t,y,y')$. C.R. Acad. Sci. Paris, Série I, L **306**, 703-706
10. Ornelas, A.: Approximation of relaxed solutions for lower semicontinuous differential inclusions. SISSA, preprint

11. Petryshyn, W.V. (1986): Solvability of various boundary value problems for the equation $x'' = f(t, x, x', x'') - y$. Pacific J. of Math., **122**, No. 1, 169-195

A Stability Analysis for a Class of Differential-Delay Equations Having Time-Varying Delay

James Louisell

Department of Mathematics, University of Southern Colorado, Pueblo, CO 81001

1 Introduction

The subject of this paper is differential-delay equations having time-varying delays. Time-varying differential-delay equations have a long history, and the literature on systems defined by such equations is vast. Within this area, the special case of systems in which the delay is itself a function of time has a much smaller literature, and results are more rare ([1]), ([2]), ([3]), ([5]), ([8]).

The specific topic of interest in this paper will be the differential-delay equation (†) $\dot{x}(t) = A_0x(t)+A_1x(t-h(t))$, where A_0, A_1 are fixed members of $\mathbb{R}^{n\times n}$, and $h(t)$ is a bounded function having domain $[0,\infty)$ and range contained in $[0,\infty)$. If $h \geq 0$ is any fixed constant for which all zeros of the characteristic function $f_h(s) = |sI - A_0 - e^{-sh}A_1|$ are contained in the open left half-plane $\{Re(s) < 0\}$, we say that h lies in the stability set, which we denote by H_s. Thus $H_s = \{h \in [0,\infty) : f_h(s) = |sI - A_0 - e^{-sh}A_1|$ is nonzero for each complex s having $Re(s) \geq 0\}$. We will assume that $h(t)$ takes its values in the set H_s, and investigate the following question: Which hypotheses on the function $h(\cdot)$ guarantee that the solutions of the above differential-delay equation (†) decay as $t \to \infty$?

In order to clarify the discussion we now fix the notations used throughout this paper, and recall those aspects of the autonomous system (*) $\dot{x}(t) = A_0x(t)+ A_1x(t-h)$ which will be useful as a point of reference in our investigation. To begin, we let $H = [0,\infty)$, and for each $h \in H$, we let σ_h denote the delay operator having duration h. A simple and useful way to form the characteristic function $f_h(s)$ for the system (*) is to first write $p(s,\sigma_h) = |sI - A_0 - A_1\sigma_h| = s^n+a_{n-1}(\sigma_h)s^{n-1}+\cdots+a_0(\sigma_h)$, where for $k = 0,\cdots,n-1$, each $a_k(\sigma_h) \in \mathbb{R}[\sigma_h]$, the ring of real polynomials in the operator σ_h. We then have $f_h(s) = p(s, e^{-sh})$. The utility of this formula is clearly seen in the following lemma.

Lemma 1.1 *Let $A_0, A_1 \in \mathbb{R}^{n\times n}$. Then there exists $\omega_0 > 0$ such that for all $h \in H$, the characteristic function $f_h(s) = |sI - A_0 - e^{-sh}A_1|$ has no zeros in $\{|s| \geq \omega_0, \mathrm{Re}(s) \geq 0\}$.*

Proof. For $z \in \mathbb{C}$, write $p(s,z) = |sI - A_0 - A_1 z| = s^n + a_{n-1}(z)s^{n-1} + \cdots + a_0(z)$. Then $p(s,z) = s^n[1 + (a_{n-1}(z)/s) + \ldots + (a_0(z)/s^n)]$, and thus $p(s,z)/s^n \to 1$ uniformly for $|z| \leq 1$ as $|s| \to \infty$. Since $|e^{-sh}| \leq 1$ for $\mathrm{Re}(s) \geq 0, h \geq 0$, we thus see that $f_h(s)/s^n \to 1$ uniformly for $(s,h) \in \{\mathrm{Re}(s) \geq 0\} \times H$ as $|s| \to \infty$, and the lemma is now apparent. □

The following lemma can be proven with the aid of Lemma 1.1 by applying the principle of the argument to the function $f_h(s)$ over any curve $c(\cdot)$ parametrizing the set $\{i\omega : |\omega| \leq \omega_0\} \cup \{|s| = \omega_0, \mathrm{Re}(s) \geq 0\}$.

Lemma 1.2 *Let $h_0 \in H$, and suppose $f_{h_0}(s)$ has no zeros in $\{\mathrm{Re}(s) \geq 0\}$. Then there is an open set $U = H \cap \{|h - h_0| < r\}$, having the property that if $h \in U$, then $f_h(s)$ has no zeros in $\{\mathrm{Re}(s) \geq 0\}$.*

2 A Lyapunov functional

The purpose of this section is to present the basic facts found necessary in the construction and use of the Lyapunov functional employed in Section 3.

We begin by taking any two matrices $A_0, A_1 \in \mathbb{R}^{n \times n}$. We set $H = [0, \infty)$, and for each $h \in H$, we define the matrix functions $T_h(s) = sI - A_0 - A_1 e^{-sh}$, $M_h(s) = T_h(s)^{-1}$, and the scalar function $f_h(s) = |T_h(s)|$. It will frequently be useful to denote $T_h(s)$ by $T(h,s)$, and to denote $M_h(s)$ by $M(h,s)$. Finally, we remind the reader that here H_s is defined by $H_s = \{h \in H : f_h(s)$ is nonzero for each complex s having $\mathrm{Re}(s) \geq 0\}$.

The simple formula $T_h(s) = s(I - \frac{1}{s}A_0 - \frac{1}{s}e^{-sh}A_1)$, valid for $s \in \mathbb{C} - \{0\}$, immediately yields $M_h(s) = s^{-1}(I - \frac{1}{s}A_0 - \frac{1}{s}e^{-sh}A_1)^{-1}$, valid throughout $\{s \neq 0, f_h(s) \neq 0\}$. Setting $F(h, i\omega) = (I - \frac{1}{i\omega}A_0 - \frac{1}{i\omega}e^{-i\omega h}A_1)^{-1}$, we immediately see that $F(h, i\omega) \to I$ uniformly for $h \in H$ as $|\omega| \to \infty$. Writing $M_h(i\omega) = \frac{1}{i\omega}F(h, i\omega)$, we have $(M_h)^*(i\omega)M_h(i\omega) = \frac{1}{\omega^2}F^*(h, i\omega)F(h, i\omega)$, and thus we see that if $f_h(i\omega)$ is nonzero for each $\omega \in \mathbb{R}$, then each of the entries of the matrix $(M_h)^*(i\omega)M_h(i\omega)$ will be absolutely integrable over the unbounded interval $(-\infty, \infty)$. For each $h \in H_s$, we now form the matrix $Q(h, \alpha)$, defined for every $\alpha \in \mathbb{R}$ as

$$Q(h, \alpha) = \frac{1}{2\pi} \int_{-\infty}^{\infty} (M_h)^*(i\omega) M_h(i\omega) e^{-i\omega\alpha} d\omega.$$

There are several basic formulas which simplify the analysis of the matrix function $Q(h, \alpha)$. First among these is the formula for $f_h(s)$ given in Section 1. Recall that we there wrote $f_h(s) = p(s, e^{-sh})$, where $p(s, \sigma_h) = |sI - A(\sigma_h)| = s^n + a_{n-1}(\sigma_h)s^{n-1} + \cdots + a_0(\sigma_h)$, and σ_h is the delay operator of duration h, with $A(\sigma_h) = A_0 + A_1\sigma_h$. We used this formula to deduce the fact that there is $\omega_0 > 0$ having the property that for each $h \in H$, $f_h(s)$ is nonzero for every complex s having $\mathrm{Re}(s) \geq 0$, $|s| \geq \omega_0$. From this formula it is also easily seen that if K is any compact subset of $\mathbb{C}$, and h_0 is any member of H_s, then $f_h(s) \to f_{h_0}(s)$ uniformly for $s \in K$ as $h \to h_0$.

Another formula useful for analyzing the matrix $Q(h, \alpha)$ is the formula for $M_h(s)$ in terms of the matrix adjugate to $T_h(s)$, i.e. $M_h(s) = T_h(s)^{-1} = \left(\frac{1}{f_h(s)}\right)(\text{adj } T_h(s))$ for each complex s having $f_h(s) \neq 0$. From this formula it is seen that, given any $h_0 \in H$, if $f_{h_0}(s)$ is nonzero for every complex s lying in some compact subset K of $\mathbb{C}$, then $M_h(s)$ is defined throughout K for each h lying in some relatively open set $U = H \cap \{|h - h_0| < r\}$, and in fact $M_h(s) \to M_{h_0}(s)$ uniformly for $s \in K$ as $h \to h_0$. With these comments as background, we can now present the first of several lemmas dealing with the matrix $Q(h, \alpha)$ found useful in Section 3.

Lemma 2.1 *Let $h_0 \in H_s$. Then there is a neighborhood $U = H \cap \{|h - h_0| < r\}$, contained in H_s, for which $Q(h, \alpha)$ is defined and continuous throughout $U \times \mathbb{R}$. Furthermore, $Q(h, \alpha) \to Q(h_0, \alpha)$ uniformly throughout $\mathbb{R}$ as $h \to h_0$.*

Proof. Existence: For any $h_0 \in H_s$, we know from Lemma 1.2 that there is a neighborhood $U = H \cap \{|h - h_0| < r\}$ with $H_s \supset U$. For any $h \in U$, since $h \in H_s$, we know that $Q(h, \alpha)$ is defined for each $\alpha \in \mathbb{R}$.

Continuity: Let U be as immediately above, and for each $h \in U$, set $R(h, \omega) = \frac{1}{2\pi}(M_h)^*(i\omega)M_h(i\omega)$, and $\tilde{R}(h, \omega) = R(h, \omega) - R(h_0, \omega)$. For any fixed $\omega_1 > 0$, we write

$$Q(h, \alpha) - Q(h_0, \alpha) = \int_{|\omega| \leq \omega_1} \tilde{R}(h, \omega) e^{-i\omega\alpha} d\omega + \int_{|\omega| \geq \omega_1} \tilde{R}(h, \omega) e^{-i\omega\alpha} d\omega.$$

Since $h_0 \in H_s$, we know from the comments preceding this lemma that for any $\omega_1 > 0$, $\tilde{R}(h, \omega) \to 0$ uniformly for $|\omega| \leq \omega_1$ as $h \to h_0$. Thus, for any $\omega_1 > 0$, we have

$$\int_{|\omega| \leq \omega_1} ||\tilde{R}(h, \omega)|| d\omega \to 0 \quad \text{as} \quad h \to h_0.$$

Again referring to the comments preceding this lemma, we write $2\pi\tilde{R}(h, \omega) = \frac{1}{\omega^2}[F^*(h, i\omega)F(h, i\omega) - F^*(h_0, i\omega)F(h_0, i\omega)]$ and recall that $F(h, i\omega) \to I$ uniformly for $h \in H$ as $|\omega| \to \infty$. We thus see that

$$\int_{|\omega| \geq \omega_1} ||\tilde{R}(h, \omega)|| d\omega \to 0 \text{ uniformly for } h \in H \text{ as } \omega_1 \to \infty.$$

Now writing

$$||Q(h, \alpha) - Q(h_0, \alpha)|| \leq \int_{|\omega| \leq \omega_1} ||\tilde{R}(h, \omega)|| d\omega + \int_{|\omega| \geq \omega_1} ||\tilde{R}(h, \omega)|| d\omega,$$

and recalling that for any $\omega_1 > 0$, $\int_{|\omega| \leq \omega_1} ||\tilde{R}(h, \omega)|| d\omega \to 0$ as $h \to h_0$, we conclude that $||Q(h, \alpha) - Q(h_0, \alpha)|| \to 0$ uniformly for $\alpha \in \mathbb{R}$ as $h \to h_0$.

If we now set $S(h, \omega, \alpha) = R(h, \omega)e^{-i\omega\alpha}$, and if for fixed $\alpha_0 \in \mathbb{R}$ we set $\tilde{S}(h, \omega, \alpha) = S(h, \omega, \alpha) - S(h_0, \omega, \alpha_0)$, then by applying an argument similar to the above, using the functions $S(h,\omega,\alpha)$ and $\tilde{S}(h,\omega,\alpha)$, one will find that $||Q(h,\alpha)-Q(h_0,\alpha_0)|| \to 0$ as $(h,\alpha)\to(h_0,\alpha_0)$. Since the original choice of $h_0 \in H_s$ was arbitrary, the proof is complete. □

The next lemma, which is actually a basic fact of real analysis, is included for use in the two lemmas which immediately follow this lemma. The proof is derived from standard real analysis techniques, and is not given here.

Lemma 2.2 *Let U_1 be any real interval, and let U_2 be either of $(-\infty,\infty),[\tau,\infty)$, where τ is any member of $\mathbb{R}$. Let $f: U_1 \times U_2 \to \mathbb{C}^{n\times n}$, where both f and $D_1 f$ are continuous throughout $U_1\times U_2$. For any $\gamma \geq 0$, set $J(\gamma)=U_2 \cap \{|x_2|\geq \gamma\}$, and now suppose that there exist $\gamma_0>0$ and a real function $\Phi : J(\gamma_0) \to [0,\infty)$ having the properties a), b) written below:*

a) $\displaystyle\int_{J(\gamma_0)} \phi(x_2)dx_2 < \infty$

b) for each $(x_1,x_2)\in U_1\times U_2$ having $|x_2|\geq\gamma_0$, both $||f(x_1,x_2)||\leq\phi(x_2)$ and $||D_1 f(x_1,x_2)||\leq\phi(x_2)$.

Then for $F(x_1)=\displaystyle\int_{U_2} f(x_1,x_2)dx_2$, we know that $F(x_1)$ is defined and finite for all $x_1\in U_1$, and in fact the derivative $F'(x_1)$ exists and is continuous throughout U_1, with $F'(x_1)=\displaystyle\int_{U_2} D_1 f(x_1,x_2)dx_2$.

In the next lemma, we again examine the behavior of the function $R(h,\omega) = \frac{1}{2\pi}M^*(h,i\omega)M(h,i\omega)$. Here we employ two formulas for $\frac{\partial R}{\partial h}(h,\omega)$ to prove existence and continuity of $\frac{\partial Q}{\partial h}(h,\alpha)$. To obtain the first of these formulas we recall that $M(h,i\omega) = T^{-1}(h,i\omega)$, where $T(h,i\omega) = i\omega I - A_0 - e^{-i\omega h}A_1$. Setting $N_1(h,\omega) = -\frac{i\omega}{2\pi}e^{-i\omega h}M^*(h,i\omega)M(h,i\omega)A_1M(h,i\omega)$, we can give the first of the two formulas for $\frac{\partial R}{\partial h}(h,\omega)$:

$$\begin{aligned}
2\pi\frac{\partial R}{\partial h}(h,\omega) &= \left(\frac{\partial M}{\partial h}(h,i\omega)\right)^* M(h,i\omega) + M^*(h,i\omega)\left(\frac{\partial M}{\partial h}(h,i\omega)\right) \\
&= (-M(h,i\omega)A_1M(h,i\omega)i\omega e^{-i\omega h})^* M(h,i\omega) \\
&\quad + M^*(h,i\omega)(-M(h,i\omega)A_1M(h,i\omega)i\omega e^{-i\omega h}) \\
&= 2\pi[N_1^*(h,\omega)+N_1(h,\omega)],
\end{aligned}$$

i.e. $\frac{\partial R}{\partial h}(h,\omega) = N_1^*(h,\omega)+N_1(h,\omega)$ for any (h,ω) having $|T(h,i\omega)|\neq 0$.

We now return to the formula $R(h,\omega) = \frac{1}{2\pi\omega^2}F^*(h,i\omega)F(h,i\omega)$, where $F(h,i\omega) = \left(I - \frac{1}{i\omega}A_0 - \frac{1}{i\omega}e^{-i\omega h}A_1\right)^{-1}$. Setting $N_2(h,\omega) = -\frac{1}{2\pi}e^{-i\omega h}F^*(h,i\omega)$ $F(h,i\omega)A_1F(h,i\omega)$, we can give the second of the formulas for $\frac{\partial R}{\partial h}(h,\omega)$:

$$2\pi\omega^2 \frac{\partial R}{\partial h}(h,\omega) = \left(\frac{\partial F}{\partial h}(h,i\omega)\right)^* F(h,i\omega) + F^*(h,i\omega)\left(\frac{\partial F}{\partial h}(h,i\omega)\right)$$
$$= (-F(h,i\omega)A_1F(h,i\omega)e^{-i\omega h})^* F(h,i\omega)$$
$$+ F^*(h,i\omega)(-F(h,i\omega)A_1F(h,i\omega)e^{-i\omega h})$$
$$= 2\pi[N_2^*(h,\omega) + N_2(h,\omega)],$$

i.e. $\frac{\partial R}{\partial h}(h,\omega) = \frac{1}{\omega^2}[N_2^*(h,\omega) + N_2(h,\omega)]$ for (h,ω) having $\omega \neq 0$, $|T(h,i\omega)| \neq 0$.

Lemma 2.3 *Let $h_0 \in H_s$. Then there is a neighborhood $U = H \cap \{|h-h_0| < r\}$, contained in H_s, for which $\frac{\partial Q}{\partial h}(h,\alpha)$ is defined and continuous throughout $U \times \mathbb{R}$, with $\frac{\partial Q}{\partial h}(h,\alpha) = \int_{-\infty}^{\infty} \frac{\partial R}{\partial h}(h,\omega)e^{-i\omega\alpha}d\omega$. Furthermore, $\frac{\partial Q}{\partial h}(h,\alpha) \to \frac{\partial Q}{\partial h}(h_0,\alpha)$ uniformly throughout $\mathbb{R}$ as $h \to h_0$.*

Proof. Let $h_0 \in H_s$, and let U be the neighborhood of Lemma 2.1, i.e., $U = H \cap \{|h-h_0| < r\}$, with $H_s \supset U$, and with $Q(h,\alpha)$ defined and continuous throughout $U \times \mathbb{R}$. Noting the formulas $2\pi R(h,\omega) = M^*(h,i\omega)M(h,i\omega)$, and $\frac{\partial R}{\partial h}(h,\omega) = N_1^*(h,\omega) + N_1(h,\omega)$, both valid throughout $U \times \mathbb{R}$, with $-2\pi N_1(h,\omega) = i\omega e^{-i\omega h}M^*(h,i\omega)M(h,i\omega)A_1M(h,i\omega)$, we see that both $R(h,\omega)$, $\frac{\partial R}{\partial h}(h,\omega)$ are continuous throughout $U \times \mathbb{R}$.

Now note the formulas $2\pi R(h,\omega) = \frac{1}{\omega^2}F^*(h,i\omega)F(h,i\omega)$, and $\frac{\partial R}{\partial h}(h\omega) = \frac{1}{\omega^2}[N_2^*(h,\omega) + N_2(h,\omega)]$, both valid for $h \in U, \omega \neq 0$, where $2\pi N_2(h,\omega) = -e^{-i\omega h}F^*(h,i\omega)F(h,i\omega)A_1F(h,i\omega)$. Recall that $F(h,i\omega) \to I$ uniformly for $h \in H$ as $|\omega| \to \infty$, and set $K = \max\left\{1 + \frac{1}{2\pi}, 1 + \frac{\|A_1\|}{2\pi}\right\}$. Then there is some $\gamma_0 > 0$ such that for $|\omega| \geq \gamma_0$ and $h \in H$, both $\|R(h,\omega)\| \leq K\omega^{-2}$, and $\|\frac{\partial R}{\partial h}(h,\omega)\| \leq K\omega^{-2}$. Writing $Q(h,\alpha) = \int_{-\infty}^{\infty} R(h,\omega)e^{-i\omega\alpha}d\omega$, we can apply Lemma 2.2 with $U_1 = U, U_2 = (-\infty,\infty)$, and $\phi(\omega) = K\omega^{-2}$ for $|\omega| \geq \gamma_0$, and conclude that for each $\alpha \in \mathbb{R}$, $\frac{\partial Q}{\partial h}(h,\alpha)$ exists and is continuous in h throughout U, with $\frac{\partial Q}{\partial h}(h,\alpha) = \int_{-\infty}^{\infty} \frac{\partial R}{\partial h}(h,\omega)e^{-i\omega\alpha}d\omega$.

Finally, noting the formulas $\frac{\partial R}{\partial h}(h,\omega) = N_1^*(h,\omega)+N_1(h,\omega)$, and $\frac{\partial R}{\partial h}(h,\omega) = \frac{1}{\omega^2}[N_2^*(h,\omega) + N_2(h,\omega)]$, one can employ a technique comparable to that used in Lemma 2.1 in proving that $Q(h,\alpha) \to Q(h_0,\alpha)$ uniformly throughout $\mathbb{R}$ as $h \to h_0$, and likewise prove that $\frac{\partial Q}{\partial h}(h,\alpha) \to \frac{\partial Q}{\partial h}(h_0,\alpha)$ uniformly throughout $\mathbb{R}$ as $h \to h_0$. Similarly, using a technique comparable to that mentioned in Lemma 2.1 for proving that $Q(h,\alpha) \to Q(h_0,\alpha_0)$ as $(h,\alpha) \to (h_0,\alpha_0)$, again noting the above formulas for $\frac{\partial R}{\partial h}(h,\omega)$, one can prove that $\frac{\partial Q}{\partial h}(h,\alpha) \to \frac{\partial Q}{\partial h}(h_0,\alpha_0)$ as $(h,\alpha) \to (h_0,\alpha_0)$. Since the choice of $h_0 \in H_s$ was arbitrary, we conclude that $\frac{\partial Q}{\partial h}(h,\alpha)$ is continuous throughout $U \times \mathbb{R}$. □

Before introducing a formula for use in dealing with $Q(h,\alpha)$ throughout this paper, we first recall some basic facts from Fourier transform theory ([9]). Consider any matrix function $f(t)$ having domain $(-\infty,\infty)$ and range in $\mathbb{C}^{n\times n}$. If $f \in L^2(-\infty,\infty)$, the Fourier transform $\hat{f}$ of f is defined as $\hat{f}(\omega) = \frac{1}{\sqrt{2\pi}}\int_{-\infty}^{\infty} f(t)e^{-i\omega t}dt$. The function $\hat{f}$ will lie in $L^2(-\infty,\infty)$, and in fact $f(t) = \frac{1}{\sqrt{2\pi}}\int_{-\infty}^{\infty} \hat{f}(\omega)e^{i\omega t}dt$. We let $\mathcal{F}$ denote the operator $\mathcal{F} : L^2(-\infty,\infty) \to$

$L^2(-\infty,\infty)$ defined by $\mathcal{F}(f) = \hat{f}$, and we let $\mathcal{F}^{-1}$ denote the inverse operator. Defining the convolution of any two members f, g of $L^2(-\infty,\infty)$ by $(f*g)(t) = \int_{-\infty}^{\infty} f(u)g(t-u)du$, we recall that $\mathcal{F}(f*g) = (\sqrt{2\pi})(\hat{f})(\hat{g})$ provided $f, g \in L^1(-\infty,\infty)$. Finally, we shall use the formula $\int_{-\infty}^{\infty} f^T(u)g(u-t)du = \int_{-\infty}^{\infty}(\hat{f})^*(\omega)\hat{g}(\omega)e^{-i\omega t}d\omega$, valid for $f, g \in (L^1 \cap L^2)(-\infty,\infty)$ and $t \in \mathbb{R}$ if f, g have range in $\mathbb{R}^{n\times n}$. This formula is readily derived by noting first that $(\hat{f})^* = \hat{f}_0$, where $f_0(t) = f^T(-t)$, by next writing $\frac{1}{\sqrt{2\pi}} \int_{-\infty}^{\infty} \hat{f}_0(\omega)\hat{g}(\omega)e^{-i\omega t}d\omega = (\mathcal{F}^{-1}\{(\hat{f}_0)(\hat{g})\})(-t) = \frac{1}{\sqrt{2\pi}}(f_0 * g)(-t)$, and by then using the variable substitution $\tilde{u} = -u$ in the expression $(f_0 * g)(-t) = \int_{-\infty}^{\infty} f^T(-u)g(-t-u)du$.

The next formula, which proves extremely useful as a characterization of the matrix $Q(h,\alpha)$, arises from the link between $M_h(s)$ and its inverse Laplace transform. To explore this further, for any $h \in H_s$, we denote the inverse Laplace transform of $M_h(s)$ by $X_h(t)$, or, when h is fixed and understood from the context, merely by $X(t)$. Then $X(\cdot)$ is the solution to the differential-delay equation $\dot{X}(t) = X(t)A_0 + X(t-h)A_1$ having initial data $X(u) = \Phi(u)$ for $-h \leq u \leq 0$, where $\Phi(u) = 0$ for $-h \leq u < 0$, and $\Phi(0) = I$. Noting that $X(t) = 0$ for $t < 0$, we write $M_h(s) = \int_0^{\infty} X(t)e^{-st}dt = \int_{-\infty}^{\infty} X(t)e^{-st}dt$. Since $h \in H_s$, we know that there exist $C \geq 1, \beta > 0$ such that $\|X(t)\| \leq Ce^{-\beta t}$ for all $t \geq 0$. Thus $X(\cdot) \in (L^1 \cap L^2)(-\infty,\infty)$, and with $\hat{X} = \mathcal{F}\{X\}$, we have $\hat{X}(\omega) = \frac{1}{\sqrt{2\pi}} M_h(i\omega)$. If we now set $f = g = X$ in the previously given formula $\int_{-\infty}^{\infty} f^T(u)g(u-\alpha)du = \int_{-\infty}^{\infty}(\hat{f})^*(\omega)\hat{g}(\omega)e^{-i\omega\alpha}d\omega$, we obtain the important formula $\int_0^{\infty} X^T(u)X(u-\alpha)du = \frac{1}{2\pi}\int_{-\infty}^{\infty}(M_h)^*(i\omega)M_h(i\omega)e^{-i\omega\alpha}d\omega$, i.e. $Q(h,\alpha) = \int_0^{\infty} X^T(t)X(t-\alpha)dt$ for each $\alpha \in \mathbb{R}$.

From this formula it is obvious that $Q(h,\alpha) \in \mathbb{R}^{n\times n}$ for $h \in H_s, \alpha \in \mathbb{R}$. Noting the definition $Q(h,\alpha) = \frac{1}{2\pi}\int_{-\infty}^{\infty}(M_h)^*(i\omega)M_h(i\omega)e^{-i\omega\alpha}d\omega$, one easily sees that $Q^*(h,\alpha) = Q(h,-\alpha)$. Thus for $h \in H_s$, $\alpha \in \mathbb{R}$, we have $Q^T(h,\alpha) = Q(h,-\alpha)$. Based also on the formula for $Q(h,\alpha)$ just derived, one may now refer to Infante and Castelan ([6]), or to Datko ([4]), and deduce the formulas for $\frac{\partial Q}{\partial\alpha}(h,\alpha)$ given in the following lemma. Alternatively, after some elementary analysis, one may use the formula derived for $Q(h,\alpha)$ to directly apply Lemma 2.2, and differentiate the integral for $Q(h,\alpha)$ with respect to α.

Lemma 2.4 *Let $h_0 \in H_s$. Then there is a neighborhood $U = H \cap \{|h-h_0| < r\}$, contained in H_s, for which $\frac{\partial Q}{\partial\alpha}(h,\alpha)$ is defined and continuous throughout $U \times (\mathbb{R} - \{0\})$. In fact, the following formulas hold for $\frac{\partial Q}{\partial\alpha}(h,\alpha)$:*

a) $\dfrac{\partial Q}{\partial\alpha}(h,\alpha) = -Q(h,\alpha)A_0 - Q(h,\alpha+h)A_1$ *for* $(h,\alpha) \in U \times (-\infty,0)$

b) $\dfrac{\partial Q}{\partial\alpha}(h,\alpha) = A_0^T Q^T(h,-\alpha) + A_1^T Q^T(h,-\alpha+h)$ *for* $(h,\alpha) \in U \times (0,\infty)$.

Corollary 2.5 *Let V_1 be any compact subset of H_s, and let V_2 be any bounded subset of $\mathbb{R} - \{0\}$. Then $\|\frac{\partial Q}{\partial\alpha}(h,\alpha)\|$ is bounded over $V_1 \times V_2$.*

Proof. Choose $\alpha_1, \alpha_2 \in \mathbb{R}^+$ such that $[-\alpha_1, 0) \cup (0, \alpha_2] \supseteq V_2$. Set $P_1(h,\alpha) = -Q(h,\alpha)A_0 - Q(h,\alpha+h)A_1$, and note that for $h \in H_s, \alpha < 0$, we have $\frac{\partial Q}{\partial \alpha}(h,\alpha) = P_1(h,\alpha)$. Noting continuity of $P_1(h,\alpha)$ and compactness of V_1, we see that $\|P_1(h,\alpha)\|$ is bounded over $V_1 \times [-\alpha_1, 0]$, and hence $\|\frac{\partial Q}{\partial \alpha}(h,\alpha)\|$ is bounded over $V_1 \times [-\alpha_1, 0)$. Setting $P_2(h,\alpha) = A_0^T Q^T(h,-\alpha) + A_1^T Q^T(h,-\alpha+h)$, and noting that $\frac{\partial Q}{\partial \alpha}(h,\alpha) = P_2(h,\alpha)$ for $h \in H_s, \alpha > 0$, we similarly find that $\|\frac{\partial Q}{\partial \alpha}(h,\alpha)\|$ is bounded over $V_1 \times (0, \alpha_2]$. We now conclude that $\|\frac{\partial Q}{\partial \alpha}(h,\alpha)\|$ is bounded over $V_1 \times V_2$. □

It is worth noting that for any $h \in H_s$, the matrix function $Q(h,\cdot)$ is differentiable from the right at $\alpha = 0$, with right derivative given by the formula

$$\left(\frac{\partial Q}{\partial \alpha}(h,\alpha)|_{\alpha=0+}\right) = -I - Q(h,\alpha)A_0 - Q(h,h)A_1.$$

In fact, the matrix function $Q(h,\cdot)$ is also differentiable from the left at $\alpha = 0$, with left derivative given by

$$\left(\frac{\partial Q}{\partial \alpha}(h,\alpha)|_{\alpha=0-}\right) = -Q(h,0)A_0 - Q(h,h)A_1.$$

Since these formulas will not be used in any of the lemmas or theorems in this paper, the proofs of these formulas will not be given here. The interested reader will find several more formulas for $\frac{\partial Q}{\partial \alpha}(h,\alpha)$ in a paper of Datko on autonomous differential-delay equations in Hilbert space ([4]).

3 Stability

In this section we consider the system $(\dagger)\dot{x}(t) = A_0 x(t) + A_1 x(t - h(t))$, having time-varying delay $h(t)$. Recalling that H_s is that subset of members h of H for which the characteristic function $f_h(s) = |sI - A_0 - A_1 e^{-sh}|$ has no zeros in $\{\text{Re}(s) \geq 0\}$, we examine the stability of the system $(\dagger)$ for certain functions $h(t)$ taking values in H_s. Thus we relate the stability of the system $(\dagger)$ to the stability of the system $(*)\dot{x}(t) = A_0 x(t) + A_1 x(t-h)$.

As is frequently the case in stability theory, we approach the system $(\dagger)$ with the aid of a Lyapunov functional. This functional is a generalization for time-varying systems of a Lyapunov functional first used to analyze autonomous systems by Infante ([7]) and Datko([4]). In order to construct this functional and identify its salient properties, we now fix the notation which will be used throughout the remainder of this section.

We begin, as throughout this paper, by taking any fixed matrices $A_0, A_1 \in \mathbb{R}^{n \times n}$. We then consider the differential-delay equations $(*)\dot{x}(t) = A_0 x(t) + A_1 x(t-h)$, and $(\dagger)\dot{x}(t) = A_0 x(t) + A_1 x(t - h(t))$. As in Section 1, we write $A(\sigma_h) = A_0 + A_1 \sigma_h$, obtaining the differential-delay equations $(*)\dot{x}(t) = A(\sigma_h)x(t)$, and $(\dagger)\dot{x}(t) = A(\sigma_{h(t)})x(t)$. We now let $C[-h, 0]$ denote the space of continuous functions Φ mapping the interval $[-h, 0]$ into $\mathbb{R}^n$. As in Infante ([7]), for each $h \in H_s$, we use the matrix function $Q(h,\alpha)$ to define the following functional V_h, taking members Φ of $C[-h, 0]$ into $\mathbb{R}$:

$$V_h(\Phi) = \Phi^T(0)Q(h,0)\Phi(0) + 2\Phi^T(0)\int_{-h}^{0} Q(h,u+h)A_1\Phi(u)du$$
$$+ \int_{-h}^{0}\int_{-h}^{0} \Phi^T(u)(A_1)^T Q(h,v-u)A_1\Phi(v)dvdu.$$

If we let $x(\cdot)$ be any trajectory of the system $(*)\dot{x}(t) = A(\sigma_h)x(t)$, with corresponding sections $x_t \in C[-h,0]$ defined by $x_t(u) = x(t+u)$ for $-h \leq u \leq 0$, we may write the functional V_h as below:

$$V_h(x_t) = x^T(t)Q(h,0)x(t) + 2x^T(t)\int_{-h}^{0} Q(h,u+h)A_1x(t+u)du$$
$$+ \int_{-h}^{0}\int_{-h}^{0} x^T(t+u)(A_1)^T Q(h,v-u)A_1x(t+v)dvdu.$$

Finally, making use of the elementary variable transformation $u' = t+u, v' = t+v$, we obtain the form for V_h given below, which is often convenient for calculating the time derivative of $V_h(x_t)$:

$$V_h(x_t) = x^T(t)Q(h,0)x(t) + 2x^T(t)\int_{t-h}^{t} Q(h,u'+h-t)A_1x(u')du'$$
$$+ \int_{t-h}^{t}\int_{t-h}^{t} x^T(u')(A_1)^T Q(h,v'-u')A_1x(v')dv'du'.$$

Now for any $\Phi \in C[-h,0]$, let $\hat{x}(\omega) = \hat{x}(\Phi,\omega)$ denote $\mathcal{F}\{\text{x(t)}\}$, where $x(t) = x(\Phi,t)$ is the solution, defined for $0 \leq t < \infty$, to the differential-delay equation $(*)$ $\dot{x}(t) = A(\sigma_h)x(t)$ having initial data Φ on $[-h,0]$. It is a routine, though somewhat lengthy exercise ([7]) to prove the formula $V_h(\Phi) = \int_{-\infty}^{\infty}(\hat{x})^*(\omega)\hat{x}(\omega)d\omega$. If one now applies Parseval's equality to this formula $V_h(\Phi) = \int_{-\infty}^{\infty}(\hat{x})^*(\omega)\hat{x}(\omega)d\omega$, one will immediately see that $V_h(\Phi) = \int_0^{\infty} x^T(\tau)x(\tau)d\tau$. Expressed in terms of the sections x_t of the solution $x(\cdot)$, this becomes $V_h(x_t) = \int_t^{\infty} x^T(\tau)x(\tau)d\tau$. From this we see that $\dot{V}_h(x_t) = \frac{dV_h(x_t)}{dt} = -x^T(t)x(t)$. As in Infante ([7]), it is now seen that if $x(\cdot)$ is any solution to the differential-delay equation $(*)\dot{x}(t) = A_0x(t) + A_1x(t-h)$, then for $t \geq 0$, we have $\dfrac{dV_h(x_t)}{dt} = -x^T(t)x(t)$, i.e. $\dot{V}_h(\Phi) = -\Phi^T(0)\Phi(0)$.

Naturally, the motivation for the definition of V_h was the expression for $V_h(\Phi)$ as the energy of the trajectory $x(\Phi,t)$, along with the important corresponding expression $\dot{V}_h(\Phi) = -\Phi^T(0)\Phi(0)$. The reason for the form of the original choice for the definition of $V_h(\Phi)$ was to emphasize the fact that, via the formula for $V_h(\Phi)$ as an integral functional in Φ, one may obtain explicit information on the behavior of $V_h(\Phi)$ as h varies throughout H_s.

Before showing how to modify the functional V_h for use in time-varying systems, we give the following lemma. This lemma is included for use in a technical detail occurring in Lemma 3.2 and in Lemma 3.3 . Since the following lemma is a basic fact of real analysis, the proof will not be given here. However, it should not be assumed that the proof is routine.

Lemma 3.1 *Let $f : (a,b) \times [c-\delta, d] \to \mathbb{R}^{n \times n}$ be continuous, where δ is a positive real number. Set $V = (a,b) \times \{c - \delta \le x_2 \le d, x_2 \ne c\}$, and suppose first that $D_1 f(x_1, x_2)$ is defined and continuous on V, and also that there is a constant M such that $\|D_1 f(x_1, x_2)\| \le M$ for $(x_1, x_2) \in V$. Now, for each $x_1 \in (a,b)$, define $F(x_1) = \int_c^d f(x_1, x_2) dx_2$. Then the derivative $F'(x_1)$ exists for each $x_1 \in (a,b)$, and in fact $F'(x_1) = \int_c^d D_1 f(x_1, x_2) dx_2$.*

We are now equipped to modify the functional $V_h(\Phi)$ for use in the case where the delay is a function of t. To begin, we let S be the class of all functions $h(t)$ with domain $[0, \infty)$ and range in H, having the following properties:

i) $h(\cdot)$ is bounded and continuous
ii) There is a compact subset D of H_s and a real number $v \ge 0$ such that $h(t) \in D$ for $t \ge v$
iii) $h(\cdot)$ is differentiable at t for $t \ge v$.

Now, for any $h(t) \in S$, set $\widetilde{h} = \sup\{h(t) : t \ge 0\}$, and define the time-varying functional $G(t, \Phi)$ for $t \ge v$ and $\Phi \in C[-\widetilde{h}, 0]$, as $G(t, \Phi) = V_{h(t)}(\Phi)$.

As is the case with $V_h(\Phi)$, it will be useful to have the functional $G(t, \Phi)$ written out explicitly for the case that $\Phi = x_t$. Letting $x(\cdot)$ be any trajectory of the system (†) $\dot{x}(t) = A(\sigma_{h(t)}) x(t)$, with corresponding sections $x_t \in [-\widetilde{h}, 0]$ defined as usual by $x_t(u) = x(t+u)$, and again setting $u' = t + u, v' = t + v$, we write the functional $G(t, x_t)$ as below:

$$
\begin{aligned}
G(t, x_t) &= x^T(t) Q(h(t), 0) x(t) \\
&+ 2x^T(t) \int_{t-h(t)}^{t} Q(h(t), u' + h(t) - t) A_1 x(u') du' \\
&+ \int_{t-h(t)}^{t} \int_{t-h(t)}^{t} x^T(u')(A_1)^T Q(h(t), v' - u') A_1 x(v') dv' du'.
\end{aligned}
$$

The next lemmas provide formulas for $\dot{V}_h$, along trajectories of the system (∗) $\dot{x}(t) = A(\sigma_h) x(t)$, and for $\dot{G}$ along trajectories of the system (†)$\dot{x}(t) = A(\sigma_{h(t)}) x(t)$. The relationship between these two quantities will be clearly displayed in these formulas, and this relationship will be the basis of the analysis relating the stability of the system (†) to the stability of (∗).

Lemma 3.2 *Let h be any member of H_s, and let $E_1(h, \cdot), E_2(h, \cdot), E_3(h, \cdot)$ be the functionals taking members Φ of $C[-h, 0]$ into $\mathbb{R}$ as given in the formulas below:*

$$E_1(h,\Phi) = 2[\Phi^T(0)(A_0)^T + \Phi^T(-h)(A_1)^T]Q(h,0)\Phi(0)$$

$$\begin{aligned} E_2(h,\Phi) = 2\Bigg[&[\Phi^T(0)(A_0)^T + \Phi^T(-h)(A_1)^T]\int_{-h}^{0} Q(h,u+h)A_1\Phi(u)du \\ &+ \Phi^T(0)Q(h,h)A_1\Phi(0) - \Phi^T(0)Q(h,0)A_1\Phi(-h) \\ &- \Phi^T(0)\int_{-h}^{0}\left(\frac{\partial Q}{\partial \alpha}(h,\alpha)|_{\alpha=u+h}\right)A_1\Phi(u)du\Bigg] \end{aligned}$$

$$\begin{aligned} E_3(h,\Phi) = \Phi^T(0)&\int_{-h}^{0}(A_1)^T Q(h,u)A_1\Phi(u)du \\ &- \Phi^T(-h)\int_{-h}^{0}(A_1)^T Q(h,u+h)A_1\Phi(u)du \\ &+ \left[\int_{-h}^{0}\Phi^T(u)(A_1)^T Q(h,-u)A_1 du\right]\Phi(0) \\ &- \left[\int_{-h}^{0}\Phi^T(u)(A_1)^T Q(h,-u-h)A_1 du\right]\Phi(-h). \end{aligned}$$

Now consider $\dot{V}_h(x_t)$, the time derivative of $V_h(x_t)$ along solutions of (∗) $\dot{x}(t) = A(\sigma_h)x(t)$, the differential equation with constant delay h. We have:

(a) $\dot{V}_h(x_t) = \sum_{i=1}^{3} E_i(h,x_t)$

(b) $-x^T(t)x(t) = \sum_{i=1}^{3} E_i(h,x_t)$.

Proof. The formula in b) follows from the formula in a) and the fact that $\dot{V}_h(x_t) = -x^T(t)x(t)$.

To prove the formula in a) for $\dot{V}_h(x_t) = \frac{dV_h(x_t)}{dt}$, consider the third of the formulas given in the definition of V_h, displayed for convenience immediately below:

$$\begin{aligned} V_h(x_t) = x^T(t)Q(h,0)x(t) + 2x^T(t)&\int_{t-h}^{t} Q(h,u'+h-t)A_1 x(u')du' \\ + \int_{t-h}^{t}\int_{t-h}^{t} &x^T(u')(A_1)^T Q(h,v'-u')A_1 x(v')dv'du'. \end{aligned}$$

Noting Lemma 2.1, one can make direct use of the chain rule and the product rule to calculate the time derivative of the first and third terms in the above expression for $V_h(x_t)$. One thus arrives at the functionals $E_1(h,\cdot)$ and $E_3(h,\cdot)$ in the expression for $\dot{V}_h(x_t)$.

The technical detail mentioned prior to Lemma 3.1 occurs in finding the time derivative of the second term in the above expression for $V_h(x_t)$. Here one begins by setting $J(\alpha_1,\alpha_2,\alpha_3) = \int_{\alpha_1}^{\alpha_2} Q(h,u'+h-\alpha_3)A_1x(u')du'$, and noting that the term being examined is equal to $2x^T(t)J(t-h,t,t)$. After noting Lemma 2.4, Corollary 2.5, and Lemma 3.1, one can calculate $\frac{\partial J}{\partial \alpha_3}(t-h,t,\alpha_3)|_{\alpha_3=t}$. One then makes routine use of the chain rule to find the derivative with respect to time of $J(t-h,t,t)$. The differentiation is completed by applying the product rule to the expression $2x^T(t)J(t-h,t,t)$, and thus one arrives at the functional $E_2(h,\cdot)$ in the expression for $\dot{V}_h(x_t)$. □

Corollary 3.3 *For any* $\Phi \in C[-h,0]$, *we have* $-\Phi^T(0)\Phi(0) = \sum_{i=1}^{3} E_i(h,\Phi)$.

Proof. This follows immediately if we let Φ be the initial data on $[-h,0]$ for the differential-delay equation $(*)$ $\dot{x}(t) = A_0x(t) + A_1x(t-h)$, and then apply the second part of the above lemma with $t = 0+$. □

Lemma 3.4 *For each* $h \in H_s$, *let* $F_1(h,\cdot), F_2(h,\cdot), F_3(h,\cdot)$ *be the functionals taking members* Φ *of* $C[-h,0]$ *into* $\mathbb{R}$ *as given in the formulas below:*

$$F_1(h,\Phi) = \Phi^T(0)\frac{\partial Q}{\partial h}(h,0)\Phi(0)$$

$$\begin{aligned} F_2(h,\Phi) = 2\Bigg[&\Phi^T(0)Q(h,0)A_1\Phi(-h) \\ &+ \Phi^T(0)\int_{-h}^{0}\left(\frac{\partial Q}{\partial \gamma}(\gamma,u+h)|_{\gamma=h}\right)A_1\Phi(u)du \\ &+ \Phi^T(0)\int_{-h}^{0}\left(\frac{\partial Q}{\partial \alpha}(h,\alpha)|_{\alpha=u+h}\right)A_1\Phi(u)du\Bigg] \end{aligned}$$

$$\begin{aligned} F_3(h,\Phi) = &\Phi^T(-h)\int_{-h}^{0}(A_1)^TQ(h,u+h)A_1\Phi(u)du \\ &+ \left[\int_{-h}^{0}\Phi^T(u)(A_1)^TQ(h,-u-h)A_1du\right]\Phi(-h) \\ &+ \int_{-h}^{0}\int_{-h}^{0}\Phi^T(u)\mathcal{N}\Phi(v)dvdu, \end{aligned}$$

where

$$\mathcal{N} = (A_1)^T\left(\frac{\partial Q}{\partial h}(h,v-u)\right)A_1.$$

Now let $h(t)$ *be any member of the class* S, *with* $h(t) \in H_s$ *for* $t \geq v$. *Consider the functional* $G(t,\Phi) = V_{h(t)}(\Phi)$, *and let* $\dot{G}(x_t) = \frac{dG(t,x_t)}{dt}$ *be the derivative of* $G(t,x_t)$ *along solutions of* $(\dagger)\dot{x}(t) = A_0x(t) + A_1x(t-h(t))$, *the differential equation with time-varying delay* $h(t)$. *For* $t \geq v$, *we have:*

(a) $\dot{G}(x_t) = \sum_{i=1}^{3} E_i(h(t),x_t) + h'(t)\left[\sum_{i=1}^{3} F_i(h(t),x_t)\right]$

(b) $\dot{G}(x_t) = -x^T(t)x(t) + h'(t)\left[\sum_{i=1}^{3} F_i(h(t),x_t)\right]$.

Proof. The formula in b) is obtained by first noting the formula in a), by then noting the relation $-\Phi^T(0)\Phi(0) = \sum_{i=1}^{3} E_i(h,\Phi)$, true for any $h \in H_s$ and $\Phi \in C[-h,0]$, and finally, by setting $h = h(t)$ and $\Phi = x_t$ in this relation.

To prove the formula in a) for $\dot{G}(x_t)$, we use the formula given in the definition of G, displayed for convenience immediately below:

$$\begin{aligned} G(t,x_t) &= x^T(t)Q(h(t),0)x(t) \\ &+ 2x^T(t)\int_{t-h(t)}^{t} Q(h(t),u'+h(t)-t)A_1x(u')du' \\ &+ \int_{t-h(t)}^{t}\int_{t-h(t)}^{t} x^T(u')(A_1)^TQ(h(t),v'-u')A_1x(v')dv'du'. \end{aligned}$$

Note that this formula is merely $V_{h(t)}(x_t)$, where the formula used for V_h is the one displayed in the preceding lemma. Now noting Lemma 2.1, Lemma 2.3, Lemma 2.4, Corollary 2.5, and Lemma 3.1, the formula in a) is obtained by a technique similar to the one used in the preceding lemma. □

It will simplify the proof of the following theorem to first present the useful lemma below.

Lemma 3.5 *Let $h(t) \in S$, with $v \geq 0$ and compact D contained in H_s both as in the definition of members of S, and recall that $\tilde{h} = \sup\{h(t) : t \geq 0\}$. Then each of the quantities defined below is finite:*

$$\begin{aligned} b_1 &= \sup\{\|\frac{\partial Q}{\partial \gamma}(\gamma,0)\| : \gamma \in D\}; \\ b_{21} &= \|A_1\|\sup\{\|Q(\gamma,0)\| : \gamma \in D\}; \\ b_{22} &= \|A_1\|\sup\{\|\frac{\partial Q}{\partial \gamma}(\gamma,\alpha)\| : \gamma \in D, 0 \leq \alpha \leq \tilde{h}\}; \\ b_{23} &= \|A_1\|\sup\{\|\frac{\partial Q}{\partial \alpha}(\gamma,\alpha)\| : \gamma \in D, 0 < \alpha \leq \tilde{h}\}; \\ b_2 &= 2(b_{21} + b_{22}\tilde{h} + b_{23}\tilde{h}); \\ b_{31} &= \|A_1\|^2\sup\{\|Q(\gamma,\alpha)\| : \gamma \in D, 0 \leq \alpha \leq \tilde{h}\}; \\ b_{32} &= \|A_1\|^2\sup\{\|Q(\gamma,\alpha)\| : \gamma \in D, -\tilde{h} \leq \alpha \leq 0\}; \\ b_{33} &= \|A_1\|^2\sup\{\|\frac{\partial Q}{\partial \gamma}(\gamma,\alpha)\| : \gamma \in D, -\tilde{h} \leq \alpha \leq \tilde{h}\}; \\ b_3 &= \tilde{h}(b_{31} + b_{32} + b_{33}\tilde{h}); \\ B &= b_1 + b_2 + b_3. \end{aligned}$$

Proof. From Lemma 2.1, Lemma 2.3, and Lemma 2.4, we know that $Q(\gamma,\alpha)$, $\frac{\partial Q}{\partial \gamma}(\gamma,\alpha)$ are both defined and continuous on $D \times \mathbb{R}$, and $\frac{\partial Q}{\partial \alpha}(\gamma,\alpha)$ is defined and continuous on $D \times (\mathbb{R} - \{0\})$. Noting continuity of $Q(\gamma,\alpha)$ and $\frac{\partial Q}{\partial \alpha}(\gamma,\alpha)$, we see that both $\|Q(\gamma,\alpha)\|, \|\frac{\partial Q}{\partial \gamma}(\gamma,\alpha)\|$ are bounded over $D \times [-\tilde{h},\tilde{h}]$. Likewise, noting Corollary 2.5, we know that $\|\frac{\partial Q}{\partial \alpha}(\gamma,\alpha)\|$ is bounded over $D \times ([-\tilde{h},0) \cup (0,\tilde{h}])$; and the lemma is proven. □

Now introducing the notation $|E|$ to denote the Lebesgue measure of any Lebesgue measurable subset E of $\mathbb{R}$, we give a theorem providing a bound on the quantity $|x(t_1)|^2 + \epsilon^2|\{v \le t < t_1, |x(t)| \ge \epsilon\}|$ in terms of the behavior of $|h'(t)|$ over the interval $[v, t_1)$.

Theorem 3.1 *Let $h(t) \in S$, with v as in the definition of S and B as in Lemma 3.5, and consider the differential-delay system $(\dagger)\dot{x}(t) = A_0x(t) + A_1x(t - h(t))$. If $x(\cdot)$ is any bounded trajectory of the system $(\dagger)$, with $|x(t)| \le \eta$ for all $t \ge 0$, then for any $\epsilon > 0$, the following inequalities hold for all $t_1 \ge v$:*

$$(a)\ G(t_1, x_{t_1}) + \epsilon^2|\{v \le t < t_1, |x(t)| \ge \epsilon\}| \le G(v, x_v) + B\eta^2 \int_v^{t_1} |h'(t)|dt$$

$$(b)\ |x(t_1)|^2 + \epsilon^2|\{v \le t < t_1, |x(t)| \ge \epsilon\}| \le G(v, x_v) + B\eta^2 \int_v^{t_1} |h'(t)|dt.$$

Proof. The formula in (b) follows from the formula in (a) and the fact that $|\Phi(0)|^2 \le G(t, \Phi)$. To prove the formula in (a), we let $x(\cdot)$ be a bounded solution of the system $(\dagger)$, with $|x(t)| \le \eta$ for $t \ge 0$. Noting the quantities defined in Lemma 3.5, we apply the Cauchy-Schwartz inequality to obtain the inequalities below for the functions $F_1(h(t), x_t)$, $F_2(h(t), x_t)$, and $F_3(h(t), x_t)$ occuring in the expression for $\dot{G}(x_t)$ in Lemma 3.4:

$$|F_1(h(t), x_t)| \le b_1\eta^2;$$
$$|F_2(h(t), x_t)| \le 2(b_{21} + b_{22}\tilde{h} + b_{23}\tilde{h})\eta^2; \text{ i.e. } |F_2(h(t), x_t)| \le b_2\eta^2;$$
$$|F_3(h(t), x_t)| \le \tilde{h}(b_{31} + b_{32} + b_{33}\tilde{h})\eta^2; \text{ i.e. } |F_3(h(t), x_t)| \le b_3\eta^2.$$

For $t \ge v$, we thus see that $|\sum_{i=1}^3 F_i(h(t), x_t)| \le B\eta^2$, with $B = b_1 + b_2 + b_3$.

Now take any $\epsilon > 0$, and for each $t_1 \ge v$, define the set E_{v,t_1} as $E_{v,t_1} = E_{v,t_1}(\epsilon) = \{v \le t < t_1, |x(t)| \ge \epsilon\}$. We examine the quantity $G(t_1, x_{t_1})$:

$$\begin{aligned} G(t_1, x_{t_1}) - G(v, x_v) &= \int_v^{t_1} \dot{G}(x_t)dt \\ &= -\int_v^{t_1} x^T(t)x(t)dt + \int_v^{t_1} h'(t)\left[\sum_{i=1}^3 F_i(h(t), x_t)\right]dt \\ &\le -\int_{E_{v,t_1}} x^T(t)x(t)dt + \int_v^{t_1} |h'(t)|\left|\sum_{i=1}^3 F_i(h(t), x_t)\right|dt \\ &\le -\epsilon^2|E_{v,t_1}| + B\eta^2\int_v^{t_1} |h'(t)|dt, \end{aligned}$$

i.e. $G(t_1, x_{t_1}) - G(v, x_v) \leq -\epsilon^2 |E_{v,t_1}(\epsilon)| + B\eta^2 \int_v^{t_1} |h'(t)|dt$, and the theorem is proven. □

One may desire to use this theorem to find hypotheses on $h(\cdot)$ which imply for any given $\epsilon > 0$ that sup $\{t_1 : |x(t_1)| < \epsilon\} = \infty$, or which yield assigned asymptotic bounds for the quantity $(|E_{v,t_1}(\epsilon)|)/(t_1 - v)$ as $t_1 \to \infty$. For the remainder of this paper, however, we investigate the consequences of the special hypothesis that $h'(\cdot) \in L^1[0, \infty)$.

Theorem 3.2 *Again let $h(t) \in S$, with v as in the definition of S, and consider the differential-delay system* $(\dagger)\dot{x}(t) = A_0 x(t) + A_1 x(t - h(t))$. *If* $\int_v^\infty |h'(t)|dt < \infty$, *then for any bounded trajectory $x(\cdot)$ of the system* $(\dagger)$, *and for any* $\epsilon > 0$, *we have* $|\{t \geq t_1, |x(t)| \geq \epsilon\}| \to 0$ *as* $t_1 \to \infty$.

Proof. Take any $\epsilon > 0$, and for each $t_1 \geq v$, define the sets E_{v,t_1} and $E_{t_1,\infty}$ as $E_{v,t_1} = E_{v,t_1}(\epsilon) = \{v \leq t < t_1, |x(t)| \geq \epsilon\}$, and $E_{t_1,\infty} = E_{t_1,\infty}(\epsilon) = \{t \geq t_1, |x(t)| \geq \epsilon\}$. Now note that if we had $|E_{v,\infty}| = \infty$, then by the Monotone Convergence Theorem, we would have $|E_{v,t_1}| \to \infty$ as $t_1 \to \infty$. Noting that $\int_v^\infty |h'(t)|dt < \infty$, there would thus be some t_1 making $B\eta^2 \int_v^\infty |h'(t)|dt + G(v, x_v) < \epsilon^2 |E_{v,t_1}(\epsilon)|$, i.e. $-\epsilon^2 |E_{v,t_1}| + B\eta^2 \int_v^\infty |h'(t)|dt + G(v, x_v) < 0$. Noting the above theorem, this would immediately yield $G(t_1, x_{t_1}) < 0$. Put concisely, if we had $|E_{v,\infty}| = \infty$, then there would be some $t_1 > v$ with $G(t_1, x_{t_1}) < 0$. Since $G(t, \Phi) = V_{h(t)}(\Phi) \geq 0$ for $t \geq 0, \Phi \in C[-\tilde{h}, 0]$, we conclude that $|E_{v,\infty}| < \infty$. Now, since $|E_{v,\infty}| < \infty$, we can again apply the Monotone Convergence Theorem to see that $|E_{v,\infty}| - |E_{v,t_1}| \downarrow 0$ as $t_1 \uparrow \infty$, i.e. $|E_{v,\infty} - E_{v,t_1}| \downarrow 0$ as $t_1 \uparrow \infty$. Since $E_{v,\infty} - E_{v,t_1} = E_{t_1,\infty}$, we have shown that $|E_{t_1,\infty}| \downarrow 0$ as $t_1 \uparrow \infty$, i.e. $|\{t \geq t_1, |x(t)| \geq \epsilon\}| \downarrow 0$ as $t_1 \uparrow \infty$. □

Theorem 3.3 *Again consider the differential-delay system* $(\dagger)\dot{x}(t) = A_0 x(t) + A_1 x(t - h(t))$, *with* $h(t) \in S$ *and with* v *as in the definition of* S. *If* $\int_v^\infty |h'(t)|dt < \infty$, *then for any bounded trajectory* $x(\cdot)$ *of the system* $(\dagger)$, *we have* $x(t) \to 0$ *as* $t \to \infty$.

Proof. Let $x(\cdot)$ be a bounded solution of the system $(\dagger)$, with $|x(t)| \leq \eta$ for $t \geq 0$. Recall the definition $\tilde{h} = \sup\{h(t) : t \geq 0\}$, and for $v < t_1, t_1 + \tilde{h} \leq \zeta_1 < \zeta_2 \leq t_1 + 2\tilde{h}$, examine the vector $x(\zeta_2) - x(\zeta_1)$:

$$
\begin{aligned}
x(\zeta_2) - x(\zeta_1) &= \int_{\zeta_1}^{\zeta_2} \dot{x}(t)dt \\
&= A_0 \int_{\zeta_1}^{\zeta_2} x(t)dt + A_1 \int_{\zeta_1}^{\zeta_2} x(t - h(t))dt.
\end{aligned}
$$

Setting $\delta = \delta(t_1, \epsilon) = |E_{t_1,\infty}(\epsilon)|$, and setting $L = ||A_0|| + ||A_1||$, this yields

$$
\begin{aligned}
|x(\zeta_2) - x(\zeta_1)| &\leq ||A_0||[(\zeta_2 - \zeta_1)\epsilon + \delta\eta] + ||A_1||[(\zeta_2 - \zeta_1)\epsilon + \delta\eta] \\
&\leq (||A_0|| + ||A_1||)(\tilde{h}\epsilon + \delta\eta),
\end{aligned}
$$

i.e. for $v < t_1, t_1 + \tilde{h} \leq \zeta_1 < \zeta_2 \leq t_1 + 2\tilde{h}$, we have $|x(\zeta_2) - x(\zeta_1)| \leq L(\tilde{h}\epsilon + \delta(t_1, \epsilon)\eta)$.

If we note from Theorem 3.2 that $\delta(t_1, \epsilon) \to 0$ as $t_1 \to \infty$, we see that there is $\tau = \tau(\epsilon) > v$ with $\delta(t_1, \epsilon) < \tilde{h}$ for each $t_1 \geq \tau$. We thus know that for any $t_1 \geq \tau$, there is $\zeta_0 \in (t_1 + \tilde{h}, t_1 + 2\tilde{h})$ with $\zeta_0 \notin E_{t_1,\infty}(\epsilon)$, and hence with $|x(\zeta_0)| < \epsilon$. For $\tau \leq t_1, t_1 + \tilde{h} \leq t \leq t_1 + 2\tilde{h}$, we now obtain: $|x(t)| \leq |x(t) - x(\zeta_0)| + |x(\zeta_0)| \leq L(\tilde{h}\epsilon + \delta\eta) + \epsilon = (1 + L\tilde{h})\epsilon + L\delta\eta$. It is now seen that for $\tau(\epsilon) + \tilde{h} < t$, $t - \tilde{h} < t_1 < t$, one has $|x(t)| \leq (1 + L\tilde{h})\epsilon + L\eta\delta(t_1, \epsilon)$.

Using the above inequality, the conclusion of the theorem follows easily. In fact, since $\delta(t_1, \epsilon) \to 0$ as $t_1 \to \infty$, we immediately see that $\limsup_{t\to\infty} |x(t)| \leq (1 + L\tilde{h})\epsilon$. Since the initial choice of $\epsilon > 0$ was arbitrary, we conclude that $\limsup_{t\to\infty} |x(t)| = 0$, i.e. $x(t) \to 0$ as $t \to \infty$. □

It is instructive to note the effect of adding terms to V_h to construct a strictly positive functional. For this purpose, we first take any real vector $k = (k_1, k_2, k_3)$ having $k_i > 0$ for each $i = 1, 2, 3$. For $h \in H_s$ and $\Phi \in C[-h, 0]$, we then define the functional W_h as

$$W_h(\Phi) = k_1 \Phi^T(0)\Phi(0) + k_2 \int_{-h}^{0} \Phi^T(u)\Phi(u)du + k_3 V_h(\Phi).$$

If one now introduces the functional $n(h, \Phi)$, defined for $h \geq 0$ and $\Phi \in C[-h, 0]$ by $n(h, \Phi)^2 = \Phi^T(0)\Phi(0) + \int_{-h}^{0} \Phi^T(u)\Phi(u)du$, it is readily seen that for $c_1 = \min(k_1, k_2)$, one has $c_1 n(h, \Phi)^2 \leq W_h(\Phi)$ for all $h \in H_s$. This inequality will prove quite useful when we modify the functional W_h for use in systems having time-varying delays. Finally, to complete our introductory comments on the functional W_h, we note that after applying Lemma 3.2 to the functional V_h, a direct calculation will then yield the following formula for $\dot{W}_h$ along the trajectories of the system $(*)\dot{x}(t) = A_0 x(t) + A_1 x(t - h)$:

$$\frac{dW_h(x_t)}{dt} = -x^T(t)[k_3 I - 2k_1 A_0^T - 2k_2 I]x(t) - [x^T(t) \\ - x^T(t-h)]\begin{pmatrix} k_2 I & k_1 A_1 \\ k_1 A_1^T & k_2 I \end{pmatrix}[x^T(t) \ - x^T(t-h)]^T.$$

As before, a simple modification of the functional W_h can be used to examine the system $(\dagger)\dot{x}(t) = A(\sigma_{h(t)})x(t)$. Here we define the time-varying functional $Y(t, \Phi)$, for any $h(t)$ in the class S and $\Phi \in C[-\tilde{h}, 0]$, as $Y(t, \Phi) = W_{h(t)}(\Phi)$. After applying Lemma 3.4, a straightforward calculation then yields the following formula for $\dot{Y}(x_t) = \frac{dY(t, x_t)}{dt}$, the time derivative of the functional Y along solutions of the differential-delay equation $(\dagger)$: $\dot{Y}(x_t) = D(h(t), h'(t), x_t) + h'(t)k_3 F(h(t), x_t)$, where $F(h(t), x_t) = \sum_{i=1}^{3} F_i(h(t), x_t)$ as in Lemma 3.4, and

$$D(h, h', \Phi) = -\Phi^T(0)[k_3 I - 2k_1 A_0^T - 2k_2 I]\Phi(0) \\ - [\Phi^T(0) \ - \Phi^T(-h)]\mathcal{M}[\Phi^T(0) \ - \Phi^T(-h)]^T,$$

where

$$\mathcal{M} = \begin{pmatrix} k_2 I & (k_1 + h'k_3 Q(h,0))A_1 \\ A_1^T(k_1 + h'k_3 Q^T(h,0)) & (1+h')k_2 I \end{pmatrix}.$$

If we again write $F(h(t), x_t) = \sum_{i=1}^{3} F_i(h(t), x_t)$ with the $F_i(h(t), x_t)$ as in Lemma 3.4, then for $h \in H_s$, we can define the functional $F_0(h, \Phi)$ by

$$F_0(h,\Phi) = 2\Phi^T(0)Q(h,0)A_1\Phi(-h) + \Phi^T(-h)\int_{-h}^{0} A_1^T Q(h, u+h)A_1\Phi(u)du$$
$$+ \left[\int_{-h}^{0} \Phi^T(u)A_1^T Q(h, -u-h)A_1 du\right]\Phi(-h).$$

Setting $\widetilde{F}_0 = F - F_0$, and noting the quantities b_i, b_{ij} defined in Lemma 3.5, we can set $\beta_0 = 2b_{21} + b_{31} + b_{32}$, and $\widetilde{\beta}_0 = b_1 + 2(b_{22} + b_{23}) + b_{33}$. Recalling that $n(h,\Phi)^2 = \Phi^T(0)\Phi(0) + \int_{-h}^{0} \Phi^T(u)\Phi(u)du$, and applying the Cauchy-Schwarz inequality separately to the terms of F_0 and $\widetilde{F}_0$, one will obtain the following inequalities for F_0 and $\widetilde{F}_0$:

$$|F_0(h,\Phi)| \le |\Phi^T(-h)|\beta_0 n(h,\Phi); \ |\widetilde{F}_0(h,\Phi)| \le \widetilde{\beta}_0 n(h,\Phi)^2.$$

Noting for $c_1 = \min(k_1, k_2)$ that $c_1 n(h,\Phi)^2 \le W_h(\Phi)$, we immediately find that $|\widetilde{F}_0(h,\Phi)| \le (\widetilde{\beta}_0/c_1)W_h(\Phi)$. If one now separately examines the cases a) $|\Phi(-h)| \le |\Phi(0)|$, b) $|\Phi(0)| < |\Phi(-h)| \le n(h,\Phi)$, and c) $n(h,\Phi) < |\Phi(-h)|$, then lengthy but relatively straightforward applications of the Cauchy-Schwarz inequality will show the following:
There exist constants μ_1, μ_2 with $-1 < \mu_1 < 0 < \mu_2$, and constants $\zeta_0, k_1, k_2, k_3 > 0$, such that for $\mu_1 < h' < \mu_2$ and $\Phi \in C[-h, 0]$, one has

$$D(h, h', \Phi) + h'k_3 F_0(h,\Phi) \le |h'|k_3\zeta_0 n(h,\Phi)^2.$$

Here ζ_0 does not depend on the choice of h, h' having $h \in [0, \widetilde{h}]$, $h' \in (\mu_1, \mu_2)$. From this one can immediately write

$$D(h, h', \Phi) + h'k_3 F_0(h,\Phi) \le |h'|k_3(\zeta_0/c_1)W_h(\Phi) \text{ for } \mu_1 < h' < \mu_2.$$

Setting $\beta_0' = \widetilde{\beta}_0/c_1, \zeta_0' = \zeta_0/c_1$, we now have

$$D(h, h', \Phi) + h'k_3(F_0(h,\Phi) + \widetilde{F}_0(h,\Phi)) \le |h'|k_3(\beta_0' + \zeta_0')W_h(\Phi),$$

i.e.

$$D(h, h', \Phi) + h'k_3 F(h,\Phi) \le |h'|k_3(\beta_0' + \zeta_0')W_h(\Phi) \text{ for } \mu_1 < h' < \mu_2.$$

Finally, setting $\eta_0 = k_3(\beta_0' + \zeta_0')$, we summarize this analysis in the following lemma:

Lemma 3.6 *Let $A_0, A_1 \in \mathbb{R}^{n \times n}$. Then there exist constants μ_1, μ_2, with $-1 < \mu_1 < 0 < \mu_2$, having the following property: If $h(\cdot)$ is any member of the class*

S satisfying $\mu_1 < h'(t) < \mu_2$ for all $t \geq v$, then there is a constant $\eta_0 > 0$ and a strictly positive time-varying functional $Y(t, \Phi) = W_{h(t)}(\Phi)$ having $\frac{dY(t,x_t)}{dt} \leq |h'(t)|\eta_0 Y(t, x_t)$ for all $t \geq v$ along each solution $x(\cdot)$ of the differential-delay equation $(\dagger)\dot{x}(t) = A_0 x(t) + A_1 x(t - h(t))$.

Noting this inequality $\dot{Y}(x_t) \leq |h'(t)|\eta_0 Y(t, x_t)$ for the function $Y(t, x_t)$, and recalling elementary analysis, one now obtains the following inequality for $Y(t, x_t)$, valid along the solutions of the differential-delay equation $(\dagger)$ if $h'(\cdot)$ satisfies $\mu_1 < h'(t) < \mu_2$ for all $t \geq v$: $Y(t, x_t) \leq Y(0, x_0)e^{\eta_0 i(t)}$ for all $t \geq v$, where $i(t) = \int_v^t |h'(\tau)| d\tau$. This inequality yields a simple proof of the final theorem for this paper.

Theorem 3.4 *Let $A_0, A_1 \in \mathbb{R}^{n \times n}$. Let μ_1, μ_2 be as in Lemma 3.6, and let $h(\cdot)$ be any member of the class S satisfying $\mu_1 < h'(t) < \mu_2$ for all $t \geq v$. If $\int_v^\infty |h'(t)| dt < \infty$, then for each solution $x(\cdot)$ of the differential-delay equation $(\dagger)\dot{x}(t) = A_0 x(t) + A_1 x(t - h(t))$, we have $x(t) \to 0$ as $t \to \infty$.*

Proof. Set $I = \int_v^t |h'(t)| dt$, and let $i(t), \eta_0$ be as above. Then for each solution $x(\cdot)$ of the differential-delay equation $(\dagger)$, one has $|x(t)|^2 \leq Y(t, x_t) \leq Y(0, x_0)e^{\eta_0 i(t)} \leq Y(0, x_0)e^{\eta_0 I}$ for all $t \geq v$. Thus all trajectories $x(\cdot)$ of the system $(\dagger)$ are bounded, and now the theorem follows directly from Theorem 3.3. □

In a paper now in preparation, we examine a Lyapunov functional for the case where the characteristic function has no zeros in $\{\mathrm{Re}(s) \geq \gamma\}$, where γ is any real number ([6]). By modifying this functional to apply to differential equations of the form $(\dagger)\dot{x}(t) = A_0 x(t) + A_1 x(t - h(t))$, we will find insights having particular significance when $\gamma < 0$. In fact, we will be able to replace the hypothesis that $h(\cdot)$ is differentiable with $h'(\cdot) \in L^1[v, \infty)$, by the hypothesis that $h(\cdot)$ is absolutely continuous with $h'(\cdot)$ satisfying $\mu_1 < h'(t) < \mu_2$ for all $t \geq v$, where μ_1, μ_2 are determined by the matrices A_0, A_1. Our theorems will then come as inequalities of the form $|x(t)|^2 \leq C_0 e^{t(2\gamma + C_1 a(t))}$, where $a(t) = \frac{1}{t}\int_0^t |h'(\tau)| d\tau$ is the average value of the magnitude of $h'(\tau)$ over the interval $[0, t]$. For $\gamma < 0$ and $C_1 \limsup_{t \to \infty} a(t) < 2|\gamma|$, we will then have exponential asymptotic stability for the system $(\dagger)$.

References

1. Burton, T.A. (1983): Volterra Integral and Differential Equations. Academic Press, New York
2. Cooke, K.L. (1970): Linear functional differential equations of asymptotically autonomous type. Journal of Differential Equations, **7**, 154-174
3. Cooke, K.L. (1984): Stability of nonautonomous delay differential equations by Lyapunov functionals. Infinite Dimensional Systems (Lecture Notes in Mathematics, **1076**), 41-52, Springer Verlag, Berlin-Heidelberg-New York
4. Datko, R. (1980): Lyapunov functionals for certain linear delay differential equations in a Hilbert space. Journal of Mathematical Analysis and Applications, **76**, 37-57

5. Hale, J.K. (1977): Theory of Functional Differential Equations. Springer Verlag, Berlin-Heidelberg-New York
6. Infante, E.F., Castelan, W.B. (1978): A Lyapunov functional for a matrix difference-differential equation. Journal of Differential Equations, **29**, 439-451
7. Infante, E.F. (1982): Some results on the Lyapunov stability of functional equations. Volterra and Functional Differential Equations, Lecture Notes in Pure and Applied Mathematics, **81**, Marcel Dekker, New York
8. Parrott, M.E. (1982): The limiting behavior of solutions of infinite delay differential equations. Journal of Mathematical Analysis and Applications, **87**, 603-627
9. Wiener, N. (1933): The Fourier Integral and Certain of its Applications. Cambridge University Press, London

On an Interval Map Associated with a Delay Logistic Equation with Discontinuous Delays

George Seifert

Iowa State University, Ames, Iowa 50011

We consider a delay logistic equation of the form

$$N'(t) = N(t)(b - N([t])),\ t \geq 0,\ N(0) > 0. \tag{1}$$

where b is a positive constant, and $[t]$ denotes the greatest integer in t. Such an equation could model a feedback control problem where the feedback term $N([t])$ depends only on the discrete values $N(k)$, $k = 0,\ 1,\ 2, \ldots$. By a solution of (1), we mean a function $N(t)$ continuous for $t \geq 0$, and satisfying (1) in each interval $(k,\ k+1)$, $k = 0,\ 1,\ 2, \ldots.$, with $N(0)$ given. Clearly, these solutions $N(t)$ are to a great extent determined by the values $N(k)$, $k = 0,\ 1,\ 2, \ldots$, and it is this set of values we propose to investigate. It is easy to see that these values satisfy

$$N(k+1) = aN(k)exp(-N(k)),\ k = 0,\ 1,\ 2, \ldots.$$

where $a = e^b > 1$. Hence, we study the function $F_a(x) = axe^{-x}$, and in particular, the semi-dynamical system generated by the iterates of F_a.

Although interval maps of a large class of functions, including quadratic functions of the form $ax(1-x)$, have been studied in detail (*cf.* for example [1], [2]), the specific details in proving our results for F_a do not seem to be in the standard literature. While many of these results can be derived from more general results appearing in this literature, it is felt that the non-specialist might appreciate seeing sufficiently detailed proofs of our results, even though in many cases, shorter arguments are possible; cf., for example, the remark appearing after the proof of Prop. 2.

Eq.(1) as well as the associated interval function $F_a(x)$ are discussed to some extent in [5]. Some estimates are given there for values of the parameter a for which certain periodic orbits of this semi-dynamical system appear but no details are given. We note that the function F_a has also been considered in a model for certain types of population variations [3].

We make the following definitions for any map F of an interval J into itself; for x in J:

(i) $F^n(x) = F(F^{n-1}(x)),\ n = 1, 2, \ldots;\ F^0(x) = x;$

(ii) If $F(x) = x$, x is called a fixed point, or a 1-periodic point, of F.
(iii) If $n > 1$ and $F^n(x) = x$, $F^m(x) \neq x$ for $m = 1, 2, \ldots, n-1$, x is called a n-periodic point of F.
(iv) The fixed point x is globally attracting if $F^n(y) \to x$ as $n \to \infty$ for any y in J.
(v) The n-periodic point x is stable (locally attracting) if there exists a $\delta > 0$ such that for any y in J with $|y - x| < \delta$, $F^{nm}(y) \to x$ as $m \to \infty$.

We shall use the following stability theorem for n-periodic points, cf. [1].

Theorem. *Let F be continuously differentiable on J, and x be an n-periodic point of F. Then x is stable if $|F^{n\prime}(x)| < 1$ and unstable if $|F^{n\prime}(x)| > 1$; here $F^{n\prime}(x) = d(F^n(x))/dx$.*

We establish the following properties of $F_a : (0,\infty) \to (0,\infty)$ which show that as a increases from 1, F_a becomes in a sense increasingly chaotic. We always assume $a > 1$ and $x > 0$.

Proposition 1. *$F_a(x) \leq a/e$, and $F_a(x) = x$ if and only if $x = \log a$.*

This follows easily; we omit the proof.

Proposition 2. *If $a \leq e^2$, F_a has no n-periodic points for $n > 1$, and fixed point $\log a$ is globally attracting.*

Proof. To show that for $a \leq e^2$, $\log a$ is a global attractor for $F_a(x)$ we can use an argument in Devaney [1]: pp. 71-72, which applies to an arbitrary continuous map $F(x)$ with a locally stable fixed point p; if $W(p)$ denotes the maximal connected component containing p of points attracted to p, $W(\mathrm{p})$ is open and invariant under F. Thus its end points, if finite, are fixed points of F or of F^2. But by the first part of our proof, $F_a^2(x) = x$ only if $x = \log a$, and since also $F_a(x) = x$ only if $x = \log a$, $W(p)$ in our case must be $(0,\infty)$. We get the stability of $\log a$ for $a < e^2$, from the fact that $|F_a'(\log a)| < 1$, and for $a = e^2$, from the fact that $\log a = 2$ is an inflection point for the graph of $y = F_a(x)$; we omit the details.

Proposition 3. *If $a > e^2$, there exist $x_1(a)$, $x_2(a)$ such that $0 < x_1(a) < \log a < x_2(a) < 2\log a$, and*

$$F_a(x_1(a)) = x_2(a), \; F_a(x_2(a)) = x_1(a).$$

Also $x_1(a) \to 0$ and $x_2(a) \to \infty$ as $a \to \infty$, and $x_1(a) \to \log a$ and $x_2(a) \to \log a$ as $a \to e^2$.

Finally, the fixed point $\log a$ is unstable.

Proof. From (2) we find that $F_a^2(x) = x$ is equivalent to

$$xe^{-x} = (2\log a - x)/a. \tag{2}$$

Since $(xe^x)'_{x=\log a} = (1-\log a)/a < -1/a$, it follows easily that $x_1(a)$ and $(x_2(a)$ exist and have the asserted properties; cf. fig. 1. Note that $x_2(a) - \log a = \log a - x_1(a)$.

The instability of $\log a$ follows from the fact that

$$F_a'(\log a) = 1 - \log a < -1,$$

and the stability theorem mentioned earlier.

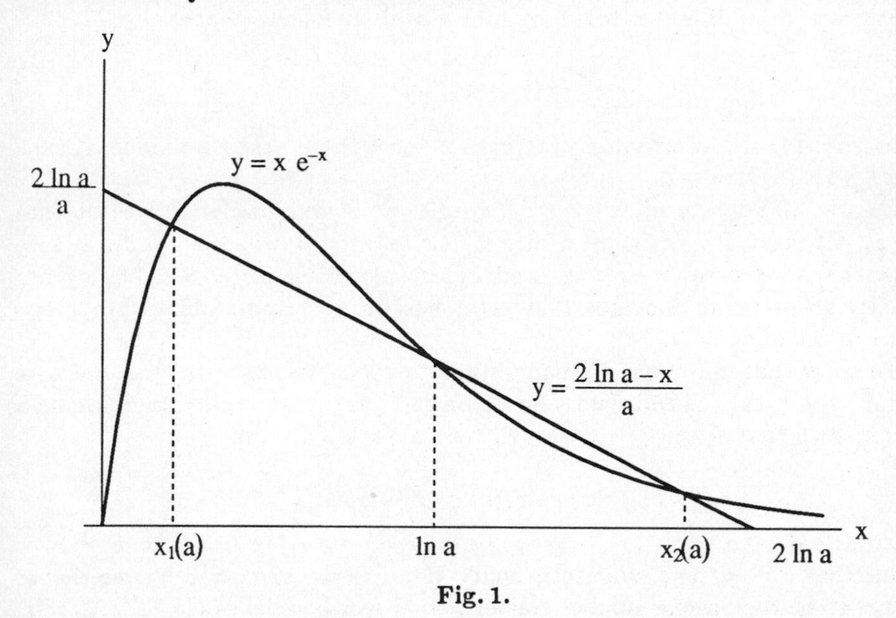

Fig. 1.

Proposition 4. *There exists an $a_o > e^2$ such that if $e^2 < a < a_o$, the 2-periodic points $x_1(a)$ and $x_2(a)$ are stable, while for $a > a_o$ they are unstable. Also $a_o < e^{1+\sqrt{3}}$.*

Proof. Let a_1 be the unique solution of $a/e = 2\log a - 1$ such that $a_1 > e^2$; note that $2\log e^2 - 1 > e^2/e$, while $2\log a - 1 < a/e$ for $a > e^2$ and sufficiently large, and since $2\log x - 1$ has decreasing derivative, the existence and uniqueness of such a a_1 follows. Then $x_1(a_1) = 1$, and $x_1(a) < 1$ for $a > a_1$. By direct calculations, we get

$$F_a^{2\prime}(x) = (1-x)(1-axe^{-x}) \quad \text{for } x = x_1(a),\ x = x_2(a), \text{ and } x = \log a. \tag{3}$$

Using (2) in (3) we see that

$$F_a^{2\prime}(x) = (1-x)(1+x-2\log a). \tag{4}$$

for $x = x_1(a)$, $x = x_2(a)$, and $x = \log a$. Using this with $x = x_1(a)$, we get

$$F_a^{2\prime}(x_1(a)) = (1 - x_1(a))\ (1 + x_1(a) - 2\log a)$$

and since $0 < x_1(a) < 1$ for $a > a_1$, and $x_1(a) < 2\log a_1$ we have $F_a^{2\prime}(x_1(a)) < 1$ for $e^2 < a < a_1$.

Now from (4), with $x = x_1(a)$ we obtain

$$\frac{d}{da} F_a^{2\prime}(x_1(a)) = 2x_1'(a)(\log a - x_1(a)) - 2/a$$

and since $x_1'(a) < 0$ and $x_1(a) < \log a$ for $a > e^2$, it follows that

$$\frac{d}{da} F_a^{2\prime}(x_1(a)) < 0 \text{ for } a > e^2.$$

Since from (4) we also have $F_a^{2\prime}(x_1(a)) \to -\infty$, there exists a unique a_o such that $F_{ao}^{2}{}'(x_1(a_o)) = -1$.

Clearly also $a_o > a_1$, since $F_{a1}^{2}{}'(x_1(a_1)) = 0$. Summarizing, we see that $F_a^{2\prime}(x_1(a)) < 1$ for $e^2 < a < a_1$, $F_a^{2\prime}(x_1(a)) = 0$ when $a = a_1$, $-1 < F_a^{2\prime}(x_1(a)) < 0$ for $a_1 < a < a_o$, and $F_a^{2\prime}(x_1(a)) < -1$ for $a > a_o$. Using our stability theorem we conclude that $x_1(a)$ has the asserted stability properties with respect to a_o.

To show that $x_2(a)$ has the same stability properties as $x_1(a)$ has, we note that if $\mid x - x_2(a) \mid$ is sufficiently small, since $F_a(x_1(a)) = x_2(a)$, there exists a $F_a^{-1}(x)$ such that $F_a(F_a^{-1}(x)) = x$, $F_a^{-1}(x_2(a)) = x_1(a)$, and

$$F_a^{-1}(x) \to x_1(a) \text{ as } x \to x_2(a).$$

In fact, if $x_1(a) \neq 1$; i.e., $a \neq a_1$, we have $F_a^1(x_1(a)) \neq 0$ and so $F_a^{-1}(x)$ is unique for $\mid x - x_1(a) \mid$ sufficiently small. If $x_1(a) = 1$, i.e., $a = a_1$, we choose $F_a^{-1}(x)$ to be the unique number greater than 1 which satisfies $F_a(F_a^{-1}(x)) = x$. By the stability of $x_1(a)$, we find that for such x,

$$F_a^{2n}(F_a^{-1}(x)) \to x_1(a) \text{ as } n \to \infty; \text{ so}$$
$$F_a(F_a^{2n}(F_a^{-1}(x))) \to F_a(x_1(a)) \text{ as } n \to \infty; \text{ i.e.,}$$
$$F_a^{2n}(x) \to x_2(a) \text{ as } n \to \infty$$

.

The fact that $\log a$ is unstable for $a > e^2$ follows easily from (4) with $x = \log a$; i.e., from $F_a^{2\prime}(x) = (1 - \log a)^2 > 1$.

To show that $a_o < e^{1+\sqrt{3}}$, we note that a_o satisfies

$$F_{ao}^{2}{}'(x_1(a_o)) = 1.$$

Using (4) and the fact that $x_1(a_o) < \log a_o$, we have

$$x_1(a_o) = \log a_o - \sqrt{(\log a_o - 1)^2 + 1}$$

If we put $v = \sqrt{(\log a_o - 1)^2 + 1}$, we find that $\log a_o = 1 + \sqrt{v^2 - 1}$, and so

$$x_1(a_o) = 1 + \sqrt{v^2 - 1 - v}.$$

Since $x_1(a_o)$ satisfies (2) with $a = a_o$, we have eventually

$$(1+\sqrt{v^2-1}-v)e^v = v+1+\sqrt{v^2-1}.$$

This last equation can be written as

$$\tanh\frac{v}{2} = \frac{v}{1+\sqrt{v^2-1}}$$

But $\tanh\frac{v}{2} < \frac{\sqrt{2}}{2}$, while $\tanh\frac{2}{2} > \frac{2}{1+\sqrt{3}}$ which show that $\sqrt{2} < v < 2$; i.e., $2 < \log a_o < 1+\sqrt{3}$. This shows $a_o < e^{1+\sqrt{3}}$ and completes the proof of Prop. 4.

Much smaller upper bounds in a_o can be obtained by using more detailed computations; however, our aim here is not toward that purpose.

Proposition 5. *Let $a_2 > 3e$ and satisfy $a_2^3e^{-a_2/e} = e$. Then if $a > a_2$, F_a has a 3-periodic point and so by Sarkovskii's theorem [1], or the result due to Li and Yorke [4], F_a has a n-periodic points for each integer $n \geq 1$. Also if $a > a_2$, the 2-periodic points are unstable.*

Proof. Let x_1 be unique solution of $axe^{-x} = 1$, such that $x_1 < 1$; since $a > 3e$ such an x_1 exists. So

$$F_a^2(x_1) = F_a(ax_1e^{-x_1}) = F_a(1) = a/e,$$

and so

$$F_a^3(x_1) = F_a(a/e) = a^2e^{-a/e}/e.$$

But since $a > a_2$, we have

$$(a^2/e)e^{-a/e} < 1/a = x_1e^{-x_1} < x_1;$$

i.e., $F_a^3(x_1) < x_1$.

Next, let x_2 be the unique solution of $axe^{-x} = \log a$ such that $x_2 < 1$; such an x_2 exists since $a > e$, and $x_2 > x_1$. So $F_a(x_2) = \log a$, and $F_a^3(x_2) = \log a > x_2$; i.e., $F_a^3(x_2) > x_2$.

By the intermediate value theorem, the existence of an $\tilde{x}$ such that $F_a^3(\tilde{x}) = \tilde{x}$, where $x_1 < \tilde{x} < x_2$ follows.

If $F_a(\tilde{x}) = \tilde{x}$, then $\tilde{x} = \log a > 1$, a contradiction since $\tilde{x} < x_2 < 1$. If $F_a^2(\tilde{x}) = \tilde{x}$, by (2) we have

$$a\tilde{x}e^{-\tilde{x}} = 2\log a - \tilde{x}.$$

But $a\tilde{x}e^{-\tilde{x}} < ax_2e^{-x_2} = \log a$, and since also

$$2\log a - \tilde{x} > \log a,$$

we have a contradiction. So $\tilde{x}$ is a 3-periodic point.

To show $x_1(a)$ and $x_2(a)$ unstable, we use Prop. 4. Since it is easy to verify that $8e < a_2 < 9e$, and since $a_o < e^{1+\sqrt{3}}$, it follows that $a_2 > e^{1+\sqrt{3}} > a_o$, and so $x_1(a)$ and $x_2(a)$ are unstable for $a > a_2$, and our proof is complete.

Proposition 6. *If $a > e$, then for any $x > 0$ there exists a $N(x) > 0$ such that*

$$(a^2/e)e^{-\frac{a}{e}} \leq F_a^n(x) \leq a/e \text{ for } n > N(x).$$

Proof. Note that $(a^2/e)e^{-a/e} = F_a(a/e) \leq a/e$ from Prop.1.

We also have $(a^2/e)e^{-a/e} < \log a$ since it easy to show that $a/e > \log a$ for $a > e$. For any x_1, $F_a(a/e) \leq x_1 \leq \log a$, we have

$$F_a(a/e) \leq x_1 \leq F_a(x_1) \leq a/e, \text{ since } F_a(x) \geq x \text{ for } 0 < x \leq \log a \tag{5}$$

If x_1 satisfies $\log a \leq x_1 \leq a/e$, we have

$$F_a(a/e) \leq F_a(x_1) \leq a/e, \text{ since } F_a(x) \leq x \text{ for } x \geq \log a. \tag{6}$$

From (5) and (6) we see that the interval $F_a(a/e) \leq x \leq a/e$ is mapped into itself by F_a.

Suppose now $0 < x_1 < F_a(a/e)$. If $F_a^n(x_1) \leq 1$ for all $n > 0$, then using a previous argument, we find that $F_a^n(x_1) \to \log a$ as $n \to \infty$, a contradiction since $\log a > 1$. So there exists a $n_1 > 0$ such that $y_1 \equiv F_a^{n1}(x_1) > 1$; i.e., $1 < y_1 \leq a/e$. But since $F_a(x)$ is decreasing for $x \geq 1$, we have $F_a(y_1) \geq F_a(a/e)$; i.e., $F_a(a/e) \leq F_a(y_1) \leq a/e$. So from the previous case, $F_a^{n_1+n}(x_1) \in [F_a(a/e),\ a/e]$ for all $n > 1$.

Finally if $x_1 > a/e > 1$, we have $F_a(x_1) < F_a(a/e)$ and by the preceding argument $F_a^n(x_1) \in [F_a(a/e),\ a/e]$, for n sufficiently large. This completes the proof.

Proposition 7. *If $a > a_2$, a_2 as in Prop. 5, F_a has at most one stable periodic orbit.*

Proof. A simple computation shows that the Schwarzian derivative

$$\frac{F_a'''(x)}{F_a'(x)} - \frac{3}{2}\left(\frac{F_a''(a)}{F_a'(a)}\right)^2$$

of $F_a(x)$ is negative for $a > a_2$. Since the fixed point $\log a$ of $F_a(x)$ is unstable, it follows immediately from Corollary II.4.2, on p. 52 in [2] that F_a has at most one stable periodic orbit.

The following questions for F_a with $a > a_2$ seem to be open and nontrivial. If $I_a = [F_a(a/e),\ a/e]$,

(i) Is the set of periodic points dense in I_a?

(ii) Does there exists x in I_a such that $F_a^n(x) : n = 0,\ 1,\ 2, ...$ is dense in I_a?

(iii) Does there exist a $\delta > 0$ such that for each x in I_a there exists a sequence $x_n \to x$ as $n \to \infty$ such that for each integer $n \geq 1$, there exists an integer m such that $[F_a^m(x_n) - F_a^m(x) |> \delta$?

If these questions all have affirmative answers, then F_a is chaotic in I_a in the sense of, for example, Devaney [1].

References

1. Devaney, R.L. (1986): An Introduction to Chaotic Dynamical Systems. Benjamin-Cummings, Inc.
2. Collet, P., Eckmann, J.P. (1980): Iterated Maps of the Interval as Dynamical Systems. Birkhauser, Boston
3. Hassel, M.P., Comins H.N. (1971): Discrete time models for 2-species competition. Theoretical Pop. Biology **9**, 202-222
4. Li, T-Y, Yorke, J.A. (1975): Period three implies chaos. Amer. Math. Monthly **82**(10), 985-992
5. Carvalho, L.A.V., Cooke, K.L. (1988): A nonlinear equation with piecewise continuous argument. Diff. and Int. Eq., **1**(3), 359-367

Vol. 1394: T.L. Gill, W.W. Zachary (Eds.), Nonlinear Semigroups, Partial Differential Equations and Attractors. Proceedings, 1987. IX, 233 pages. 1989.

Vol. 1395: K. Alladi (Ed.), Number Theory, Madras 1987. Proceedings. Vll, 234 pages. 1989.

Vol. 1396: L. Accardi, W. von Waldenfels (Eds.), Quantum Probability and Applications IV. Proceedings, 1987. Vl, 355 pages. 1989.

Vol. 1397: P.R. Turner (Ed.), Numerical Analysis and Parallel Processing. Seminar, 1987. Vl, 264 pages. 1989.

Vol. 1398: A.C. Kim, B.H. Neumann (Eds.), Groups – Korea 1988. Proceedings. V, 189 pages. 1989.

Vol. 1399: W.-P. Barth, H. Lange (Eds.), Arithmetic of Complex Manifolds. Proceedings, 1988. V, 171 pages. 1989.

Vol. 1400: U. Jannsen. Mixed Motives and Algebraic K-Theory. Xlll, 246 pages. 1990.

Vol. 1401: J. Steprans, S. Watson (Eds.), Set Theory and its Applications. Proceedings, 1987. V, 227 pages. 1989.

Vol. 1402: C. Carasso, P. Charrier, B. Hanouzet, J.-L. Joly (Eds.), Nonlinear Hyperbolic Problems. Proceedings, 1988. V, 249 pages. 1989.

Vol. 1403: B. Simeone (Ed.), Combinatorial Optimization. Seminar, 1986. V, 314 pages. 1989.

Vol. 1404: M.-P. Malliavin (Ed.), Séminaire d´AIgèbre Paul Dubreil et Marie-Paul Malliavin. Proceedings, 1987–1988. IV, 410 pages. 1989.

Vol. 1405: S. Dolecki (Ed.), Optimization. Proceedings, 1988. V, 223 pages. 1989. Vol. 1406: L. Jacobsen (Ed.), Analytic Theory of Continued Fractions III. Proceedings, 1988. Vl, 142 pages. 1989.

Vol. 1407: W. Pohlers, Proof Theory. Vl, 213 pages. 1989.

Vol. 1408: W. Lück, Transformation Groups and Algebraic K-Theory. Xll, 443 pages. 1989.

Vol. 1409: E. Hairer, Ch. Lubich, M. Roche. The Numerical Solution of Differential-Algebraic Systems by Runge-Kutta Methods. Vll, 139 pages. 1989.

Vol. 1410: F.J. Carreras, O. Gil-Medrano, A.M. Naveira (Eds.), Differential Geometry. Proceedings, 1988. V, 308 pages. 1989.

Vol. 1411: B. Jiang (Ed.), Topological Fixed Point Theory and Applications. Proceedings. 1988. Vl, 203 pages. 1989.

Vol. 1412: V.V. Kalashnikov, V.M. Zolotarev (Eds.), Stability Problems for Stochastic Models. Proceedings, 1987. X, 380 pages. 1989.

Vol. 1413: S. Wright, Uniqueness of the Injective lll_1 Factor. III, 108 pages. 1989.

Vol. 1414: E. Ramirez de Arellano (Ed.), Algebraic Geometry and Complex Analysis. Proceedings, 1987. Vl, 180 pages. 1989.

Vol. 1415: M. Langevin, M. Waldschmidt (Eds.), Cinquante Ans de Polynômes. Fifty Years of Polynomials. Proceedings, 1988. IX, 235 pages.1990.

Vol. 1416: C. Albert (Ed.), Géométrie Symplectique et Mécanique. Proceedings, 1988. V, 289 pages. 1990.

Vol. 1417: A.J. Sommese, A. Biancofiore, E.L. Livorni (Eds.), Algebraic Geometry. Proceedings, 1988. V, 320 pages. 1990.

Vol. 1418: M. Mimura (Ed.), Homotopy Theory and Related Topics. Proceedings, 1988. V, 241 pages. 1990.

Vol. 1419: P.S. Bullen, P.Y. Lee, J.L. Mawhin, P. Muldowney, W.F. Pfeffer (Eds.), New Integrals. Proceedings, 1988. V, 202 pages. 1990.

Vol. 1420: M. Galbiati, A. Tognoli (Eds.), Real Analytic Geometry. Proceedings, 1988. IV, 366 pages. 1990.

Vol. 1421: H.A. Biagioni, A Nonlinear Theory of Generalized Functions, Xll, 214 pages. 1990.

Vol. 1422: V. Villani (Ed.), Complex Geometry and Analysis. Proceedings, 1988. V, 109 pages. 1990.

Vol. 1423: S.O. Kochman, Stable Homotopy Groups of Spheres: A Computer-Assisted Approach. Vlll, 330 pages. 1990.

Vol. 1424: F.E. Burstall, J.H. Rawnsley, Twistor Theory for Riemannian Symmetric Spaces. III, 112 pages. 1990.

Vol. 1425: R.A. Piccinini (Ed.), Groups of Self-Equivalences and Related Topics. Proceedings, 1988. V, 214 pages. 1990.

Vol. 1426: J. Azéma, P.A. Meyer, M. Yor (Eds.), Séminaire de Probabilités XXIV, 1988/89. V, 490 pages. 1990.

Vol. 1427: A. Ancona, D. Geman, N. Ikeda, École d'Eté de Probabilités de Saint Flour

XVIII, 1988. Ed.: P.L. Hennequin. Vll, 330 pages. 1990.

Vol. 1428: K. Erdmann, Blocks of Tame Representation Type and Related Algebras. XV. 312 pages. 1990.

Vol. 1429: S. Homer, A. Nerode, R.A. Platek, G.E. Sacks, A. Scedrov, Logic and Computer Science. Seminar, 1988. Editor: P. Odifreddi. V, 162 pages. 1990.

Vol. 1430: W. Bruns, A. Simis (Eds.), Commutative Algebra. Proceedings. 1988. V, 160 pages. 1990.

Vol. 1431: J.G. Heywood, K. Masuda, R. Rautmann, V.A. Solonnikov (Eds.), The Navier-Stokes Equations – Theory and Numerical Methods. Proceedings, 1988. Vll, 238 pages. 1990.

Vol. 1432: K. Ambos-Spies, G.H. Müller, G.E. Sacks (Eds.), Recursion Theory Week. Proceedings, 1989. Vl, 393 pages. 1990.

Vol. 1433: S. Lang, W. Cherry, Topics in Nevanlinna Theory. II, 174 pages.1990.

Vol. 1434: K. Nagasaka, E. Fouvry (Eds.), Analytic Number Theory. Proceedings, 1988. VI, 218 pages. 1990.

Vol. 1435: St. Ruscheweyh, E.B. Saff, L.C. Salinas, R.S. Varga (Eds.), Computational Methods and Function Theory. Proceedings, 1989. Vl, 211 pages. 1990.

Vol. 1436: S. Xambó-Descamps (Ed.), Enumerative Geometry. Proceedings, 1987. V, 303 pages. 1990.

Vol. 1437: H. Inassaridze (Ed.), K-theory and Homological Algebra. Seminar, 1987–88. V, 313 pages. 1990.

Vol. 1438: P.G. Lemarié (Ed.) Les Ondelettes en 1989. Seminar. IV, 212 pages. 1990.

Vol. 1439: E. Bujalance, J.J. Etayo, J.M. Gamboa, G. Gromadzki. Automorphism Groups of Compact Bordered Klein Surfaces: A Combinatorial Approach. Xlll, 201 pages. 1990.

Vol. 1440: P. Latiolais (Ed.), Topology and Combinatorial Groups Theory. Seminar, 1985–1988. Vl, 207 pages. 1990.

Vol. 1441: M. Coornaert, T. Delzant, A. Papadopoulos. Géométrie et théorie des groupes. X, 165 pages. 1990.

Vol. 1442: L. Accardi, M. von Waldenfels (Eds.), Quantum Probability and Applications V. Proceedings, 1988. Vl, 413 pages. 1990.

Vol. 1443: K.H. Dovermann, R. Schultz, Equivariant Surgery Theories and Their Periodicity Properties. Vl, 227 pages. 1990.

Vol. 1444: H. Korezlioglu, A.S. Ustunel (Eds.), Stochastic Analysis and Related Topics Vl. Proceedings, 1988. V, 268 pages. 1990.

Vol. 1445: F. Schulz, Regularity Theory for Quasilinear Elliptic Systems and – Monge Ampère Equations in Two Dimensions. XV, 123 pages. 1990.

Vol. 1446: Methods of Nonconvex Analysis. Seminar, 1989. Editor: A. Cellina. V, 206 pages. 1990.

Vol. 1447: J.-G. Labesse, J. Schwermer (Eds), Cohomology of Arithmetic Groups and Automorphic Forms. Proceedings, 1989. V, 358 pages. 1990.

Vol. 1448: S.K. Jain, S.R. López-Permouth (Eds.), Non-Commutative Ring Theory. Proceedings, 1989. V, 166 pages. 1990.

Vol. 1449: W. Odyniec, G. Lewicki, Minimal Projections in Banach Spaces. VIII, 168 pages. 1990.

Vol. 1450: H. Fujita, T. Ikebe, S.T. Kuroda (Eds.), Functional-Analytic Methods for Partial Differential Equations. Proceedings, 1989. VII, 252 pages. 1990.

Vol. 1451: L. Alvarez-Gaumé, E. Arbarello, C. De Concini, N.J. Hitchin, Global Geometry and Mathematical Physics. Montecatini Terme 1988. Seminar. Editors: M. Francaviglia, F. Gherardelli. IX, 197 pages. 1990.

Vol. 1452: E. Hlawka, R.F. Tichy (Eds.), Number-Theoretic Analysis. Seminar, 1988–89. V, 220 pages. 1990.

Vol. 1453: Yu.G. Borisovich, Yu.E. Gliklikh (Eds.), Global Analysis – Studies and Applications IV. V, 320 pages. 1990.

Vol. 1454: F. Baldassari, S. Bosch, B. Dwork (Eds.), p-adic Analysis. Proceedings, 1989. V, 382 pages. 1990.

Vol. 1455: J.-P. Françoise, R. Roussarie (Eds.), Bifurcations of Planar Vector Fields. Proceedings, 1989. VI, 396 pages. 1990.

Vol. 1456: L.G. Kovács (Ed.), Groups – Canberra 1989. Proceedings. XII, 198 pages. 1990.

Vol. 1457: O. Axelsson, L.Yu. Kolotilina (Eds.), Preconditioned Conjugate Gradient Methods. Proceedings, 1989. V, 196 pages. 1990.

Vol. 1458: R. Schaaf, Global Solution Branches of Two Point Boundary Value Problems. XIX, 141 pages. 1990.

Vol. 1459: D. Tiba, Optimal Control of Nonsmooth Distributed Parameter Systems. VII, 159 pages. 1990.

Vol. 1460: G. Toscani, V. Boffi, S. Rionero (Eds.), Mathematical Aspects of Fluid Plasma Dynamics. Proceedings, 1988. V, 221 pages. 1991.

Vol. 1461: R. Gorenflo, S. Vessella, Abel Integral Equations. VII, 215 pages. 1991.

Vol. 1462: D. Mond, J. Montaldi (Eds.), Singularity Theory and its Applications. Warwick 1989, Part I. VIII, 405 pages. 1991.

Vol. 1463: R. Roberts, I. Stewart (Eds.), Singularity Theory and its Applications. Warwick 1989, Part II. VIII, 322 pages. 1991.

Vol. 1464: D. L. Burkholder, E. Pardoux, A. Sznitman, Ecole d'Eté de Probabilités de Saint- Flour XIX-1989. Edited by P. L. Hennequin. VI, 259 pages. 1991.

Vol. 1465: G. David, Wavelets and Singular Integrals on Curves and Surfaces. X, 107 pages. 1991.

Vol. 1466: W. Banaszczyk, Additive Subgroups of Topological Vector Spaces. VII, 178 pages. 1991.

Vol. 1467: W. M. Schmidt, Diophantine Approximations and Equations. VIII, 217 pages. 1991.

Vol. 1468: J. Noguchi, T. Ohsawa (Eds.), Prospects in Complex Geometry. Proceedings, 1989. VII, 421 pages. 1991.

Vol. 1469: J. Lindenstrauss, V. D. Milman (Eds.), Geometric Aspects of Functional Analysis. Seminar 1989-90. XI, 191 pages. 1991.

Vol. 1470: E. Odell, H. Rosenthal (Eds.), Functional Analysis. Proceedings, 1987-89. VII, 199 pages. 1991.

Vol. 1471: A. A. Panchishkin, Non-Archimedean L-Functions of Siegel and Hilbert Modular Forms. V, 157 pages. 1991.

Vol. 1472: T. T. Nielsen, Bose Algebras: The Complex and Real Wave Representations. V, 132 pages. 1991.

Vol. 1473: Y. Hino, S. Murakami, T. Naito, Functional Differential Equations with Infinite Delay. X, 317 pages. 1991.

Vol. 1474: S. Jackowski, B. Oliver, K. Pawałowski (Eds.), Algebraic Topology, Poznań 1989. Proceedings. VIII, 397 pages. 1991.

Vol. 1475: S. Busenberg, M. Martelli (Eds.), Delay Differential Equations and Dynamical Systems. Proceedings, 1990. VIII, 249 pages. 1991.